A History of the Sciences

STEPHEN F. MASON, M.A., D.Phil.

NEW REVISED EDITION

COLLIER BOOKS

A Division of Macmillan Publishing Co., Inc.

NEW YORK

First Collier Books Edition 1962
20   19   18   17   16   15   14
A History of the Sciences originally appeared under the title Main
Currents of Scientific Thought
This revised Collier Books Edition is published by arrangement with
Abelard-Schuman Limited
Macmillan Publishing Co., Inc.
866 Third Avenue, New York, N.Y. 10022
Collier Macmillan Canada, Ltd.

PRINTED IN THE UNITED STATES OF AMERICA

Contents

Part Six / Twentieth-Century Science:
New Fields and New Powers

Acknowledgements

THANKS are gratefully offered to all those friends, students, and colleagues who discussed with me the contents of this book at the preliminary note stage, and who read through parts in manuscript and proof, providing not a few criticisms and corrections. The errors, ineptitudes, and misinterpretations that remain are, however, entirely the author's responsibility. I am indebted especially to Mr. J. E. C. Hill of Balliol College, Oxford, for the provision of numerous bibliographical references, and for many illuminating discussions, with regard to all aspects of the history of the seventeenth century, and also to the following who made available to me the results of some of their researches which were unpublished at the time of writing: Dr. J. Needham of Gonville and Caius College, Cambridge, part of whose book on *Science and Civilization in China* I read in manuscript, and Dr. S. Lilley of Birmingham University Extra-Mural Department who provided me with new material relating to the British geologists of the late eighteenth and early nineteenth centuries.

A History of the Sciences

Chapter 1

Introduction

SCIENCE, as we know it today, was a comparatively late product of the general development of human civilization. Prior to the modern period of history, we cannot say that there was much of a scientific tradition, distinct from the tradition of the philosophers on the one hand, and that of the craftsmen on the other. The roots of science, however, ran deep, stretching back to the period before the appearance of civilization. No matter how far back in history we go there were always some techniques, facts, and conceptions, known to craftsmen or scholars, which were scientific in character, though before modern times such knowledge in general was subordinate to the requirements of either the philosophical or the craft tradition. Philosophical considerations, for example, limited the important scientific achievement of the ancient Greeks, so that both of their two main astronomical systems conflicted with observations known in antiquity.

Science had its historical roots in two primary sources. Firstly, the technical tradition, in which practical experiences and skills were handed on and developed from one generation to another; and secondly, the spiritual tradition, in which human aspirations and ideas were passed on and augmented. Such traditions existed before civilization appeared, if we are to judge by the continuity in the development of the tools used by the men of the stone age, and by their burial practices and cave paintings. In the bronze age civilizations, the two traditions appear to have been largely separate, perpetuated on the one hand by craftsmen, and on the other by corporations of priestly scribes, though the latter had some important utilitarian techniques of their own.

In the subsequent civilizations, the two traditions remained separate for the most part, though both became differentiated, the philosopher separating off from the priest and the scribe, and the artisans of one trade from those of another. There were occasional rapprochements, notably those in ancient

Greece, but, in general, it was not until the late middle ages and early modern times that elements from the two traditions began to converge, and then combine, producing a new tradition, that of science. The development of science then became more autonomous, and, containing both practical and theoretical elements, science produced results which had both technical and philosophical implications. Thus, science reacted back and influenced its sources, and indeed it had an effect ultimately upon domains far removed from its immediate origins. With these things, and with the inner development of the scientific movement, we shall be concerned in this book.

PART ONE

ANCIENT SCIENCE

Chapter 2

Science in the Ancient Civilizations of Babylonia and Egypt

BEFORE THE appearance of the first urban civilizations, mankind had already brought into being a considerable array of techniques, instruments, and skills. The men of the old stone age had developed a wide variety of tools for the fashioning of materials and of weapons for the catching of prey, whilst the men of the new stone age were responsible for the most important innovation of settled agriculture, perhaps as recently as the sixth millennium B.C. The agriculture of the stone age quickly exhausted the fertility of the land, a factor which seems to have limited the size and stability of most of the neolithic communities. Such a limitation obtained to a lesser degree in the valleys of the Indus, the Tigris-Euphrates, and the Nile, where the natural flooding of the rivers deposited a fresh layer of fertile silt annually. More settled communities flourished in these valleys and, by the drainage of marshland and the irrigation of desert, they extended considerably the areas under permanent cultivation. These communities grew in size from villages to towns, and from towns to cities, whilst an administrative system governed by priests was evolved to organize the complex activities of their way of life. Thus, within a century or so before 3000 B.C., appeared the first urban civilizations in the valleys of the Tigris-Euphrates and the Nile.

The craft techniques of the first Middle East civilizations were markedly in advance of those of the earlier neolithic communities. During the fourth millennium B.C. the Sumerians of the Tigris-Euphrates valley harnessed domestic animals to the newly discovered plough, passing from the plot cultivation of neolithic man to large-scale field agriculture. They constructed animal-drawn wheeled vehicles, built ships, and used the potter's wheel to make baked ceramic ware. By about 3000 B.C. the Sumerians had already reached the peak achievement of bronze-age metallurgy. They were aware that copper could be obtained by the reduction of certain mineral ores in a fire, that it could be melted and cast into various shapes, and that copper could be alloyed with tin to give the harder and more fusible bronze. The equipment of the Egyptians was similar, though they did not use bronze or wheeled vehicles until they were invaded by the barbarian Hyksos about 1750 B.C.

The produce of these technical arts was controlled and distributed through an organization governed by priestly scribes. The resources with which the priests dealt were large and varied, and it seems that they could not keep account of them by memory alone. Accordingly, the priests made permanent records of the produce that passed through their hands by making imprints on clay tablets which were then baked and preserved for future reference. The imprints consisted of numbers and of abbreviated pictures of the produce enumerated, the tablets on which they were inscribed giving us the first recorded numeral system and pictographic script. The earliest records of the Sumerians, dating from about 3000 B.C. deal only with accounts of produce taken into the temple storerooms or handed out. Later, both the numeral system and the pictographic script became conventionalized, and a literature was developed dealing with mathematics, astronomy, medicine, history, mythology, and religion.

In the bronze-age civilization the picture-signs of the early records were simplified into ideogram forms, and arbitrary representations were made of things which could not be depicted. Such a script still exists in China, the number of Chinese ideograms multiplying with the development of the language. In Sumer the number of symbols was reduced by making an ideogram stand not only for the object depicted but also for the sound of its name. In this way a combination of ideograms came to stand for a complex word, or a phrase, rendering a number of symbols superfluous. Some two thousand signs were used in the earliest Sumerian records, but by 2500 B.C. the number had been cut down to about six hundred. At the same time the signs were further simplified, and ultimately they were made into combinations of wedge-shaped impressions, giving the cuneiform script. The Semitic Akkadians to the north transcribed their language into the cuneiform script phonetically, a practice which was imitated by every race which established its rule in Mesopotamia until the time of the Greeks.

In Sumer numbers were represented at first by imprints in clay made with a reed. The numerals up to ten were denoted by the requisite number of imprints made with the reed held obliquely; tens and multiples of ten were represented by imprints made with the reed held vertically. Side by side with this decimal system there was a notation based upon the number sixty. A small reed was used to designate units and tens, while a large reed was used obliquely to represent units of sixty, and vertically to represent units of six hundred. By about 2500 B.C. the decimal system was no longer in use, and the reeds were replaced by the wedge-shaped stylus used in writing the cuneiform script. A single vertical imprint was used to designate any power of sixty—1, 60, 3,600 and so on—whilst two such imprints made at an angle to form an arrow-head impression represented 10, 600, 36,000, and so on. The particular value which these signs denoted was determined by their position or place in a given number, as in our Indo-Arabic numeral system.

Before 2500 B.C. the Sumerians had drawn up multiplication tables which they used for the determination of the areas of fields, by multiplying length times breadth, and the estimation of the volumes of such things as brick stacks, by multiplying together the length, breadth, and height. In calculating the area of a circle and the volume of a cylinder they took the value of π as three, probably determining the value by direct measurement and rounding it off to a whole number for simplicity.

By 2000 B.C. the Sumerians had been conquered and had disappeared from history, only their language and their script living on as the vehicles of learning and religious ceremony, like Latin in the middle ages. The ruling power in Mesopotamia was then the Semitic Hammurabi dynasty of Babylon which set up temple schools for the training of priestly administrators. Here mathematics were developed further, fractions being designated in the same way as the Sumerians had represented integers. The vertical imprint made with a stylus now stood not only for 1, 60, 3,600, and so on, but also for $\frac{1}{60}$, $\frac{1}{3,600}$, etc., whilst the arrow-head impression stood for $\frac{1}{10}$, $\frac{1}{600}$, and so on, as well as 10, 600, etc. Other fractions were resolved into these sexigesimal fraction units, fractions, such as $\frac{1}{7}$ or $\frac{1}{11}$ which could not be so resolved, being either approximated to or ignored.

With such fractions, the operation of division was now carried out by multiplying the number to be divided by the reciprocal of the divisor. Tables of reciprocals were drawn up for the purpose, though the places for awkward numbers like $\frac{1}{7}$ were left blank. Tables of the squares of numbers, their square roots, and cube roots, were also drawn up, and the tables were used to solve problems involving quadratic and cubic equations. In the field of geometry the Babylonians were aware that all triangles inscribed in a semi-circle were right-angled, and they knew of the so-termed theorem of Pythagoras concerning right-angled triangles, not only for particular cases, but in full generality. The geometry of the Babylonians, like their arithmetic, possessed a marked algebraic character and was invariably expressed in the form of particular examples. Their methods indicate that the Babylonians were aware of several general algebraic rules, but they formulated mathematical problems only with specific numerical values for the coefficients of the equations.

Generally speaking the Egyptians achieved less in the field of mathematics than the Babylonians. The Egyptians possessed a better value for π, namely $256/81$, but they could solve only simple linear equations, and they were not acquainted with the properties of right-angled triangles known to the Babylonians. The Egyptians did not discover the device of making the position of a numeral determine its value in a number nor did they reduce fractions to conventional bases, a procedure which went with the place-value system of the Babylonians. From about 3000 B.C. the Egyptians used a numeral system based upon the number ten, units being represented by a stroke which was repeated to denote numbers up to nine, whilst ten, a hundred, and a thousand, were designated by further symbols which were repeated to represent multiples of these numbers. With such a notation, computations were difficult, as with the similar Roman numeral system.

In the field of astronomy, the Egyptians again accomplished less than the Babylonians. Perhaps because of their astrological beliefs, the Babylonians were close observers of the heavens, leaving numerous astronomical records behind them. No records of Egyptian observational astronomy survive, but from inscriptions and from pictures of the heavens painted on coffin lids, we find that the Egyptians divided the stars of the equatorial belt of the skies into thirty-six groups. Each star group, when it rose above the horizon just before dawn, indicated the beginning of one of the ten-day periods into which the year was divided. The Egyptians were aware that a year was approximately three hundred and sixty-five days as they added an extra five days to the thirty-six ten-day periods. Such a discovery may have been made by averaging out over several years the number of days separating the regular annual flood of the Nile, with which the Egyptian new year started. About 2700 B.C. the Egyptians regulated their calendar by the rising of the star Sotkis, our Sirius, which rose just before dawn about the time of the Nile flood.

The Babylonians did not have an official year. They based their measurement of time upon the lunar month, adding extra months from time to time to keep their calendar in step with the seasonal agricultural festivals. About 2000 B.C. the Babylonian year consisted of three hundred and sixty days, divided into twelve months of thirty days each. Besides the month, the Babylonians gave us the week as another unit of time, naming the days after the sun, the moon, and the five planets. They were responsible also for the division of the day into twelve double hours, and of the hour into sexigesimal minutes and seconds. Furthermore the Babylonians gave the names we use to the constellations of the stars, and they mapped those in the equatorial belt through which the sun passes, the zodiac, into twelve groups corresponding to the months.

The most accurate of the astronomical observations made in Mesopotamia concerned the motions of the planets. As early as 2000 B.C. it was noted that Venus returned to the same position five times in eight years. From about 1000 B.C. the observations of the Mesopotamians became comparatively accurate, and from 700 B.C. such observations were systematically recorded. They were then able to calculate the correct average values of the main periodic phenomena of the heavens, such as the periods of the planetary revolutions, and to make quite good predictions of astronomical events. The

Mesopotamians found, for example, that lunar eclipses occur every eighteen years, the so-termed Saronic cycle. Moreover, in the fourth century B.C. they developed an algebraic method of analysing the complex periodic phenomena of the heavens into a number of simple periodic effects. They found, for example, that on the average the lunar month was twenty-nine and a quarter days, and that the deviations from this average were regular and periodic too. This method, when put into a geometrical form by the Greeks, served as the main method of analysing the motions of the heavenly bodies down to moddern times.

The inhabitants of Mesopotamia did not use geometrical methods to interpret their astonomical observations until Greek times, so that their cosmological views as to the spatial features of the universe remained separate from their science. At first the Mesopotamians considered that the earth and the heavens were two flat discs supported by water, though later the heavens were thought of as a hemispherical vault resting on the waters surrounding the flat disc of the earth. Above the vault were more waters, and beyond the waters was the dwelling of the gods. The sun and the other heavenly bodies were such gods who emerged from their dwelling daily and traced finite orbits over the immobile vault. The gods controlled terrestrial affairs, and thus the motions of the heavenly bodies were taken to be indications of the destiny that the gods were to mete out to man on earth.

Conceptions in ancient Egypt as to the structure of the universe were not dissimilar. Here the world appeared to be a rectangular box, the earth at the bottom being slightly concave, and the sky at the top, supported by the peaks of four mountains at the corners of the earth, being flat or somewhat vaulted. The Nile flowed down the middle of the earth, branching off at the south from a universal river that flowed round the earth. The level of the universal river was not much below that of the mountain peaks supporting the sky, and thus it served to carry the boat of the sun-god on his daily journey across the sky. The boat always kept as near as possible to the earth bank of the universal river, and thus, in the time of the Nile flood, it could come nearer to the earth than in the winter, hence accounting for the seasonal changes in the position of the sun.

These worlds were thought to have come into existence from a primeval chaos of waters. The heavens, earth, air, and other natural objects and forces personalized as gods, were considered to have come into being through the union of the male and the female gods of the chaos. The younger natural forces, or gods, then continued the work of ordering the universe by magic spells or words of command. In the Mesopotamian creation myths there is a later stage in which the younger gods used physical force to tame nature, battling against the older gods of the chaos. Such creation myths appear to have been bardic memories of the creation of the ancient civilizations. First, the reclamation of land from water by primitive tribal communities in which the power of procreation seemed all important. Then the continuance of the work with a more elaborate organization controlled by priests who governed by word of command. Finally in Mesopotamia there arose the city governors, or battle kings, who governed by force of arms in the period when the cities fought for land. The Egyptians did not conceive of battle or physical force as a world-ordering process; their gods were powerful without being violent. The rule of kings and the control of the forces of nature were more certain in Egypt than in Mesopotamia. The Nile rose with predictable regularity each year, whilst the floods of the Tigris-Euphrates were uncertain, and were feared rather than welcomed; they were chaos reasserting itself. The dynasties of Egypt were long-lived, their government following a definite legal code, whilst the dynasties of Mesopotamia were transient, their rule being uncertain and arbitrary. In Egypt the future seemed certain and predictable; in Mesopotamia the future was less sure, requiring occult methods of divination to ascertain its course, such as the practice of astrology and the inspection of sacrificial livers.

Whilst the Mesopotamians excelled in the field of astronomy, perhaps because of their preoccupation with astrology, the Egyptians were the more able in the field of medicine. There are no cuneiform medical texts earlier than the seventh century B.C., but the Egyptian medical papyri go back to 2000 B.C., and contain earlier material of the time of Imhotep, who was

physician and vizier to King Zoser c. 2980 B.C. Imhotep traditionally was the founder of Egyptian medicine and in later
times he was accorded divine status, being regarded as the
patron deity of medicine. A medical profession presumably
existed in ancient Mesopotamia, for a legal code promulgated
by Hammurabi of Babylon about 2000 B.C. prescribed a fee
of 2-10 shekels for a successful surgical operation (a craftsman receiving about 10 shekels a year at the time), whilst an
unsuccessful surgeon was to have his hands cut off.

In the medicals texts of both the Egyptians and the later Mesopotamians the 'demon' theory of disease was prevalent. The
illness itself was personified as an evil spirit, which the physician sought to expel from the patient by the use of emetics
and purges or revolting medicaments to put the demon to flight.
The earliest Egyptian medical texts consist largely of lists of
drug recipes for the practising physician, different diseases
being vaguely referred to rather than described in any detail.
The Ebers medical papyrus, dating from about 1600 B.C.,
differs from others in that it gives descriptions of some forty-
seven diseases, giving the symptoms of the case, followed by a
diagnosis and a prescription. Later texts are more magical in
character. They describe how the cause of an illness was ascertained by the examination of omens, and how the patient
was treated by drawing off the demon of the disease into a
statuette of dough, which was then burnt, or into an animal, an
ointment, or amulet.

There are no Egyptian or Mesopotamian treatises on anatomy or physiology, though the Egyptians must have acquired
a knowledge of anatomy from the practice of mummification.
Even so the Egyptian hieroglyphic signs for bodily organs are
derived from animal, not human, anatomy, indicating that
there was little contact between the physicians and the em-
balmers. Surgery too seems to have been a separate art. The
Edwin Smith surgical papyrus, dating from about 1700 B.C.,
consists of a series of descriptions of injuries, starting with the
head and moving downwards. The instructions it gives are entirely practical, apart from some magical formulae added
later. The cases are classified under one of three verdicts,
favourable, uncertain, or unfavourable. The last type of case

was incurable, and was 'not to be treated', a judgement found in no other ancient medical text.

A notable feature of the ancient literature is the absence of chemical texts until comparatively late times. Mineral compounds are mentioned as drugs in the medical papyri, but we do not find specifically chemical papyri until the third and fourth centuries A.D. when alchemy appeared at Alexandria. Mesopotamian chemical texts go back to the Assyrian period, about the seventh century B.C., and there is one cuneiform tablet of the seventeenth century B.C. dealing with the making of a lead glaze coloured with copper. It is a very cryptic text, using Sumerian ideograms as puns upon the equivalent Semitic words, which has been taken as indicating that the lead glaze was a recent and unusual innovation which had to be preserved yet kept a secret. The later Assyrian chemical texts are comparatively straightforward, though rituals involving blood and embryos are mentioned as requisite to the metallurgist's art, indicating perhaps that the production of metals was considered to be a kind of birth.

It would seem that the priestly scribes of Mesopotamia and Egypt recorded mainly those disciplines which they had developed themselves during the course of their duties: mathematics for the purpose of keeping accounts and surveying, astronomy for calendar making and astrological prognostication, medicine for curing disease and driving away evil spirits. Until later times they rarely recorded a knowledge of the chemical arts, metallurgy, dyeing, and so on, which belonged to another tradition, that of the craftsmen who handed on their experience orally. The rift between the clerical and the craft traditions was noted at the time. In an Egyptian papyrus of about 1100 B.C. a father advises his son: 'Put writing in your heart that you may protect yourself from hard labour of any kind, and be a magistrate of high repute. The scribe is released from all manual tasks; it is he who commands. I have seen the metal worker at his task at the mouth of his furnace, with fingers like a crocodile. He stank worse than fish-spawn. I have not seen a blacksmith on a commission, a founder who goes on an embassy.'

Perhaps because of the lack of contact between them, both the craft and the clerical traditions became stationary, producing little that was new during the high bronze age. The priestly scribes came to depend more and more upon the written word of their predecessors, valuing their texts the more highly the older they were. Such a scholastic attitude of mind was not conducive to the development of new discoveries, and in fact few inventions appeared during the high bronze age. From the scribes of Babylonia came the extension of place-value to fractions about 2000 B.C., and from the craftsmen of Egypt the discovery of glassmaking about 1600 B.C., but the important innovations of the period, the discovery of the smelting of iron, and the evolution of an alphabetic script, came from peoples at the periphery of bronze-age civilization. An efficient method of smelting iron was developed by the Kizwanda tribe in the Armenian mountains during the second millennium B.C., the method beginning to spread after 1400 B.C., and becoming generally known after 1100 B.C. An alphabet, or rather two alphabets, were evolved by the Phoenicians about 1300 B.C., one from the Babylonian cuneiform characters, and the other from the Egyptian hieroglyphic script. The second alphabet was the more convenient for writing on papyrus, and it appears to have been the parent of the subsequent Indo-European and Semitic scripts.

The widespread use of iron and the alphabetic scripts were solvent forces in the ancient bronze-age civilizations. With the alphabet, men outside of the clerical corporations could read and write, for craftsmen have left their names on their wares. Iron was more plentiful than bronze, and now even ploughshares could be made of iron where formerly they were made of wood. Iron weapons gave the barbarians, like the Greek tribes, the force to conquer the bronze-age cultures and set up new civilizations in their stead.

Bronze-age civilization had spread from the valleys of the Nile and the Tigris-Euphrates until it had covered the whole of Asia Minor and much of the Middle East from mainland Greece to northern Iran. Babylonia was occupied by a succession of invaders during the first millennium B.C., while Egypt fell to its own mercenaries, and finally both were incorporated into the empire of the Persians by 500 B.C. Similarly, on the mainland of Greece the bronze-age Mycenaean civilization fell to the then barbarian tribes of classical Greece.

Chapter 3

The Natural Philosophies of the Pre-Socratic Greeks

THE NEW opportunities afforded by the introduction of iron and alphabetic writing were exploited to the greatest advantage by those communities who took to sea-going commerce, or those who emerged directly from barbarism into the iron age and thus by-passed to some degree the traditions of bronze-age civilization. The Etruscans and the Phoenicians, who pushed westwards from Asia Minor and the Middle East to Italy and North Africa, were seafarers, but they perpetuated some of the bronze-age traditions of their homeland, such as the practice of inspecting the livers of sacrificial animals for divination purposes. The Romans and the Hebrews came to civilization during the iron age, but they were primarily agriculturalists, not seafarers, and they did not make notable contributions in the field of science.

The Greeks alone were a people who came to iron-age civilization directly from barbarism and who took to sea-going commerce from the start. The Greeks had the traveller's feel for space, the sense of geometry, which was lacking in the settled agricultural communities of pre-Greek times and later ones, such as the civilization of China, which were cut off from Greek thought. They possessed also the traveller's knowledge of a variety of cultures and traditions, which allowed them to pick out what was valuable from each without rigidly following any particular one.

The achievements attributed to the first of the Greek natural philosophers, Thales of Miletus, c. 625-545 B.C., exemplify these things. He is said to have been a merchant who travelled to Egypt, where he obtained a knowledge of geometry, and to Mesopotamia, where he studied astronomy. He is credited with the prediction of a solar eclipse, though such forecasts were not possible at that time, and with having proved that a circle is bisected by its diameter, a proposition long known as a fact. Thales doubtless came across the creation stories of the

Babylonians and the Egyptians, in both of which water featured as the primeval chaos, for he supposed that all things came from water in the beginning. The earth, he supposed, was a cylinder or a disc with waters below, on which it floated, and with waters above, from which the rains came.

In the philosophies of Thales and the other Ionian Greeks nature became more impersonal than it had been in the bronze-age cosmologies. The pre-Socratic Greek philosophers tended to remove the gods from nature, supposing that the heavenly bodies were solid material objects, not powerful personalized beings. In a complementary way, their approximate contemporaries, the Hebrew, Amos, the Persian, Zoroaster, and the Indian, Buddha, separated their gods from nature. These religious reformers minimized the roles assigned to the gods in the bronze-age civilizations, the tasks of making rain and providing a good harvest, and indicated that the gods were concerned primarily with the spiritual welfare of man. Hence the old gods became more abstract and spiritual, just as the world of the Greeks became more impersonal and material.

The Babylonians and the Egyptians had conceived of water, then air and earth, as the primary constituents of the world. The second Miletian philosopher, Anaximander, c. 611-547 B.C., added a fourth element, that of fire, and conceived of a primal substance anterior to the elements. After their formation from the primal substance, the four elements had separated out as strata in the order, earth, water, air, and fire. The fire evaporated the water, producing dry land, and the water vapour arose to enclose the fire in circular tubes of mist. What appeared to be the heavenly bodies were holes in these tubes, the holes permitting us to see the inner fires. The diameter of the tube containing the sun was some twenty-seven times as large as the diameter of the earth and that of the tube containing the moon was eighteen times as large. The earth itself was a cylinder some three times higher than it was broad. The heavens were concentric round the earth, 'like the bark of a tree', the earth being poised at the centre, 'because of the equal distance from everything'. Anaximander believed that living organisms had arisen from elemental water and that the higher animals had developed from the lower: 'Living creatures rose from the moist element as it was evaporated by the sun. Man was like another animal, namely a fish, in the beginning.'

The third of the Miletian philosophers, Anaximenes, c. 550-475 B.C., took mist or air as his primordial substance and derived the other elements from it. By rarefaction mist became fire, for, he argued, when air is puffed from the mouth it appears to be warm whilst it seems cold when blown out under pressure. Similarly, by a process of condensation, mist became first water and then earth. The differences between the elements were therefore quantitative, the elements being mist condensed or rarefied to varying degrees.

The analogies used by the Miletian philosophers to explain the structure and the workings of the world differ markedly from those employed in the creation stories of the Egyptians and the Babylonians. The Greeks regarded neither organic procreation nor the magic word of command as world-building principles, relying more upon analogies based upon craft processes. Anaximenes compared the process of element formation to that of felt-making: 'Cloud is formed from air by felting, and this still further condensed becomes water.' Anaximander regarded the formation of the world as a sort of cooking process with fire as the active agent. The inclusion of fire amongst the four elements in itself indicates that the philosophers were interested in the arts, for the use of fire was the prerogative of the cook, metallurgist, and potter.

The pre-Socratic philosophers used one analogy in common with the Babylonians, namely, that of human trespass and retribution. The Babylonians had explained the flaking properties of flint as a retribution imposed by the gods upon flint-stones for committing a transgression against them. In the same way Anaximander regarded the interconversion of the elements and the production of objects from them as processes of trespass and retribution: 'And into that from which things take their rise, they pass away once more, as is meet, for they make reparation and satisfaction to one another according to the order of time.' Thus in winter, cold commits an injustice to heat, and in summer, heat exacts its retribution. All things are transient, for when an object comes into being, it commits an injustice against things already existing and a reparation must be made. The notion that there was a principle of retribution in natural processes was derived by analogy from the customs of human society in which the practice of vengeance preceded that of the due process of law. Thus the early meaning of the Greek word for *cause*, 'aitia', was *guilt*. Such a notion was replaced ultimately by the conception that nature, like human society, was governed by laws.

The idea of retribution was used extensively as a principle explaining the world order by Heraclitus of Ephesus, *c.* 550-475 B.C. He supposed that such a principle governed the motions of the heavenly bodies, the interconversion of the elements, and the processes of nature generally. Retribution was a principle of change, and it was with processes of change in nature, rather than the structural features of the world, that Heraclitus concerned himself. He was of the view that fire was the origin and image of all things, the burning flame symbolizing the universal flux and change in nature. Fire was the common substratum of all things, and so, behind the qualitative changes in nature, which were governed by the principle of retribution, there was a quantitative continuity of substance, governed by principles analogous to those employed in commercial transactions: 'All things are an exchange for fire, and fire for all things, even as wares for gold, and gold for wares.'

Such commercial analogies, with their emphasis upon the quantitative features of phenomena, contributed also to the systems of the Pythagoreans and the Atomists, who were of the view that number-units or discrete particles were the substratum of the universe, just as the quantitative units of coined money provided the basis of commerce. To the Pythagorean, Philolaus of Tarentum, *c.* 480-400 B.C., was attributed the saying: 'You can see the power of number exercising itself not only in the affairs of demons and of gods, but in all the acts and thoughts of men, in all handicrafts, and in music.'

Pythagoras, *c.* 582-500 B.C., was a native of Samos, but he left his birthplace for the Greek colony of Croton in southern Italy where he founded a brotherhood devoted to a life of mathematical speculation and religious contemplation. Men and women were admitted on equal terms to the association in which all property was held in common. Even their mathematical discoveries were considered to be common property within the brotherhood, though they were preserved as secret mysteries from outsiders. However, the Pythagoreans split up into a scientific and a religious wing in the fifth century B.C., and Philolaus of Tarentum, representing the scientific party, made known the views of the Pythagoreans.

For the Pythagoreans numbers provided a conceptual model of the universe, quantities and shapes determining the forms of all natural objects. At first they thought of numbers as geometrical, physical, and arithmetical entities made up of unit points or particles. They arranged such unit points at the corners of various geometrical figures and spoke of them as triangular numbers, square numbers, and so on. Thus, for the Pythagoreans, numbers had a geometrical shape as well as a quantitative size, and it was in this sense that they understood numbers to be the forms and images of natural objects.

In mathematics the Pythagoreans found that their views were self-contradictory. The geometrical theorem which carries their name indicated that the sum of the squares on the sides of a triangle which contained a right-angle was equal to the square on the hypotenuse. In some such triangles the length of the hypotenuse was found to be incommensurable, as in the simplest case of a right-angled isosceles triangle where the hypotenuse is $\sqrt{2}$ times the length of one of the other sides. The Pythagoreans showed that $\sqrt{2}$ could be expressed neither by a whole number nor by a combination of whole numbers, and, holding an atomic view of quantity and number, they found themselves at an impasse. The dilemma was resolved in two ways, the one mathematical, the other physical. In mathematics, arithmetic was largely abandoned and attention was concentrated upon geometry, as $\sqrt{2}$ could be represented by a line of definite length if not by a definite number of unit points. In physics, the Atomists did away with the numerical aspects of Pythagorean unit-points and speculated upon their physical properties.

According to the Pythagoreans, the universe was divided into three parts. In order of increasing nobility and perfection, these were the Uranos, or the earth and the sublunary sphere, the Cosmos, or the movable heavens bounded by the sphere of the fixed stars, and the Olympos, or the home of the gods. The earth, the heavenly bodies, and the universe as a whole, were spherical as the sphere was the most perfect of the geometrical solids. The various bodies within the universe moved with a uniform circular motion, as the circle was the perfect geometrical figure, and their motions were such that they moved the more slowly the more noble and divine their status. The axiom that the motions of the heavenly bodies must be uniform and circular, a principle which was to dominate astronomical science down to modern times, was explained by a later Greek astronomer, Geminus of Rhodes, c. 70 B.C. in the following way:

'It was the Pythagoreans, the first to approach these questions, who laid down the hypothesis of a circular and uniform motion for the sun, moon, and planets. Their view was that, in regard of divine and eternal things, a supposition of such disorder as that these bodies should move now more quickly and now more slowly, or should even stop as in what are called the stations of the planets, is inadmissible. Even in the human sphere such irregularity is incompatible with the orderly procedure of a gentleman. And even if the crude necessities of life often impose upon men occasions of haste or loitering, it is not to be supposed that such occasions inhere in the incorruptible nature of the stars. For this reason they defined their problem as the explanation of the phenomena on the hypothesis of circular and uniform motion.'

The Pythagoreans felt it aesthetically objectionable that the sun, moon, and planets should move round the heavens from west to east in their own periodic times, and that they should rotate simultaneously in the opposite direction once in every twenty-four hours. To overcome the objection, and to satisfy the principle that base bodies moved more quickly than the noble, Philolaus suggested that the earth moved from west to east in an orbit round a fire at the centre of the universe once a day. Its motions were such that the same side of the earth was always presented to the central fire, just as the same side of the moon was constantly presented to the earth. Greece was on the side of the earth away from the central fire, but even on the other side of the earth the fire could not be seen as there was another body between the earth and the fire, the counter-earth, which kept pace with the earth and always obscured the fire.

The daily motion of the earth round the central fire, postulated by the Pythagoreans, explained the apparent diurnal rotation of the heavens about the earth, and indicated that all of the movable bodies in the universe moved round the central fire in the same west-to-east direction, their periods of revolution increasing the more noble they became. The earth, the basest body in the universe, moved round the central fire once a day, the moon took a month, the sun a year, and the planets longer periods still, whilst the sphere of the fixed stars was stationary. Such a view required that, as the earth traced out a finite course daily, the fixed stars should change their apparent positions relative to one another between dusk and dawn, unless they were an infinite distance away from the earth. The Pythagoreans held that the distances between the heavenly bodies and the central fire were in the same numerical ratios as the intervals of the musical scale, a relation which placed the stars at a finite distance from the earth. However, stellar parallax, a shift in the relative apparent positions of the stars, was not observed, and so the original Pythagorean system of the world had to be modified. The absence of stellar parallax implied that the daily orbit of the earth round the central fire was much smaller than had been supposed previously. Following up the implication, two Pythagoreans, Hicetus and Ecphantus, both of Syracuse, assumed that the earth was at the centre of the universe and that it rotated daily upon its axis, a view which accounted for the absence of stellar parallax and which saved the main outlines of the Pythagorean theory.

The Pythagorean schools of southern Italy also had their biologists and anatomists. The first of note was Alcmaeon of Croton, who flourished c. 500 B.C. He is said to have discovered by dissection the optic nerves connecting the eyes with the brain and the Eustachian tubes linking the ears with the mouth. Alcmaeon held that man and the universe as a whole were built upon the same plan, man being a microcosmic copy of the macrocosm in its entirety. Such a view became a settled part of the Pythagorean teachings.

Another philosopher influenced by the Pythagoreans was Empedocles of Agrigentum in Sicily, who lived c. 500-430

B.C. Empedocles held that all things were composed of quanti-tatively different proportions of the four elements, just as all colours were compounded by the artist from different amounts of four pigments. The agencies which combined the elements and separated them from their compounds were Love and Strife, both forces being inherent in the very nature of the elements. Water was allied to earth because they had the quality of coldness in common, yet they were opposed at the same time for water possessed the quality of wetness and earth that of dryness. Similarly air and fire were both hot, but the one was moist whilst the other was dry. Again, water and air, and earth and fire, possessed both a common quality and opposing qualities.

Empedocles had a curious theory of organic evolution which illustrates how the Pythagoreans tended to think in terms of units even though such units did not necessarily have a numeri-cal significance. He supposed that in the beginning there were various unit-parts of animals and man, eyes, legs, arms, heads, etc., wandering about alone. Chance combinations of the unit parts were formed by attraction or Love, giving all kinds of monstrous creatures as well as the present forms. Some had many legs, others had the body of an ox and the head of a man, but these were not viable and only the creatures with the attributes necessary for survival were left:

'Many races of living creatures must have been unable to continue their breed: for in the case of every species that now exists, either craft, or courage, or speed, has from the begin-ning of its existence protected and preserved it.'

The Atomists carried the unit conception of nature further, extending it from the organic to the physical world. The early Atomists were Leucippus of Miletus, *fl.c.* 440 B.C., and Demo-critus of Abdera, *fl.c.* 420 B.C., their contributions being in-separable. They believed that everything in the universe was composed of atoms, which were physically indivisible. There were an infinite number of atoms, and they moved perpetually in an infinite void. They had existed from eternity, for they had not been created, and could not be destroyed. Atoms differed as to their size, shape, and perhaps their weight. By their motions in the void, the atoms had set up a whirlpool motion, which drove the larger atoms to a centre where they formed the earth. Such large earth atoms interlocked so that their motions became restricted; they could only throb or oscillate. The finer atoms of water, air, and fire, were driven out into space, where they generated a vortex motion round the earth. Large atoms outside of the earth came together to form moist lumps, which were dried and ignited by their motion through the vortex. These were the heavenly bodies. As the number of atoms, and the space in which they moved, were infinite, there were many worlds, some coming into being, others passing away. Some had no sun or moon, others had several of each, but all had a beginning and an end. Like most of the Ionian philosophers, Democritus considered the earth to be a cylinder, not a sphere as the Pythagoreans thought.

The Atomists supposed that life had developed out of a primeval slime, man as well as animals and plants. Man was a microcosm of the universe, for he contained every kind of atom. Life and the soul were akin to fire, for they all consisted of small spherical atoms. Such atoms were constantly given off from the body, and were constantly taken in with the air. Thus when respiration ceased, life departed. The cosmology of the Atomists was almost entirely mechanistic; all things were pre-determined—'By necessity were foreordained all things that were, and are, and are to be.' They did not use the analogies of human purposes, love and strife, or the principle of retribution, to explain the workings of the world. In this connection it is interesting to note that, contemporary with Democritus, Protagoras of Abdera put forward the view that wrongdoers should be punished, not to exact vengeance, but to deter further crimes.

In early Greek medicine there were three main strands. First, and perhaps earliest, there was the medicine of the temples dedicated to Aesculapius, the god of medicine. Secondly, there was the philosophical Pythagorean school of southern Italy, and thirdly, there was the more practical Ionian school of Hippocrates. The Hippocratic writings are the earliest of the Greek medical works, dating from the fourth century B.C. The writings were the product of a school rather than of any one man, though Hippocrates of Cos, c. 460-377 B.C., was an outstanding figure. The school regarded medicine as an art or technique, rather than a theoretical science as the Pythagoreans did, though they developed theories of their own. They originated the doctrine that the human body contains four humours, or juices, the melancholic, the sanguineous, the choleric and the phlegmatic, the correct proportions of which were indispensable for health, illness arising from an excess of any one of them. The theory seems to have been based on the observation that four substances may be obtained from blood, a dark clot, representing the melancholic humour, a red fluid, equivalent to the sanguine humour, a yellow serum, or the choleric humour, and fibrin, which was connected with phlegm.

Chapter 4

Natural Philosophy in Athens

ATHENS DID not flourish at so early a date as the Greek towns of Ionia and southern Italy, though her culture proved to be more stable and enduring. The Ionian towns were subjugated by Persia in 530 B.C., Miletus being entirely destroyed a few years later. Athens profited indirectly by the eclipse of the Ionians, for the trade with the Greek colonies round the coast of the Black Sea fell into her hands. Politically Athens commanded the leadership of the Greek cities against the Persians, who were defeated on land at Marathon in 490 B.C., and at sea ten years later. The crafts at Athens flourished, particularly from the time of Solon, c. 639-559 B.C., who decreed, according to Plutarch, that a son need not support his father unless the father had taught him a trade. This was the period when the Greek inventors are said to have lived; such men as Anarcharsis, the Scythian, to whom the invention of the bellows, and the improvement of the anchor and the potter's wheel were ascribed, the Theodorus of Samos, who was credited with the invention of the level, the set square, the lathe, the ruler and the key. It was a time too, when the Greek word 'sophia' still meant technical skill, and not intellectual wisdom.

After her victory over the Persians, Athens entered upon her period of prosperity and greatness. Pericles brought to Athens Anaxagoras, 488-428 B.C., a pupil of the last Miletian philosopher, Anaximenes, to add to the cultural life of the city. Anaxagoras was typically an Ionian philosopher, holding that the earth was a cylinder, and not a sphere as the Pythagoreans believed. He followed his school too in that he believed the heavenly bodies to be of much the same quality as the earth, they were not divine as the Pythagoreans held. Thales had thought that a lodestone possesses a soul because it could move a piece of iron, and Anaxagoras generalized this view, ascribing all motions to the operation of a mind or soul.

He was of the opinion that the sun was a piece of red-hot

35

stone, not much bigger than Greece. The moon and the planets too were like the earth, the moon possessing mountains and inhabitants. He was the first to suggest that the moon shone by reflected light, and the first to explain eclipses in terms of the moon's shadow falling on the earth, and the earth's shadow falling on the moon. For denying that the heavenly bodies were divine, Anaxagoras was prosecuted for impiety, and he was only saved by the intervention of Pericles.

His suggestion regarding the existence of mountains in the moon, indicates that he observed the heavens fairly closely. Towards the end of his life astronomical observations were made at Athens by Meton, who announced the discovery of the so-termed Metonic cycle at the Olympiad of 432 B.C. The Metonic cycle consisted of nineteen years, which contain a whole number of lunar months, so that the cycle could be used conveniently to regulate the calendar. The cycle was known to the contemporary inhabitants of Mesopotamia, who used it as their standard calendar cycle, but it was not adopted by the Greeks.

By the time of Meton, Athenian society had grown and differentiated in a way that had forced apart the craft and the philosophical traditions. Some two hundred years after Solon's measures to promote the crafts, Xenophon wrote that, 'What are called the mechanical arts carry a social stigma, and are rightly dishonoured in our cities.' Not only the mechanical arts, but the older natural philosophy, which used craft analogies, was dishonoured too. According to Xenophon, his teacher Socrates, 470-399 B.C., considered astronomy 'a waste of time'. During the lifetime of Socrates, Athens fell on evil days, the Peloponnesian wars of 431-404 B.C. resulting in the victory of the Spartans over the Athenians. For Socrates, therefore, the prime task of the philosopher was the ordering of man and human society, not the understanding or the control of nature. Natural philosophy he rejected, and he concerned himself primarily with problems of an ethical and political character.

The work of Socrates was continued by his pupil, Plato, 427-347 B.C., who was, however, more sophisticated than his master. Plato saw that any philosophy with a claim to generality must include a theory as to the nature of the universe. Such a theory could be subordinate to ethics, politics, and theology, and, if suitably framed, it could enhance their cogency. Plato accordingly evolved a natural philosophy that was harmonious with, and subordinate to, his political and theological views. Plutarch tells us that Plato removed the taint of atheism from astronomical studies by making 'natural laws subordinate to the authority of divine principles'.

Plato developed the view of the Pythagoreans that the heavenly bodies were divine and noble beings whose motions were perfectly uniform and circular. 'All of us Hellenes', he said, 'tell lies about those great gods, the sun and the moon. . . . We say that they, and diverse other stars, do not keep the same path, and we call them planets or wanderers.' On the contrary, he affirmed, 'each of them moves in the same path—not in many paths, but in one only, which is circular, and the varieties are only apparent.' Plato accordingly set his students the problem of finding the particular circular movements that would explain the apparent motions of the heavens. In so doing he did not wish to stimulate the observation of the heavens, on the contrary he desired only to make astronomy a branch of mathematics. 'Astronomy,' he wrote, 'like geometry, we shall pursue by the help of problems, and leave the starry heavens alone.' However, his students took to the observation of the heavens to obtain data for the calculations.

Plato, like his predecessors in Greece, Babylonia, and Egypt, held that the universe was an uncreated chaos in the beginning. The ordering of this chaos was not a mechanical process, as the Ionian philosophers had imagined, but a result of the activities of a supernatural being. Plato's God differed from the gods of the Bronze Age in that he did not order the universe by a process of organic procreation, nor by uttering words of command, but by realizing an intellectual design. The most important feature of the ordering of the universe from chaos according to Plato, was the *formulation* of a rational design for the world by the Creator; the mechanics of the process whereby the design was put into *practical operation* were ignored by Plato, or rather were taken for granted as something occurring naturally. Such a conception pervades Plato's general view of causation. Events occur primarily because rational purposes and designs are formulated by intelligent beings. The inner movement of nature, the flow of forces from cause to effect within events, was unimportant, and need only be considered in so far as it thwarted the achievement of intellectual purposes. Such a view of causality illustrates how far apart the craft and philosophical traditions now lay. The movement of everyday life appeared, to the philosopher, to depend primarily upon the purposes and designs he formulated; how these designs were carried into effect by the craftsman were hidden from him, and did not appear to contribute to the final result. However, the designs of the philosopher did not always come to fruition. The world possessed an autonomous movement of its own that sometimes thwarted human purposes. This apparent truculence of nature Plato designated 'Neccessity'. It was not the necessity of the Atomists, for its operations were chance and random, without order or design. Plato's idea of 'Necessity' was closely akin to the Greek conception of Fate, a kind of superhuman will and purpose, that thwarted lesser human purposes and designs. It was an expression of the limited degree of control that the Greeks exercised over the world in which they lived. Herodotus expressed this feeling well when he said that, 'Of all the sorrows that afflict mankind, the bitterest is this, that we should have consciousness of much, but control over nothing.'

Plato's view of the universe was essentially a mathematical one. In the beginning, he supposed there were two sorts of right-angled triangles—the one sort, half a square, and the other, half an equilateral triangle. From these triangles were rationally derived four of the regular solids which composed the particles of the four elements. The particles of fire were tetrahedra, those of air were octahedra, whilst those of water were icosahedra, and those of earth were cubes. The fifth regular solid composed of regular pentagons, the dodecahedron, formed the quintessence, the fifth element composing the material of the heavens. The universe as a whole was a sphere because the sphere is symmetrical and perfect, being the same at all points of its surface. The universe was also alive, possessing a soul throughout its space, and being alive it moved. The movement of the universe was a rotation because circular motion was the most perfect, requiring neither hands nor feet.

Each of the four elements was present in the universe to such an amount that the proportion of fire to air was the same as the proportion of air to water, and water to earth. All objects could be designated by a number expressing the proportions of the elements that they contained.

Of all the animals, man appeared first, and of all the parts of men, the head was first created, because it was the organ of the soul, and very nearly spherical. The other parts of the body appeared to prevent the head rolling about on its own, and they contained a lower soul, governing the animal desires of man. The other animals were produced by the degeneration of man, his soul transmigrating into lower bodily forms. Men who had lived badly, 'were suitably reborn as women in the second generation', whilst 'beasts who go on all fours came from men who were wholly unconversant with philosophy'. Poking fun at the Ionian philosophers, Plato held that, 'Birds sprang by a change of form from harmless but light-witted men who paid attention to things in the heavens but in their simplicity supposed that the surest evidence in these matters is that of the eye.' Specifically scoring off Anaximander, who had thought that man had come from fishes, Plato suggested that, 'The fourth kind of animal whose habitat is water came from the most utterly mindless men.' Much of this was lightly written, but it is indicative of Plato's way of thinking. In the same vein, he postulated the existence of an evil world spirit, a demon who was responsible for the views of his arch-opponents, the Atomists.

Plato's philosophy was most influential, though his important successors did not so much develop his views as diverge from them. Even his own pupils at the Academy which he founded, notably Eudoxus of Cnidus, 409-356 B.C., were forced to observe the movements of the heavenly bodies, a procedure which Plato had deplored, in order to explain those movements geometrically. Eudoxus was the first to unite quantitative astronomy with cosmological speculation, thus allowing observation to play a part in determining the configuration assigned to the universe. By the time of Eudoxus, the Babylonians had developed a method of resolving complex periodic phenomena of the heavens into a number of simple periodic movements. Eudoxus either heard of this procedure, or developed it independently, and transposed it from an arithmetical into a geometrical form. To each simple periodic movement he assigned a circle, or a spherical shell, that centred on the earth, a combination of such spheres describing reasonably well the complex periodic movement of a particular heavenly body. Each sphere accounted for a particular motion—one described the apparent daily rotation of the heavens, another the monthly, annual, or other period of revolution, and others yet further periodic phenomena. All of the spheres were homocentric, they centred on the earth, and the axes of rotation of the lower spheres were embedded in the surfaces of spheres further out. In this way Eudoxus explained the motions of the heavens using twenty-seven spheres, one for the fixed stars, three each for the sun and the moon, and four each for the five known planets. As observations multiplied and fresh periodic phenomena were discovered, the system had to be enlarged. The pupil of Eudoxus, Callipus, c. 325 B.C., gave each heavenly body an extra sphere, bringing the total up to thirty-four, whilst Aristotle added a further twenty-two spheres. However, the system of homocentric spheres entailed certain difficulties from the start. It demanded that the heavenly bodies should always remain the same distance from the earth, whilst it had long been known that such planets as Venus and Mars exhibited variations in apparent brightness which implied that these planets possessed motions of recession and approach relative to the earth. Moreover, eclipses of the sun had been observed to be sometimes total, sometimes annular, indicating that the relative distances of the sun and the moon from the earth varied. As the astronomer, Sosigenes, c. 45 B.C., observed:

'The spheres of the partisans of Eudoxus do not account for the phenomena. Not only do they not account for the phenomena which have been discovered after them, they do not account for the phenomena which were known before them, and which they themselves regarded as true.'

The achievement of the Eudoxian system was limited by the preconception that the heavenly bodies moved with uniform speed in circles which centred on the earth. An attempt was made to overcome some of the difficulties of the Eudoxian scheme by Heraclides of Pontus, *fl. c.* 373 B.C. It had been observed that the planets, Mercury and Venus, never moved far from the sun, and so Heraclides suggested that these planets moved in circular orbits round the sun, thus accounting for the changes in their apparent brightnesses. He also adopted the view of the Pythagoreans, Hicetas and Ecphantus, that the earth rotated on its axis daily in order to explain the apparent diurnal rotation of the heavens. Heraclides supposed further that the universe was infinite, each star being a world in itself, composed of an earth and other bodies. However, Heraclides found few supporters, the followers of both Plato and Aristotle adopting the Eudoxian system.

The works of Aristotle, 384-322 B.C., have been placed in a sequence which displays a growing divergence between the philosophies of Plato and Aristotle both in content and method. The early work of Aristotle on the nature of the heavens was speculative in method and Eudoxian in content whilst his later biological works were based more closely upon observation and contained much new material. In the field of astronomy Aristotle was responsible for the notion that the spheres which carried the heavenly bodies round their courses were real physical bodies, not mere geometrical constructions as Eudoxus had supposed. Each sphere passed on its motions to the sphere immediately below it so that the outermost sphere carrying the fixed stars, by rotating daily upon its axis, caused the diurnal rotation of all the spheres and the heavenly bodies attached to them with it. In order that the motions peculiar to one heavenly body should not be passed on to the body immediately below it, Aristotle inserted a number of 'unrolling spheres' between each set of spheres carrying a planet. These 'unrolling spheres' had the same axes of rotation, the same speeds, and were of the same number as the spheres which moved the planet above them, but they moved in an opposite direction so that they neutralized all the motions peculiar to that planet, only the diurnal rotation being passed on. Mathematically Aristotle's system was the same as that of Callipus, the twenty-two 'unrolling spheres' being geometrically redundant.

According to Aristotle, the outermost sphere of the fixed stars was moved by the Primum Mobile or the Unmoved Mover at the periphery of the universe which governed all the spheres and the universe as a whole. He seems to have suggested also that each of the other spheres had a lesser unmoved mover which was responsible for the particular motion of that sphere. The movers were spiritual in character, the relation of a mover to its sphere being akin to that of a soul to a body. The movers of a planet worked against the Primum Mobile so that the planets had their own west-to-east motions contrary to the diurnal rotation. The outermost planet, Saturn, had the greatest difficulty in overcoming the force of the Primum Mobile and so had the longest period of rotation, while the innermost body, the moon, had the shortest. Thus Aristotle, like Plato, arranged the heavenly bodies in order outwards from the earth according to their apparent periods of revolution, namely, the Moon, Sun, Venus, Mercury, Mars, Jupiter, and Saturn.

Aristotle supposed that there was an absolute difference in kind between the material of the heavens and terrestrial matter. All things below the sphere of the moon were made up of the four terrestrial elements, earth, water, air, and fire. The heavens were composed of a fifth and purer element, the quintessence. The heavenly bodies were incorruptible and eternal and so too were their motions which accordingly were circular and uniform. On earth there was generation and decay, and thus terrestrial motions were rectilinear, possessing a beginning and an end like all terrestrial phenomena. The heavenly bodies were always within their appointed spheres, but terrestrial bodies were not, and they strove continuously to return to their proper places. The elements of earth and water possessed gravity, tending to move towards the centre of the universe. Air and fire possessed a lightness which tended to take them upwards to their proper places in the upper atmosphere. Fire was a more noble element than air as it had a higher natural place. Similarly air was more noble than water, and water more noble than earth. All of the heavenly bodies were more noble than any terrestrial object, but they increased in perfection the further they were from the centre of the universe. The moon was the least perfect, as could be seen by its blotched appearance, whilst the sphere of the fixed stars and the Primum Mobile were the most perfect, for 'the more excellent is that which encompasses and is the limit than that which is completed'.

In the field of physics, Aristotle was of the view that a body could be maintained in motion only so long as it was in direct contact with a continuously operating mover. Should the mover cease, or lose contact the body would stop immediately. Such movers were either internal to bodies, as in the case of self-moving organisms, or external as when an object was pushed or pulled by outside forces. Homogeneous bodies could be set in motion only by external movers, for self-moving bodies were necessarily composite, consisting of a mover and that which was moved. Thus homogeneous bodies, such as a stone hurled from a catapult, never moved freely. When the stone left the catapult it was maintained in motion by the air which streamed in behind it to prevent the formation of a vacuum. According to Aristotle, a 'vacuum' could not exist as space must be filled with matter to transmit physical effects by direct contact in this way. Consequently the Atomists were wrong in supposing that the world consisted fundamentally of atoms in the void for space must be a continuum of matter. In the case of the composite self-moving entities the mover was considered to be more noble and spiritual than the body moved. Aristotle indicated that, 'In every composite thing there is always a ruling and a subject factor', an idea which linked up with Plato's view that, 'Nature ordains obedience and slavery for the body, command and mastery for the soul. The movement of a homogeneous body was akin to the motion of a cart which ceased if the cart were separated from the ox or if the ox suspended its tractive effort. The relation between the elements of a self-moving body was similar to that between the soul and the body or the relation between the philosopher and the craftsman in Athenian society at the time, the intellectual designs of the philosopher appearing to be the driving force of the craftsman's activity.

Aristotle, like Plato, considered that intellectual designs and purposes were the formative and guiding princples of all natural processes. However, Aristotle had a richer view of causality than Plato as he accepted also some of the doctrines expressed earlier upon the matter. There were, Aristotle indicated, four main types of cause. Firstly, there was the material cause of things, the primary matter out of which objects were made. Secondly, there were formal causes, the designs, patterns, and forms which were impressed upon the primary matter. There were, thirdly, efficient causes, providing the mechanisms whereby such designs were realized, and fourthly, final causes, which were the purposes for which objects were designed. Potter's clay provided the material cause of a vessel whilst the formal cause lay in its design. The potter's wheel and hands were the efficient cause, and the purposes for which the vessel was intended the final cause. Aristotle himself was mainly concerned with formal and final causes. Formal causes, he believed, were inherent within all natural objects and processes. At first they were latent, but such forms became manifest during the development of the object or the creature. Ultimately they arrived at a completion where the finished being served the purpose, or the final cause, for which it was designed.

In the biological works of Aristotle such views found their most complete expression. He classified some five hundred and forty animal species according to the gradation of their forms, and he must have dissected animals from at least fifty different species whilst studying their anatomical structures which, for Aristotle, were expressions of their formal causes or designs. Aristotle noted some correlations between the structures of animals, such as, 'No animal has, at the same time, both tusks and horns', and 'A single-hooved animal with two horns I have never seen.' Nature, he suggested, did nothing in vain, and so no animal required both horns and tusks for protection. Aristotle observed that ruminants had a multiple stomach and deficient teeth, and as 'nature invariably gives to one part what she subtracts from another', he supposed that ruminants had complex stomachs to compensate for their deficient teeth.

Aristotle investigated also the development of organic forms during the embryological growth of the chick and other animals, the degree of maturity of an animal at birth being an important criterion in his classificatory system. In this connection Aristotle pointed out that whales, which bear their young alive, are more akin to mammals than to fishes, which lay eggs. He produced other good correlations along these lines, as, for example, 'Four-footed beasts which produce their young alive have hair, four-footed beasts that lay eggs have scales.' The male and the female, according to Aristotle, contributed unequally to their offspring, the female providing the matter, the material cause, and the male the design or the formal cause. The male was the carpenter, he said, the female, the timber.

Aristotle was of the view that the various animal species formed a continuous scale of creatures increasing in perfection from plants up to man. There were eleven main grades of perfection, distinguished by embryological criteria. The highest animals were warm and moist creatures which bear their young alive. Others were moist without being warm, and they laid eggs which developed inside the female, as in the case of the shark. Warm and dry animals laid 'complete' eggs, like the birds, whilst cold and earthy animals, such as the frog, laid 'incomplete' eggs. In this way Aristotle drew up a hierarchy of creatures, 'graded according to the degree to which they are infected with potentiality', as expressed by their maturity at birth. The highest creature of one grade was directly contingent upon the lowest creature of the grade above, so that 'their continuity renders the boundary between them indistinguishable'.

Aristotle indicated that the grade of a creature's perfection was recognized by its structural form, but was not determined by that form, as the structure of an organism was governed by its habits and functions. Plants, which only grow and reproduce, possess simpler, and a smaller variety of structures than animals, which not only grow and reproduce but also move about and possess the faculty of sensation. According to Aristotle, the habits and functions of an organism, and thus its structure and its degree of perfection, were governed by the quality of its soul or souls. Plants had only a vegetative soul, responsible for growth and reproduction, whilst animals possessed in addition a sensitive soul which governed the faculties of self-movement and sensation. Man possessed not only vegetative and senstive souls but also a rational soul, the seat of which was the heart, not the brain as previous philosophers had taught.

Aristotle marks a turning point in the history of Greek

science, for he was the last to formulate a system of the world as a whole, and he was the first to embark upon extensive empirical enquiries. Previous philosophers had erected large bodies of theory upon slender empirical foundations, a trend which Aristotle had followed in his early work on astronomical questions. His later zoological works embodied a great deal of observation, and it was this trend which was developed by his successors. Aristotle had set up the Lyceum in opposition to Plato's Academy, and here he was succeeded by his pupil Theophrastus, 372-287 B.C., in 322 B.C. Theophrastus continued the biological work of his master, describing and classifying numerous species of plants. Many of the names given to plants by Theophrastus survive in modern botany, as also the technical terms which he introduced, such as *carpos* for fruit, and *pericarpion* for seed vessel. Theophrastus was aware that the reproduction of higher plants was of a sexual nature, though this knowledge was lost in later antiquity.

Reacting against the search for purposes and final causes in nature, Theophrastus maintained that efficient causes only were the concern of science. He suggested that the scientist should explain natural phenomena in terms of the processes observed in the mechanical arts, laying it down that 'we must in general proceed by making reference to the crafts, and drawing analogies between natural and artificial processes'. In this way Theophrastus explained the phenomenon of lightning by analogy with the sparks which could be struck from stones, and he suggested that the red colour of the sun at dawn and sunset was due to causes similar to those which gave rise to red and smoky flames when green logs were burnt.

Following Theophrastus, Strato of Lampsacus was head of the Lyceum from 287 to 269 B.C. Strato appears to have gone beyond observation to experiment. He weighed a piece of wood before and after heating, and he found that the charcoal produced had the same volume as the wood but possessed a smaller weight. Strato presumed therefore that matter had departed from the wood, leaving small vacuous pores. In another experiment he showed that partially evacuated vessels would suck up water, an effect which he ascribed to the water filling the voids between the particles of air. Strato was of the view that bodies in general consisted of minute particles with voids between them. If there were no such voids, he argued, light would not be able to pass through water and air, nor would heat be able to flow from body to body.

Strato is said to have written Book IV of Aristotle's *Meteorology*, which is the only Greek work dealing with chemical problems before the time of the Alexandrian alchemists. The *Meteorology* contains the theory that all mineral substances derive from two exhalations which rise from the interior of the earth. The one was a smoky exhalation which possessed hot and dry qualities, and the other was a steamy or vaporous exhalation which had cold and wet qualities. The interaction of the two exhalations produced all the varied mineral substances, infusible stones, red pigments, and sulphur containing more of the smoky exhalation and metals more of the vaporous exhalation.

After Strato very little work of scientific importance came from Athens: the main centres of Greek science shifted elsewhere, notably to Alexandria. The atomic theory was revived at Athens by Epicurus of Samos, 342-270 B.C., but he used the theory mainly to combat religion. The question as to whether the moon shone by its own light or by reflected light was not important, he suggested, but it was essential to stress that the moon was earthly and not divine. In opposition, the Stoics, notably Zeno, and Cleanthes of Assus, 300-225 B.C., emphasized the doctrine of the divinity of the heavenly bodies, supposing that the heavens controlled the destinies of man on earth. The Stoics adopted the notion that man was a microcosmic copy of the whole universe, and they supposed that both the lesser and the greater worlds were subject to the government of an absolute power. The sun was the ruling power of the universe, not the Primum Mobile as the Aristotelians believed, as the sun was the organ of the macrocosm equivalent to the heart, the governing power of the microcosm. Such a view, deriving from the ancient despotisms of Mesopotamia, had a considerable appeal in the period of the newer imperialisms of first the Greeks and then the Romans.

Chapter 5

Greek Science of the Alexandrian Period

AFTER THE conquests of Alexander the Great, the main centre of Greek science shifted from Athens to Alexandria. The Athenians had become either superstitious or cynical, as we see in the Stoics and the Epicureans. They had been defeated by the Spartans in 404 B.C., and again by Philip of Macedon in 338 B.C., and, losing their early vigour, they did little more than preserve their earlier achievements. The last of their great philosophers, Aristotle, had been a Macedonian, and his pupil, Alexander the Great, continued the conquests of his father, Philip. In 334 B.C. Alexander crossed over to Asia Minor, and after defeating a Persian army, he entered Egypt. Here in 332 B.C. he founded the city of Alexandria, which prospered on the trade diverted from the Phoenician cities of Tyre and Sidon. In the following year Alexander marched out of Egypt to conquer Mesopotamia and the whole of central Asia as far as the Indus and the Punjab.

On all of his campaigns, Alexander took with him engineers, geographers, and surveyors. Such men mapped out the conquered countries, and noted their resources, collecting a vast store of observations on natural history and geography. Theophrastus made use of the observations on plants in his botanical works, whilst another of Aristotle's pupils, Dicaearchus, c. 355-285 B.C., made use of the geographical information to make a map of the known world. Dicaearchus was the first to draw a line of latitude across a map, the line running from the Straits of Gibraltar, along the Taurus and Himalaya ranges, to the Pacific ocean. Thus the information collected by the armies of Alexander provided the means, and perhaps the stimulus, for the change in Greek science from the speculative to the empirical which occurred in the life time of Aristotle. We shall note later a similar reorientation from the theoretical to the practical in French science with the conquests of Napoleon.

When the Greeks took over Mesopotamia, Babylonian astronomy and mathematics became known to them in detail. The Greeks now adopted the sexigesimal numeral system, though using letters to represent numbers, they lost the Babylonian discovery of place-value. They took over Mesopotamian algebra. In the solution of quadratic equations the Greeks used distinctively Babylonian methods, multiplying the equation by the coefficient of the square, instead of dividing as we do. A fresh wave of Babylonian astrology also passed to the Greeks at this time, finding an expression in Stoic philosophy as the Fate ordained to men by the stars. This was one of the factors that made the Stoic philosophy so congenial to the Romans, for they were already versed in Babylonian astrology and liver divination through the Etruscans, who had originated in Asia Minor. From the Babylonians too came a knowledge of the correct order of the heavenly bodies outwards from the earth. The earlier Greeks thought that the sun came immediately after the moon in distance outwards from the earth, and then the planets. The later Greeks knew that after the moon came Mercury, then Venus, the sun, Mars, Jupiter, Saturn, and finally the fixed stars. Cicero tells us that the Stoic, Diogenes of Babylon, c. 160 B.C., first taught the latter order, bringing it from Mesopotamia. It is probable also that Hipparchus, 190-120 B.C., used Babylonian observations to measure the precession of the equinoxes, which had been discovered earlier by the Babylonian, Kidenas, c. 340 B.C.

Upon the death of Alexander the Great, in 323 B.C., his empire fell apart. Egypt was taken over by one of his generals, Ptolemy, who like Alexander himself had studied under Aristotle. Ptolemy employed Strato, later head of the Lyceum, as tutor to his son, and founded the Museum at Alexandria, a research and teaching institute modelled on the Lyceum, but much vaster in scale. The Museum was staffed by some one hundred professors, who were given salaries by the state. It was equipped with a library of about half a million rolls, and possessed a zoo, botanical gardens, an astronomical observatory, and dissecting rooms. The Museum lasted for about six hundred years, though the first two hundred years of its existence were the most important for science. As the Ptolemies became more Egyptianized, they favoured science less and less, Ptolemy VII, 146-117 B.C., going so far as to persecute the

Greek element in Alexandria. Thus Hipparchus of Nicaea, who has been described as the greatest astronomer of antiquity, lived and worked at Rhodes during this period. Another centre was Pergamum, from which Galen came, where animal skins, parchment, were developed for written works, as the Ptolemies prohibited the export of papyrus. Last, but not least, there was Syracuse, where Archimedes lived and worked.

An expression of the new empirical and practical trend in Greek science was the rise of a group of literate engineers in Alexandria. The earlier Greek engineers had reached a high level of technical proficiency, but they left no written records. In the time of Pythagoras, the engineer Eupalinus of Megara constructed a subterranean aqueduct for Polycrates, the tyrant of Samos, starting at both ends and meeting in the middle with an error of only two feet. The geometrical construction required for this operation was first described by the engineer, Hero of Alexandria, c. A.D. 100, though it must have been known in the time of Pythagoras. The Alexandrian school of engineers is said to have been founded by Ctesibus, fl. 285-222 B.C., the son of a barber of Alexandria. None of his works survive, though they were described by his younger contemporary, Philo of Byzantium. He is said to have invented the force-pump, and to have constructed a water-organ, and a water-clock with a mechanical action. Ctesibus and Philo suggested that the elastic force of compressed air or metal springs could be used to make siege catapults, instead of twisted ropes or leather thongs, which were sensitive to damp. However, reconstructed models have shown these suggestions to be unworkable. The works of Philo and Hero, who lived sometime between 100 B.C. and A.D. 150, deal with three main topics, military engineering, scientific instruments, and mechanical toys. They do not touch upon civil engineering, nor craft problems, to any considerable extent, except those of surveying. The instruments described are the water-clock for time keeping at nights, the Dioptra, a precursor of the theodolite, used in surveying, and the hodometer for measuring distances.

There were Greeks of this period who combined engineering with science, the most notable being Archimedes of Syracuse, 287-212 B.C. Archimedes visited Alexandria, inventing, it is said, in the course of his stay, the Archimedian screw for raising water, which is still used in Egypt. He must have been a mechanic of considerable skill, for, according to Cicero, he made a planetarium, a model of the sun, moon, earth and planets, that reproduced the apparent movements of the heavenly bodies in some detail, showing even eclipses. Archimedes developed the cross-staff for astronomical work, and made an apparatus for measuring the angle subtended by the sun on the earth. This instrument consisted of a disc sliding along a rule at right angles to it. At dawn, when the sun may be viewed with the naked eye, he moved the disc along the rule until it just covered the circle of the sun. The relation between the diameter of the disc and its distance from the eye along the rule, then gave the angle subtended by the sun. Archimedes' discovery of the principle of buoyancy and relative densities is another indication of the practical nature of his investigations. Tradition has it that he found King Hiero's crown to displace more water than an equal weight of gold, showing the goldsmith to have debased the crown by adding a metal of smaller density. Up to the time of Archimedes, the Greeks had taken the weight of a body to be proportional to its volume. He showed that this was not the case, some bodies being denser than others.

In his works Archimedes presented scientific knowledge as a deductive system of theorems from self-evident propositions, like the geometry of Euclid. It is possible however that he first obtained his results experimentally, and then deduced them from postulated axioms, for he tells us in his work *On Method* that he made thought experiments in the investigation of areas and volumes. He measured the areas of plane figures by weighing, in imagination, their shapes cut out of uniform material, thus gaining some insight into their relationships, which he then proceeded to demonstrate mathematically. In geometry he originated a method of deducing π, the ratio of the circumference of a circle to its diameter. The ratio of the perimeter of a regular polygon to the distance between its centre and a corner was readily ascertained, and by considering such polygons inscribed, and circumscribing a given circle, he showed that π could be obtained to any degree of accuracy desired.

In Alexandria itself geometry was systematized by the most widely known of the mathematicians of antiquity, Euclid of Athens, c. 330-260 B.C. Very little of his *Elements of Geometry* appears to have been original, but Euclid gathered together propositions and proofs from widely scattered sources, and presented them in an orderly text-book fashion. The first of the notable astronomers of Alexandria, Aristarchus of Samos, c. 310-230 B.C., on the other hand produced what was perhaps the most original scientific hypothesis of the Alexandrian period. According to Archimedes, he held that the earth rotated on its axis daily, and moved round the sun in a circular orbit once a year, the sun and the fixed stars being stationary, and the planets moving in circular orbits with the sun at the centre. Aristarchus' works describing this theory, if it was recorded at all, have not been preserved, though his theory seems to have been well known at the time, for according to Plutarch, the head of the Stoic school of philosophy, Cleanthes, said that Aristarchus ought to have been indicted for impiety. The work of Aristarchus, *On the Sizes and Distances of the Sun and the Moon*, has come down to us. This work contains the first scientific attempt to measure the relative distances of the sun and the moon from the earth. Aristarchus supposed that at half moon, the sun, moon, and earth, would form a right-angled triangle, from which the relative distances of the sun and moon could be determined by measuring the angular separation of the sun and moon from the earth. He measured the angle as eighty-seven degrees, from which he calculated that the sun was nineteen times more distant from the earth than the moon, though actually the angle, and thus the ratio of the distances of the sun and the moon from the earth, are larger. As the moon, on the average, just covered the sun during a solar eclipse, Aristarchus presumed that the diameter of the sun was nineteen times larger than that of the Moon. From lunar eclipses he estimated that the breadth of the earth's shadow, and thus the approximate diameter of the earth, was equal to three moon diameters. Thence, he argued, the diameter of the sun must be six to seven times as great as that of the earth.

To obtain estimates of the absolute dimensions of the sun and the moon, and of their distances from the earth, a measurement of the size of the earth was required. Such a measurement was first made by Eratosthenes of Cyrene, 284-192 B.C., who was chief librarian of the Museum at Alexandria. He noted that at Syene the sun was directly overhead on midsummer's day, whilst at Alexandria the sun's rays were seven degrees from the vertical, a value estimated from the length of a shadow cast by a rod of a know height. Eratosthenes estimated that Alexandria was 5,000 stades due north of Syene, so that the circumference of the earth was 250,000 stades. Estimates of the stade vary, but if ten stades be taken to the mile, his polar diameter of the earth comes out only fifty miles short of our modern value.

Eratosthenes also developed the study of mathematical and astronomical geography. He gathered together the work of his predecessors, who had evolved the conception of the earth as a globe, with two poles and an equator, and made a map of the known earth, marked with lines of latitude and longitude, and delineating five zones, two frigid, two temperate, and one torrid. He took as his fundamental meridian of longitude that of Alexandria and Syene, which passed, he thought, through Byzantium. As his fundamental parallel of latitude, he took the 36° line, passing through the Straits of Gibraltar and the island of Rhodes. Along this parallel he thought land stretched for 78,000 stades, from the Atlantic to the Pacific, the rest being sea. According to Strabo, Eratosthenes thought that, 'if it were not for the vast extent of the ocean, it would be possible to sail from Spain to India along the same parallel', because the tides were similar in the Atlantic and Indian oceans, showing them to be connected.

The heliocentric system of the world, suggested by Aristarchus, was an attempt to overcome the difficulties inherent in the Eudoxian system, which we have mentioned earlier. The views of Aristarchus were not accepted, because the Greeks, in general, found it difficult to rid themselves of the conception that the earth and the heavens were entirely different in regard both to their material constitutions, and to the laws they obeyed. Such a conception entailed the view that the base earth was stationary at the centre of the universe, whilst the more perfect heavenly bodies moved uniformly in circles through the purer regions above. All attempts made by the

Greeks, after Aristarchus, to overcome the limitations of the
Eudoxian system retained this view. Apollonius of Perga, c.
220 B.C., suggested a geometrical construction whereby the
variation in the distance of the planets from the earth could be
explained. He pointed out that if a planet moved in a circle,
the epicycle, the centre of which moved on another circle, the
deferent, which centered on the earth, then the distance of
the planet would vary, and by an appropriate choice of circles
the motions of the planet could be accounted for quantitatively. Another device was the suggestion that the heavenly
bodies moved in circles eccentric to the earth, the centres of
their orbits lying at a distance from the centre of the earth.

Both of these devices, epicycles and eccentrics, were used
by the astronomer, Hipparchus of Nicea, 190-120 B.C., who
lived and worked at Rhodes. He explained the apparent motions
of the sun by means of a fixed circular orbit eccentric to the
earth, and the motions of the moon with a moving eccentric
orbit. The motions of the planets he accounted for by means
of a system of epicycles. The greatest contribution of Hipparchus lay in the field of observational astronomy. He collected and collated the records made by previous Greek
observers, and also those of the Babylonians, whose reliable records stretched back to the seventh century B.C. In so doing
he found that the tropical year, the time taken for the sun to
return to the same equinoctial point, was a little shorter than
the sidereal year, the time taken by the sun to return to the
same position amongst the fixed stars. The difference, known
as the precession of the equinoxes, he estimated to be about
thirty-six seconds of arc a year, the modern value being about
fifty seconds. Hipparchus saw that, for observations of this
character, a catalogue of star places would be needed by future
observers. Accordingly he determined the positions of some
1,080 stars, classifying them into six magnitudes of brightness.
Hipparchus also continued the work of Aristarchus on the
determination of the sizes and distances of the sun and the
moon. By observing the altitude of the moon at two different
latitudes, he found that the moon was about thirty-six earth-
diameters away, which is a little large, but a great improvement on the value of nine earth-diameters obtained by Aris-

tarchus from the angle subtended by the moon to an observer on the earth.

The work of Hipparchus was continued at Rhodes by Posidonius of Apamea, *fl.c.* 100 B.C., and his pupil, Geminus of Rhodes, *fl.c.* 70 B.C. Posidonius made a fresh determination of the size of the earth, by measuring the distance and the latitude difference between Rhodes and Alexandria. The value he obtained, 180,000 stades, was smaller than that of Eratosthenes, but it was adopted by Ptolemy, the last notable astronomer of antiquity, and became the generally accepted value. Claudius Ptolemy, A.D. 85-165, observed at Alexandria between A.D. 127 and 151. His observations are said to be less accurate than those of Hipparchus, displaying an error of about a quarter of a degree, whilst those of Hipparchus were accurate to within one-sixth of a degree. However, his estimate of the average distance of the moon from the earth (29·5 earth diameters) was nearer to the modern value (30·2) than that of Hipparchus. Ptolemy adopted and developed the system of eccentrics and epicycles used by Hipparchus to explain the apparent motions of the heavens. The number of periodic movements in the heavens now known was large, a system of some eighty circles being required to explain them. Ptolemy himself made a discovery which showed that the system could not have a physical reality, and it seems that he himself may have regarded the scheme as a mathematical convenience. He found a second inequality in the motion of the moon, which he accounted for by adding an epicycle to the orbit of the moon. The epicycle carried the moon to distances that were in the ratio of 1/2 from the earth, which would have produced changes in the apparent area of the moon in the ratio of 1/4, which were not observed.

Ptolemy produced also the last important geographical work of antiquity. Hipparchus had suggested that the latitudes and longitudes of important cities and coastal points should be determined and collected for the making of maps. Such a project was accomplished in part by Marinus of Tyre, *c.* A.D. 150, whose work Ptolemy took over and completed. Six out of the eight books that comprise Ptolemy's geographical work consist of lists of place-positions in latitude and longitude. However, it seems that most of his positions were estimated distances from his prime meridian and parallel reduced to degrees, for none of his longitudes had been determined astronomically, and only a few of his latitudes had been so determined. Ptolemy adopted the smaller value for the circumference of the earth, determined by Posidonius, so that all of his landward distances in terms of degrees were exaggerated, as he took five hundred instead of six hundred stades to the degree. Thus the ocean distance from Europe to Asia, across the Atlantic, appeared to be much less than Eratosthenes' estimate, and it was this that ultimately led to the attempt of Columbus to reach Asia from the west. Ptolemy knew of much more of the world than his predecessors. Eratosthenes' map extended as far east as the Ganges, but Ptolemy knew of the Malay peninsula, and the silk-lands, or China.

The fortunes of the biological sciences at Alexandria paralleled those of the physical and mathematical sciences. There was a burst of activity in the medical and biological field in the third century B.C. under the early Ptolemies, and a later one in the second century A.D. under the Romans, biology flourishing elsewhere in the intervening period. The earliest medical teacher at Alexandria was Herophilus of Chalcedon, *fl.c.* 300 B.C., who was the first to perform anatomical dissections in public. He recognized the brain to be the seat of the intelligence, which Aristotle had placed in the heart, and he connected the nerves with the functions of motion and sensation. Herophilus was the first to distinguish between veins and arteries, noting that the latter pulsate, whilst the former do not. His younger contemporary, Erasistratus of Chios, *fl.* 300-260 B.C., traced the course of the veins and the arteries, and their subdivisions, all over the human body to the limits perceptible with the naked eye. He did the same for the nervous system, which he centred upon the brain, and he connected the complexity of the convolutions in the brain of man with his superior intelligence. Erasistratus, like his teacher, Strato of the Lyceum, was something of an experimentalist. He weighed a bird in a cage, and noted that it lost weight continuously between feeding times. He was also interested in the pneumatic problems studied by his teacher, attaching great importance to the role of air in his physiology. Strato had shown how a partial vacuum exerts a pull on liquids, and conversely that liquids can exert a pull on air. Erasistratus thought similarly that air was drawn into the body by the pull of the blood as it surged downwards in the body, and was expelled by the blood rising again. Normally, he thought, the arteries are full of air, or rather air transformed into a vital spirit, as he found the arteries to be empty in dead animals. In a live animal he held that the air escapes when an artery is severed, and the blood follows after it. When air entered the body, it was drawn through the lungs into the heart, where it was transformed into vital spirit. The vital spirit was relayed by the arteries all over the body, a small amount finding its way to the brain, where it was transformed into animal spirit, which was distributed by the nerves.

The school of Alexandria declined in the second century B.C., and medicine found its home elsewhere, notably on the mainland of Asia Minor. Crateuas, 120-63 B.C., physician to the king of Pontus, gathered and described plants of use in medicine; he was the first to give illustrations of the plants he collected and classified. About the same time, Apollonius of Citium, c. 100 B.C., gave sketchs of surgical operations, and methods of bandaging. Later, Dioscorides of Anazarba, c. A.D. 50, a military surgeon in the army of the emperor Nero, wrote his *Materia Medica,* a work on drugs, and the plants from which they were derived, illustrated in the style of Crateuas.

The last notable medical writer of antiquity was Galen, A.D. 129-199, the son of an architect of Pergamum. At Pergamum a large library of parchment manuscripts had been built up, and it had become an important centre of learning. Here Galen studied medicine, later visiting Alexandria and other medical centres. Finally he settled at Rome, where he became physician to the emperors Marcus Aurelius and L. Verus. Galen dissected and investigated both dead and living animals, though he did not dissect human bodies. He showed that Erasistratus was wrong on the question of the arterial fluid by tying off a section of an artery in a living animal to isolate its contents, and then opening it up to show that it contained blood, and not air. Galen's anatomical studies were based on the dissection of the Barbary Ape which is anatomically similar to man but sufficiently different to cause confusion to later students. His physiological system of man was based mainly on that proposed earlier by Erasistratus, though he brought in the Hippocratic doctrine of the four humours, and Aristotle's views on the nature of man.

Aristotle had divided terrestrial creatures into three types, the vegetable, which exhibited growth by virtue of a vegetable soul, the animal, which moved themselves by virtue of a sensitive soul, and man, who showed intelligence by virtue of a rational soul. Men possessed all three souls, animals the first two, and plants only a vegetable soul. Galen suggested that the seats of these three levels of vital activity were located in certain internal organs and that they were connected with a common source of vitality, the pneuma, or the spirit of the air. The philosophical school dominant at the time, the Stoic, held that the air was the breath and soul of the universe, the macrocosm, and that it served to sustain the life of man, the microcosm. Respiration, therefore, was the function that connected man with the cosmic spirit, and renewed his vital activities, through the intake of the spiritous part of the air, the pneuma, which in itself was half-air, half-fire.

Galen placed the seats of the three vital activities of man in the digestive, respiratory, and nervous systems. He held that the useful part of the ingested food was transported as 'chyle' from the alimentary tract, through the portal vein, to the liver where it was transformed into the dark venous blood. The useless parts of the food went to the spleen where they were converted into black bile. The liver was the seat of the vegetative life, and it was here that the natural spirit, which governed the nourishment and growth of the body, was prepared and infused into the venous blood. From the liver the venous blood was transported by its mover, the natural spirit, in a largely one-way motion to the right chamber of the heart. Galen was aware that the valves of the heart allowed the blood from the veins to enter the right chamber, but not to return, and blood from the left chamber to flow into the arteries, but not back again. However, he considered that the valves of the heart were imperfect, so that there was a small back flow of venous blood from the right chamber to the veins, and of arterial blood from the arteries to the left chamber, but, contrary to the traditional interpretation of his writing, Galen did not hold that there was a large scale ebb and flow of blood in the veins and arteries. Apart from the small amount that went back into the veins, the blood in the right chamber of the heart passed either through the dividing wall of the heart, the septum, into the left chamber, or through the pulmonary artery to the lungs. In the left chamber of the heart, the soot and waste products of the venous blood were separated off and discharged through the pulmonary vein into the lungs. By the same route air was brought from the lungs to the left chamber, where the *pneuma* of the air was separated off and infused into the blood as the vital spirit. The bright red arterial blood so prepared was then carried by its mover, the vital spirit, into the arteries, and through them to all parts of the body, which were thus fitted for their animal activities. Some of the arteries led to a network of vessels at the base of the brain, the *rete mirable*, where the vital spirit was transformed into animal spirit (from *anima*, the soul). The animal spirit was then distributed through the nerves, which were thought to be hollow tubes, conferring sensitivity upon the various parts of the body.

Technically the system of Galen was erroneous in several ways. He held that the heart was responsible for respiration, air being drawn into the body when the heart expanded and expelled when it contracted, though the heart beat is much faster than the respiration rate. The dividing wall of the heart is a solid muscle which does not permit the passage of blood, as his sixteenth-century critic, Vesalius, indicated. Again the *rete mirabile* is not found in man, though it is prominent in the ruminants which Galen studied. Intellectually Galen conformed to the dominant Greek preconception that terrestrial motions were rectilinear, circular motion being the prerogative of the celestial bodies. Such a notion held up the idea of the circulation of the blood, as also did the conception that there were two different sorts of blood, each with its own distinct function, mover, and distribution system.

Galen's theories were most influential, dominating medicine down to modern times. More of his works than of those of any other ancient author have been preserved—eighty-three of his one hundred and thirty-one medical writings having come down to us. His works were popular partly because they were felt to be useful, more useful, say, than the contemporaneous astronomical and geographical works of Ptolemy, and partly because they were permeated with a marked religious feeling, which made them more acceptable to the scholars of Islam and the medieval Church. Like Plato and Aristotle, Galen was much concerned with the cosmic purposes for which objects and organisms were intended, though, in keeping with the spirit of the age, his religion was more mystical and less intellectual than that of the Athenians.

Chapter 6

Rome and the Decline of Ancient Science

THE ROMANS, like the Greeks, came to civilization during the iron age, directly from barbarism. However, they did not get away from the traditions of the bronze age so completely as the Greeks. When the Romans expelled the Tarquins in 510 B.C. they took over the systems of astrology and liver divination that the Etruscans had brought with them from their original home in Asia Minor. Moreover, the Romans did not develop a civilization of seaboard city states like the Greeks. Rome was a warrior-agriculturalist community, like Sparta, the least intellectual of the Greek States. Commerce was forbidden to the senators of Rome, whilst her merchants submitted to the values of their society, aspiring to become the owners of farming land. The Romans therefore lacked above all the quantitative and spatial thinking of the merchant-traveller, rendering them weakest in the mathematical sciences. When the civilization of Rome was fully mature, Cicero, 106-43 B.C., could observe that, 'Greek mathematicians lead the field in pure geometry, while we limit ourselves to reckoning and measuring.'

The Romans did not add a great deal to science. Their contribution lay elsewhere, in the field of organization—the formation of a public medical service, the building of roads and aqueducts, the introduction of the Julian Calendar, and the formulation of Roman law to regulate their organizations. The Romans were in contact with the Greeks of Sicily and southern Italy from an early date, and as they subjugated the dynasties that had emerged from the empire of Alexander the Great in the second century B.C., they became increasingly aware of the superiority of Greek culture. Some, such as Cato the censor, 234-149 B.C., and Varro, 116-27 B.C., reacted against the Greek knowledge, but for the most part the Romans endeavoured to assimilate the learning of the Greeks, particularly in the first century B.C. Cato the Censor wrote a work on medicine and agriculture to show the Romans superior to the

61

Greeks. His medicine was mainly magical formulae and herb remedies; Rome, he thought, was 'healthy without doctors'. Varro endeavoured to do the same over a wider field, dealing with nine 'liberal arts', grammar, dialectic, rhetoric, geometry, arithmetic, astronomy, music, medicine and architecture. The last two were cut out by Cassiodorus, A.D. 490-585, leaving the seven liberal arts studied in the Middle Ages.

During the process of assimilation, the Stoic philosophy of the Greeks came to have the greatest influence upon the Romans, for it presented the traditional beliefs they had inherited from the Etruscans in a more sophisticated form. The opposing school of Epicurus found a notable exponent in Lucretius, c. 95-55 B.C. but his philosophy was not an important force in Rome. Lucretius, like Epicurus, preserved the content, but not the spirit, of the early Atomist's philosophy, adding nothing that was new. They used the atomic philosophy mainly to combat religion, not to extend man's understanding and control of nature.

The Romans failed to assimilate the limited degree of unity which the Greeks had achieved between theory and experimentation in science. The Greek practice of dissection in medical teaching, for example, never took root in Rome. They took over the content of Greek science without the method, and their works tended to be therefore either primarily philosophical, like that of Lucretius, *On the Nature of Things,* or largely empirical, like Pliny's *Natural History.* The latter work, by Pliny the Elder, A.D. 23-79, was a vast compilation in thirty-seven books of facts and observations derived from some two thousand previous works, written by one hundred and forty-six authors of Roman origin, and three hundred and twenty-six who were Greeks. He was quite uncritical of his sources, reporting all the things of which he had read, the unicorn and the phoenix equally with the lion and the eagle. Pliny stressed the usefulness of the things which he described, the general attitude permeating his work being that nature existed to serve the purposes of man. However, he recorded not only the things of which he had read, but the things he had observed for himself as well; in fact he was killed while observing an eruption of Vesuvius close at hand.

Not all of the content of Greek science was absorbed by the

Romans, the mathematical sciences in particular having little appeal for them. The Romans had no mathematicians or astronomers of note, and only one important geographer, Pomponius Mela, *c.* A.D. 43, who took over the qualitative aspects of Eratosthenes' geography, avoiding the mathematics and the measurements. Subsequent Latin geography displayed a marked decline. Isodore of Seville, A.D. 570-636, represented the known world as a circle divided by a T, so that Asia was represented by a semicircle and Europe and Africa by two quadrants, with seas between and all around.

Medicine, perhaps for utilitarian reasons, was assimilated more successfully by the Romans. The first notable teacher was the Greek, Asclepiades of Bithynia, died *c.* 40 B.C., who set up a medical school in Rome. One of his students, Celsus, wrote an important work *On Medical Matters* about A.D. 30, a treatise which was a good compilation from Greek sources. The teaching of medicine was extended in the time of Vespasian, A.D. 70-79, with regard to the training of army surgeons. Teachers of medicine were paid by the state, and medical centres were established in the provinces. About the time of the barbarian invasions, Roman medicine showed a marked decline. It was then, Vesalius wrote in 1543,

'that the more fashionable doctors, first in Italy, in imitation of the old Romans, despising the work of the hand, began to delegate to slaves the manual attentions they judged needful for their patients, and themselves merely to stand over them, like architects. Then, when all the rest who practised the true art of healing gradually declined the unpleasant duties of their profession, without however abating any of their claim to money, or to honour, they quickly fell away from the standard of the doctors of old. Methods of cooking, and all the preparation of food for the sick, they left to nurses; compounding of drugs they left to the apothecaries; manual operations to the barbers.'

When the Greeks were conquered by the Romans they tended to become either cynical or religious, just as had the Athenians when they were subjugated by Sparta and then by Philip of Macedon. Both tendencies were reflected in the sciences of the Greeks, though the religious trend was ultimately the stronger. In the field of astronomy, Geminus of Rhodes, *c.* 70 B.C., expressed the somewhat sceptical opinion that an astronomical system of the world was a mathematical convenience rather than a representation of physical reality. Astronomers, he suggested, were not concerned with physical questions, but with saving the phenomena mathematically. At the time such a view entailed, of course, the tacit acceptance of the physical axiom that the motions of the heavenly bodies were circular and uniform, an axiom which was based upon the idea that the matter of the heavens was superior to the terrestrial elements.

'In general', wrote Geminus, 'it is not the astronomer's business to see what by its nature is immovable, and of what kind the moved things are, but framing hypotheses as to some things being in motion, and others being fixed, he considers which hypotheses are in conformity with the phenomena of the heavens.'

Later, astronomy was to be fitted with a theological dress, but the more immediate impact of religion upon science was in the field of chemistry, with the rise of the Alexandrian alchemists about the second century A.D. The earlier Greeks did not pay much attention to chemistry, perhaps because the subject was associated with the crafts which most of the Greek philosophers regarded as degrading. Notable exceptions were the early Ionian philosophers and the later followers of Aristotle, such as Theophrastus and Strato, who used craft analogies to explain natural processes. It is not surprising to find therefore that the sole surviving purely Greek work upon chemical matters, Book IV of the *Meteorology* in the Aristotelian corpus, is ascribed to Strato, and that, according to later ancient authors, Democritus, the Atomist, is said to have written upon chemical matters.

In early times it seems that craft processes were associated with magical rituals, which were regarded as essential to the success of the operations involved. Greek potters placed masks upon their furnaces to frighten away the demons that were thought to crack the pots. The production of metals in particular was connected with quasi-organic processes, such as a birth, or a death and resurrection. We have already mentioned an Assyrian text of about 700 B.C. which associated the production of metals with a kind of birth ritual. In a Persian myth, dated to about 500 B.C., the birth of the metals was regarded as the consequence of the death of a divine primeval being. When the being was killed, lead came from his head, tin from his blood, silver from the marrow, copper from the bones, steel from the flesh, and gold from the soul.

It appears that the magical and the practical elements of chemical processes were gradually separated, for we find in Egypt during the early centuries of our era purely practical chemical recipes side by side with the magic and the mysteries of alchemy. The earliest alchemical work, that of pseudo-Democritus, c. A.D. 100, contains practical recipes and mystical speculations in different portions of the same treatise. Later, chemical works became entirely practical, like the Leiden and the Stockholm papyri compiled during the third century, or largely alchemical, such as the works of Zosimus, dating from the same period. The practical papyri contain technical recipes for the falsification of gold and silver, and for the preparation of artificial precious stones and dye-stuffs. Gold and silver were simulated by alloys of other metals, and by gilding the surface of a base metal, or again by debasing the noble metals and removing the base metal in the surface layer with corrosives. In the technical treatises the products were regarded as imitations, but in the alchemical works they were thought to be real silver and gold.

During the period of the early alchemists, Stoicism was the dominant philosophy, though the views of Plato were revived within the Stoic school, coming to maturity with the Neoplatonic system of Plotinus, A.D. 204-270. The Stoics believed, amongst other things, that all of the varied objects of nature were alive and growing. Each entity developed from a seed which contained from the beginning the form or plan determining the characteristics of the mature object. Such a form or plan was a soul or a spirit, and it was brought into activity and sustained by the universal spirit of nature, the pneuma. Plato, as we have seen, believed in the transmigration of souls, a view which implied, in the context of the Stoic philosophy and the religions of the time, that the form or characteristics of one object could be transferred to another by a process of death and resurrection.

Following such lines of thought, the alchemists believed that the metals were living organisms which were developing gradually towards the perfection of gold. The development might be urged on, or copied artificially, by isolating the form or soul of gold and transferring it to base metals, which would thence assume the form and characteristics of gold. The soul or form of a metal was regarded as a spirit or vapour, which manifested itself notably in the colour of the metal. Hence the surface gilding of a base metal was thought to be a transmutation by the alchemists. According to Zosimus:

All sublimed vapour is a spirit, and such are the tinctorial qualities. . . . The mystery of the gold tincture is to change bodies into spirits in order to tint to spirituality.'

A fairly general process adopted by the alchemists was the alloying of the four base metals, copper, tin, lead, and iron, in order to obtain an approximation to primordial matter devoid of form. The alloy was then whitened on the surface with arsenic or mercury vapour, so as to confer upon it the spirit or form of silver. Next, a little gold was added as a seed or ferment to bring the whole mass to gold, for 'just as yeast raises a great quantity of dough, so also a little quantity of silver or gold acts'. Finally the transmutation was completed by another surface process, either by etching away a surface layer of base metal to leave a gold surface, or by treating the alloy with sulphurous waters to give it a bronzed appearance. In such an attempted transmutation process the base metals were regarded as dead when they had lost their individuality in the alloy; that is, they had lost their specific form or soul. Similarly the assumption of the new forms, first of silver and then of gold, by surface tinting were thought to be resurrection processes.

Another idea to which the early alchemists gave currency was the perhaps more primitive notion that the metals were a product of sexual generation, and that they were themselves either male or female. One of the early alchemists, Mary the Jewess, who is said to have invented the water-bath, the

bain-marie as the French term it, recorded the secret of transmutation in the instruction: 'Unite the male with the female,
and you will find what you seek.' Silver, she added, does this
readily, but copper couples 'as the horse with the ass, and the
dog with the wolf'. Such an idea assumed a greater importance
in Islamic and Medieval alchemy when the theory of Aristotle's *Meteorology* was revived, and sex was assigned to the
two exhalations which were presumed to give rise to the metals
and the minerals.

The age in which the early alchemists lived, the first to the
fifth century A.D., gradually became less and less intellectual as
the Roman empire crumbled and fell apart. It is said that
Diocletian burned the books of the alchemists in A.D. 292, and
that the library at Alexandria was destroyed in a Christian riot
of A.D. 389, whilst, according to an anti-infidel account dating
from the thirteenth century, the Museum was finally obliterated by the Muslims in A.D. 640. With the rise of Christianity
came a revival of the bronze-age theory that the earth was
flat, supported by waters beneath and covered by waters in the
vault above. Such a view had an appeal for some of the early
Church Fathers, who considered that the universe was analogous in general configuration to the tabernacle. The flat-earth
theory was supported the most strongly by the Syrian Church,
notably by Cyril of Jerusalem, *c.* 360, and Diodorus, Bishop
of Tarsus, died 394, the latter going so far as to declare that
the Greek system of the world was atheistical.

In the western Church the tabernacle theory also found
favour, though in general the outlines of the Greek theory were
preserved, notably the idea of a spherical earth and a spherical
heaven. Ambrose of Milan, *c.* 397, spoke of the heavens as
spherical, and so too did his disciple, Augustine, 354-430. However, Ambrose did not consider such questions important, for
he indicated that, 'To discuss the nature and the position of the
earth does not help us in the hope of life to come.' Isodore,
570-636, Bishop of Seville, quoted the 'philosophers' to the
effect that the earth was a sphere at rest, and the heavens a
rotating sphere, without committing himself on the point, and
so too did the Venerable Bede, 673-735. The detail of the
Greek systems of the world were lost to the Church Fathers of

the west, and it was in eastern Christendom that the systems of Aristotle and Ptolemy were given a Christian theological dress.

In Byzantium science had deeper roots than in Rome, and at Athens the Academy and the Lyceum were active until A.D. 529, when they were closed down by Justinian. Proclus, 412-485, the last notable pagan philosopher of Byzantium, indicated that by his time a theory of the world had been evolved which incorporated the mathematical system of Ptolemy together with elements of Aristotle's physical cosmology. It was supposed that the heavens consisted of nine concentric spherical shells, the first carrying the moon, the second, Mercury, and so on upwards, the eighth carrying the fixed stars, and the ninth being the Primum Mobile. Each shell was of such a thickness that it accommodated the epicycles of the Ptolemaic theory, and successive shells were contingent, as Aristotle had supposed. Thus the furthest distance one heavenly body moved away from the earth was equal to the nearest distance of approach of the next higher body, that is, the greatest distance of the moon's recession equalled the smallest distance of Mercury's approach, and so on. Accordingly, the thickness of the shells carrying the heavenly bodies and the mean distances of those bodies from the earth could be calculated, giving what were thought to be the absolute dimensions of the universe.

These nine spheres required movers which, according to the Platonic and Aristotelian theories, had to be of a more noble and spiritual character than the bodies moved. Plato had postulated that there was a hierarchy of spirits in the heavens, and Dionysius, who was probably one of Proclus' pupils and who was certainly in the Neoplatonic tradition, identified these spirits with the various angelic beings mentioned in the Scriptures. Dionysius arranged the angelic beings into a hierarchy of nine orders, grouped into three subsidiary hierarchies, and subsequently these orders of angelic beings were taken to be the movers of the nine celestial spheres. First came the Seraphim, Cherubim, and Thrones, then the Dominations, Virtues, and Powers, and finally the Principalities, Archangels, and the Angels. The Seraphim were responsible for the movement of the Primum Mobile, the Cherubim for the sphere of the fixed stars, and so on, down to the Angels which propelled the sphere of the moon. Above the hierarchy of angelic beings was God in a tenth Empyrean heaven, and below were the creatures of the earth, first man, then animals, plants, and finally the very dregs of the universe. Each order of creatures was hierarchically arranged in itself, so that there was a chief Seraph and lesser Seraphim arranged in rank, just as on earth, said Dionysius, there was the Patriarch of the Church, his bishops, and so on, down a certain scale. Thus the universe appeared to be constituted of a continuous chain of creatures, stretching down from God at the periphery of the universe to the basest inhabitant of hell at the centre of the earth, for the highest creature of one order was directly contingent upon the lowest creature of the order above it.

The physical aspects of such a view of the world were opposed by John Philoponos, a sixth-century writer of Alexandria, who was named as a heretic by the Church. Philoponos denied that angelic beings moved the heavenly bodies. He held that in the beginning God had given the heavenly bodies an impetus, a motive power of their own which did not decay with time, just as He had given heavy objects the tendency to fall towards the earth. Terrestrial and celestial bodies were comparable in this respect, they did not differ essentially in kind. Philoponos suggested that in general a body in motion did not require to be in constant physical contact with a mover, for a force supplied impetus to a body and it was impetus that kept a body in motion. Thus a vacuum could exist, as the impetus theory did not require a material continuum to transmit actions by physical contact. Again an arrow could fly through such a vacuum under its own impetus: it would not require a constant pressure of air behind it as Aristotle had supposed.

The theory of impetus suggested by Philoponos was revived during the thirteenth century, and it constituted an important departure from the views of Aristotle and Dionysius, which were generally accepted during the middle ages. Dionysius had considerable influence, for it was then thought that he was Dionysius the Areopagite, the Athenian convert of St. Paul. The Byzantine emperor, Michael, sent a copy of his works to the western emperor, Louis the Pious, in 827, and they were translated from the Greek by the Irish philosopher,

John Scot, and others. The works of Aristotle came by a more devious route through Islam, and from the twelfth century they provided, together with the Scriptures and the works of the Christian Neoplatonists, such as Dionysius, the texts basic to medieval learning.

PART TWO

SCIENCE IN THE ORIENT
AND MEDIEVAL EUROPE

Chapter 7

The Science and Technology of the Chinese

IN CHINA the traditions of early civilized society were more persistent and continuous than elsewhere, the ideographic script, irrigation agriculture, and scholar bureaucracy, which we associate with the ancient bronze-age civilizations, living on in China down to modern times. Like the Babylonians and the Egyptians, the ancient Chinese failed to develop a theoretical geometry, nor did they base their theories concerning the spatial structure of the universe upon their quantitative astronomical observations. Again, the ancient Chinese did not evolve a scientific method, their philosophies and techniques remaining largely isolated from one another until modern times.

The earliest phase of Chinese civilization which may be dated with any certainty was that of the Shang dynasty which ruled at Anyang on the Yellow River about 1500 B.C. Excavations at Anyang have shown the Chinese of this period to be bronze workers, with the potter's wheel, and horse-drawn chariots, but growing rice instead of the barley cultivated in the west, and weaving silk instead of flax. They had the present ideographic script in an early pictographic form, and a sexigesimal numeral system, which has been taken as evidence of Babylonian influence. About 1000 B.C. the Shang dynasty fell to the border Chou peoples, who extended considerably the area of Chinese civilization. The peripheral fiefs of the Chou empire gradually developed into a number of autonomous feudal states with the Chou state weakened in power at the centre, and during the period 480-220 B.C. these states were constantly at war with one another.

Iron came to China about the sixth century B.C., the first reference to it being recorded in 513 B.C. The most westerly state of Ch'in, which was associated with the making of iron, gradually conquered the other states and set up the Ch'in dynasty, 221-207 B.C. To unite his empire the first Ch'in emperor extended the waterworks of China, built a network of roads, and made notable additions to the Great Wall. He also endeavoured to consolidate his position by burning the history books of all the states except Ch'in, though he retained copies of the prohibited books in the Imperial Library. His rule in general was authoritarian, and when he died one of his minor officials inaugurated the more tolerant and long-lived Han dynasty, which lasted from 202 B.C. to A.D. 220. The Han emperors founded the Imperial university in 124 B.C. and set up a bureaucracy of scholars to govern their empire. Such scholars wrote at first on bamboo slips, then upon silk, and finally upon paper. The invention of paper is ascribed to Tshai Lun, A.D. 105, surviving specimens dating back to A.D. 150.

The Han period was notable for its technical innovations. It saw not only the invention of paper, but also the discovery of the orientating effect of the magnet *c.* 100 B.C., and the first record of the casting of iron. A work of A.D. 31 described a mechanism whereby a horizontal waterwheel drove a bellows through a system of pulleys and belts, the bellows working an iron furnace for the casting of agricultural implements. Later, in A.D. 290, the vertical waterwheel appeared in China, and with it came a water-driven pestle and mortar working by a tilt-hammer action.

The period of the Warring States, 480-221 B.C., the Ch'in, 220-207 B.C., and the Han, 202 B.C. to A.D. 220, was a time in which there was much speculation on questions of a scientific and philosophical nature. There were said to be a 'Hundred Schools' of philosophers in the Warring States period, though of these only the Legalist, Logicians, Mohists, and more particularly the Taoists and Confucians were important. These schools were divided on the question of how to overcome the devastating wars between state and state. The Legalists held that the chaos could only be ordered by positive law—'Laws fixed beforehand', as they put it. They were influential under the Ch'in rulers, but for the most part disappeared with their patrons. The Mohists, traditionally the followers of Mo Ti, *c.* 479-381 B.C. taught a doctrine of universal love, though they were not pacifists, for they trained themselves in the military arts in order to help the weak states oppressed by the strong.

Such activities led the Mohists to investigate problems in physics, notably optics, mechanics, and methods of fortification. They studied the reflection of light from plane, concave, and convex mirrors, and obtained empirical rules connecting the size and position of objects and images with the curvature of the mirrors used. In mechanics they were interested in lever systems and pulleys, which again they studied empirically. They had no theory of light, nor did they employ geometrical constructions in their work. Their results were obtained experimentally and were expressed as empirical rules. The Mohists and the related school of Logicians tried to work out a scientific method of reasoning, so that the men of the Warring States period with their varying opinions could arrive at a common agreement. The two schools were opposed on this matter, the Mohists holding that sense experience was the surest foundation of human knowledge, while the Logicians held that such experience was illusory, and that logical argument was the only way of arriving at a consensus of opinion.

By the beginning of the Han dynasty the Legalists, Mohists and Logicians had become unimportant, leaving the Confucians and the Taoists as the dominant schools of thought. The Confucians, traditionally the followers of Confucius, 552-479 B.C., taught that the following of ancient custom and tradition was the solution to the problems of the Warring States period, and indeed of all periods. Confucianism was the official philosophy of the scholar bureaucracy from the time of its inception under the Han emperors. The Confucians had little or no natural philosophy until the Neo-Confucian movement of Sung times, A.D. 960-1279, but they were associated with the official astronomers of the bureaucracy, and shared their views as to the nature of the universe. The Confucians were not interested in chemical and craft problems, such questions being studied almost entirely by the Taoists.

Traditionally the Taoists were the followers of Lao Tsu, who is said to have lived some time between the sixth and the fourth century B.C. The Taoists suggested that men should abandon civilized society and return to the simple egalitarian communities of ancient times. Such was the age of perfect virtue when 'men lived in common with birds and beasts, and formed one family with all creatures'. Thus many of the early Taoists took to the wilderness, and became hermits studying nature with the magic of the earlier shaman sorcerers. The Tao was the way of nature and the way of man; it was the cosmic process. Men should follow the Tao—'Riding on the normality of the universe', as it was put by Chuang Chou, 369-286 B.C. A Taoist story of this period tells of a famous royal butcher whose chopper never grew blunt because he knew of the Tao of animals, the run of bones and joints in each kind of carcase, which he accordingly cut up with skill. The Taoists sought out the secrets of nature, the way nature worked being suggested to them by the primitive beliefs of their idealized tribal society. The world and all the objects within it were thought to have come into being by a process analogous to sexual generation, the interaction of two opposite principles.

From about the fourth century it was thought, not only by the Taoists but also by others, that the two principles which produced all things by their interaction were what were termed the Yin and the Yang. The Yin was the passive, dark, and female force, while the Yang was the active, light, and male force. These two principles came from a primordial mixture of matter and energy in the form of a fluid in gyratory motion. Such a kind of motion separated out that which was dark and heavy from that which was light and fine, the former giving rise to the earth and the Yin principle, while the latter became the heavens and the Yang principle. The interaction between the two principles then produced the five elements, water, fire, wood, metal, and earth. First came water and fire which were largely Yin and Yang respectively in composition, then came wood in which Yin was slightly predominant and metal containing a small excess of Yang, while finally came earth in which the two principles were balanced. The continued interaction of the two principles produced a further differentation into all the objects of nature, the 'ten thousand things' of the world.

The Taoists followed the way of the universe in order to control human mortality; they endeavoured to extend the span of human life and to make youth eternal. For this purpose they developed respiration techniques, imitating what they thought to be the breathing of an embryo in the uterus. They advocated that men should sunbathe and women expose themselves to the moon, so that they would absorb the Yang and the Yin essences given out by the sun and the moon respectively. They developed gymnastic and sexual exercises to build up the life-giving Yang principle in men and the Yin principle in women. But, most important of all, they endeavoured to isolate the Yin and the Yang principles chemically, developing alchemy, dietetics, and pharmaceutics in so doing.

The first Ch'in emperor is said to have consulted Taoist magicians concerning the possibility of extending his life span, but the first known reference to alchemy occurs in the *History of The Earlier Han Dynasty*. Here it is related that in 133 B.C. an alchemist came to the Emperor Han Wu-Ti and offered to demonstrate how to make gold from cinnabar, liquids imbibed from vessels made of such gold conferring immortality upon the drinker. Chinese alchemy was much more concerned with the quest for the 'pill of immortality' than with the transmutation of base metals into gold, though it was associated with gold making from the start. Gold was thought to be important because it was the yellow sun metal, filled with the life-giving Yang principle. But cinnabar ranked higher, because of its red colour, and because it gave the living metal, quicksilver, on heating. Ko Hung, of the fourth century A.D., the most famous of the Chinese alchemists, wrote:

'After grass and wood have been burnt, they become ashes, but cinnabar can be changed into mercury by heating it over a fire, and vice versa. It is far different from the ordinary vegetable substance so it can make people immortal.'

As in the west, it was thought that minerals and metals grow underground. In China the doctrine goes back to the second century B.C. Ho Ting, of the fifth century A.D., wrote that in the earth, cinnabar is fertilized by a green Yang, giving after two hundred years a green substance pregnant with the metals. Lead is born first, then silver, and finally gold. Gold was thus the son of cinnabar. Following the death and resurrection theme, Ho Ting held that for gold to be born, the Yang must die and the Yin be condensed.

As in Europe, it was thought in China that such natural processes could be copied in the laboratory. Ko Hung regarded sublimation and distillation processes as most important in this respect, for they combined the Yang action of heating with the Yin action of cooling. The two principles could be isolated in a chemical form as mercury and sulphur. Mercury was largely Yin, and sulphur mainly Yang, their combination giving cinnabar, which was the starting point for both the natural and the artificial formation of metals, and for the preparation of the pill of immortality.

Chinese medicine was also shaped by Taoism through the dietetic techniques adopted to prolong life and cure sickness. It was the custom in China on New Year's Day to swallow a hen's egg which provided enough vitalizing substance for the following year. Foods were made of animals enjoying a long life, like the tortoise, whilst mineral substances of a supposedly high Yang content, such as sulphur, and saltpetre, were ingested to promote vitality. All the anatomical features of the human body, and the diseases to which it was subject, were of a Yin or Yang character. Fevers were regarded as Yang disorders, chills as Yin. Great attention was paid to the pulse in Chinese medicine, Yang disorders strengthening the pulse, Yin disorders weakening it. Diseases diagnosed by the pulse were treated by drugs to correct for the excess of Yin or Yang. Stimulants and hot pungent drugs were Yang purgative and bitter astringent drugs were Yin.

The standard Chinese medical work was the *Canon of Medicine*, which dates from Han times. The theories of anatomy and physiology that it contains are based mainly on the analogies of man to the state, and man, the microcosm, to the universe as macrocosm. The heaven is round and the earth is square, and thus the head is round and the feet are square. There are four seasons and twelve months in a year, and thus man has four limbs and twelve joints. The heart is the prince of the body, whilst the lungs are his ministers. The liver is the general of the body, with the gall bladder as the central office. The spleen and stomach are granaries, the intestines the communication and sewage systems. The *Canon of Medicine* contains the assertion, 'The blood flows continuously in a circle and never stops', but this cannot be regarded as a discovery of the circulation of the blood, for the Chinese did not distinguish veins from arteries. It was only an analogy drawn between the movement of the blood and the cycle of nature, the succession of the seasons, and the movements of the heavenly bodies, without empirical demonstration of its actuality.

The doctors were associated with the Chinese bureaucracy through a Medical Bureau, whilst the alchemists were outside it. The scientists and technicians who were the most closely associated with the official administration were however the mathematicians, astronomers, surveyors, and calendar makers, as in Egypt and Babylonia. The earliest Chinese mathematical work is the *Arithmetic in Nine Sections* which is said to have been first written under the early Chou emperors, *c.* 1000 B.C., though existing versions date from Ch'ang Tshang, who rewrote it *c.* 200 B.C. The problems dealt with by this work were firstly those of surveying, the determination of areas of triangles, trapeziums, and circles. The value of π was taken as 3 at first, and then $\sqrt{10}$ in the first century A.D. Secondly, commercial questions of proportions, percentages, and partnerships, were discussed, and then methods of determining the volumes of figures from their sides, and the length of their sides from their volumes, involving the use of square and cube roots. Pythagorean triangles and linear simultaneous equations were dealt with, and the rule of false position was given. This rule was a method of solving a problem by guessing at the answer, and determining the true solution from the errors introduced by such guesses.

In astronomy the Chinese have been credited with the making of observations in the far distant past. The Jesuits who came to Peking in the seventeenth century thought that Chinese astronomy dated from four thousand years before their time. However, the oracle bone inscriptions of the Anyang period *c.* 1500 B.C. are the oldest records of Chinese astronomy whilst there is not a great deal that is reliable before 400 B.C. Eclipses were noted on the Anyang oracle bones, and they were regularly recorded, together with the passage of comets from the seventh century B.C., covering periods, such as the Dark Ages, when there was little observation in the west.

The most famous Chinese astronomer was Shih Shen, *c.* 350 B.C., who mapped out the relative positions of some eight hundred stars. He knew that eclipses were due to interference effects between the heavenly bodies, and he gave rules for the prediction of eclipses based on the expected positions of the sun and the moon. He divided the circle into 365¼ ° to correspond to the number of days in a year, and he was aware of the Metonic cycle of 19 years which contains almost exactly 236 lunar months. A later official astronomer, Hu Hsi, discovered the precession of the equinoxes in A.D. 336, assigning the value of 1° in 50 years to the precession.

The pole star, and the circumpolar stars which did not rise or set, were the most important of the heavenly bodies for the Chinese. The pole star was regarded as the emperor of the skies, as it never moved, whilst the circumpolar stars were princes, and the other stars were officials. Chinese measurements of the positions of the heavenly bodies accordingly differed from those made in the west before modern times, in that they were based on the celestial sphere with its stationary pole, rather than upon the terrestrial sphere with its stationary observer.

One of the most difficult problems in early astronomy was to determine the sun's position relative to the fixed stars, for in the day-time, when the sun appears, the stars are blotted out. The Babylonians, Egyptians, and the Greeks after them, solved this problem by observing the stars that arose just before the dawn, such stars determining the sun's position at that time. The Egyptians in early times knew, for example, that Sirius arose with the sun about the time of the Nile flood. Such helical risings of the stars were terrestrial-horizon phenomena, and they gave as the basic line for astronomical measurements, the sun's apparent annual orbit through a belt of fixed stars, the ecliptic. The Chinese, on the other hand, keyed in the stars which rise and set with the circumpolar stars, so that they could calculate where they were, even when they were not visible. Thus the celestial sphere, studded with fixed stars as points of reference, provided the co-ordinates for the measurement of the positions of the heavenly bodies, the basic line being the celestial equator, and not the ecliptic. By a curious irony of history, the Jesuits in the late sixteenth century introduced the Chinese to the Greek method of using ecliptical co-ordinates, at the same time as Tycho Brahe was introducing equatorial measurements into Europe.

The astronomical calculations made by the Chinese were almost entirely algebraic, and thus their astronomy did not give them a picture of the configuration of the universe. Chinese technical astronomy was therefore largely separate from their cosmological speculations, which remained qualitative in character throughout their history. In the Han period there were three main world systems. The oldest was the Ka Thien system, which held that the skies were a hemispherical vault, and the earth an inverted bowl with linear edges, so that it formed a convex square. The heaven was not a regular hemisphere. It was elevated in the south, and depressed in the north, like 'an umbrella inclined over a chess board'. Thus the sun, rotating with the hemisphere, was visible when in the south but not when in the north. The sun, moon, and planets, rotated with the heaven, but they had proper motions of their own, 'like ants on a mill wheel', as it was put. Round the rim of the earth was an ocean into which the vault of heaven dipped at its periphery, the heaven and earth being supported by virtue of the air that was caught under their bowls. The heaven was 80,000 Li above the earth (three Li equal one English mile).

This 'Hemisphere Heaven' theory had disappeared by the end of Han times, being replaced by the Hun Thien 'Celestial Sphere' theory in the official dynasty histories after the Ch'in, 207 B.C. The Hun Thien theory is said to have originated in the second century B.C., the first account of it being given by Chang Heng in the first century A.D. According to this theory, the universe was a spheroid, some two million Li in diameter, being 1,000 Li shorter in the north-south direction than in the east-west direction. Chang Heng compared the universe to an egg, the yolk being the earth, supported on water, the shell being the heavens, supported by vapours. The third, Hsuan Yeh, or 'Empty Infinite Space' theory, is said to be old, though accounts of it date from the later Han period. According to this theory, the universe had no shape, nor any substance beyond the earth and heavenly bodies. Space was infinite and empty, the heavenly bodies were attached to nothing at all, they moved freely propelled by 'hard winds'. The 'Empty Infinite Space' theory was associated with the Taoists, whilst the 'Celestial Sphere' theory was adopted by the official Confucian scholars. However, the Confucians took over the early Taoist natural philosophy, as the Taoists gradually became religious mystics. Thus elements of the 'Infinite Space' theory were imported into the official 'Celestial Sphere' theory, particularly during the Neo-Confucian movement of the twelfth century. However, as early as the fourth century, the official astronomer, Yu Hsi, who discovered the precession of the equinoxes, thought that the heavens were immeasurably high, though they had a limit, with the heavenly bodies moving freely beneath them.

The Chinese did not conceive of the heavenly bodies as divine and powerful beings who determined events on earth. They had a system of astrology, but it was peculiar in that the determinations between heaven and earth cut both ways. The appearance of a comet might portend disaster, but equally it might exemplify the cosmic disarray caused by some human departure from the customary order of things. Thus there was no Divine law-giver controlling the universe; the cosmic process was a web of interrelations between the varied objects of nature, ordered only by custom. For the Taoists, all of the elements in the cosmic process were of equal weight, none had supremacy over another. The Taoist, Chuang Chou, 369-286 B.C., wrote:

'It might seem as if there were a real Governor, but there is no evidence of his existence. One may believe that he exists, but we do not see his form. The hundred parts of the human body, with its nine orifices, and six viscera, all are complete in their places. Which should one prefer? Is there any true ruler other than themselves?'

The Confucians did not think that all things were of equal weight in the cosmic process, but they were all interrelated, and all bound by the force of custom. The Confucian, Hsun Ch'ing, c. 300-240 B.C., developed a hierarchical classification of nature, similar to Aristotle's division of terrestrial creatures into those with vegetable, animal and rational souls. All of his orders were moved and related by custom and tradition. He said:

'Custom is that whereby heaven and earth unite, whereby the sun and moon are brilliant, whereby the four seasons are ordered, whereby the stars move in their courses, whereby rivers flow, whereby all things prosper, whereby love and hate are tempered, whereby joy and anger keep their proper place. It causes the lower orders to obey, and the upper classes to be illustrious; through a myriad changes it prevents going astray. If one departs from it, one will be destroyed. Is not Custom the greatest of all principles?'

The customary behaviour of men, and of the years and the four seasons, was easily observed and ascertained, but for more detailed knowledge, methods of divination were brought in. During Shang times animal shoulder bones were cracked by the application of a hot rod, the run of the cracks being taken as a forecast of the trend of events. In the Chou period the drawing of lots was resorted to, the choice being between a long and a short bamboo stick. The range of choice was gradually extended by forming eight trigrams from such long and short sticks, the random selection of trigrams being practised in the Warring States period. Finally in Han times, a complex method of prognostication was developed whereby a ladle, a model of the Great Dipper, the most important of the circumpolar stars, was spun on a divination board of trigrams, the result being indicated by the position at which the ladle came to rest. It has been suggested that the orientating property of the magnet was discovered in this way, a ladle of lodestone magically appearing to return always to the same position.

The developments described hitherto took place during the early history of China, roughly up to the end of the Han dynasty, A.D., 220. From this time, the previous pattern of Chinese history repeated itself. The empire of the Han rulers broke up into a number of autonomous warring states, each battling for the supremacy that was finally achieved by the Sui in A.D. 581. The Sui dynasty, 581-618, like the Ch'in before it, was ruthless and short-lived, giving way to the Thang, 618-906, and the Sung, 960-1279, where the more tolerant system of government by Confucian custom prevailed. During the second period of warring states, 221-580, Buddhism came to China, and it gained a strong hold, filling the moral vacuum of the times. Taoism now became a mystical religion with its temples, monks, and a succession of popes, forming a native opposition to Buddhism. However, in Thang times Taoism revived, and the Chinese alchemists became active once more. They practised the distillation of mercury from cinnabar, and perhaps also the distillation of alcoholic liquors. The quest for the secrets of longevity was pursued afresh, indeed, seven of the twenty-two Thang emperors are said to have died from an overdose of the 'pill of immortality'.

Porcelain manufacture, which had started with the crude proto-porcelain of Han times, came to a high level of perfection under the Thang, an imperial bureau for its manufacture being set up in 621. The wheelbarrow had been invented during the fifth century, and in the seventh appeared tread-mill-operated paddle boats, fitted with water-tight bulkheads and a stern-post rudder. During the Thang period, the first block printing began in the Buddhist monasteries of China. The earliest printed book is the Diamond Sutra, dated A.D. 868, found in the Caves of the Thousand Buddhas at Kansu. The printing of books soon became general throughout China, the Confucian classics being printed from 932 and the official dynastic histories between 994 and 1063. Block printing spread to the border tribe of the Uigurs some time before 1206, the year in which they were overrun by the Mongols. The Uigurs printed the Buddhist works, with Sanskrit notes and Chinese page numbers, in their own Turkish language, using an alphabetic script that had originated in Syria. Movable type of clay was invented in China by Pi Sheng in the 1040's. Wooden movable type came into use some time later, specimens dating from c. 1300 being found in the Caves of the Thousand Buddhas. Finally, movable type of cast metal was developed, specimens of fonts dating from 1403 being found in Korea, and books printed from such type dating from 1409.

By the end of the Thang period, gunpowder had been developed in China, whilst firearms appeared before the end of the Sung. Saltpetre occurs as a natural efflorescence of the soil in China and India. It is first mentioned in Chinese texts of the first century B.C. The Chinese alchemists of the third century A.D. mixed sulphur and saltpetre in the correct proportions for gunpowder, and exposed the mixture to high temperatures. Such experiments may have been the origin of the fireworks mentioned in seventh-century texts. In the wars of the Thang period, fire arrows were used, but they were probably only burning pitch attached to an arrow head. In 969 a new type of fire arrow appeared, which seems to have been a kind of rocket. A record of 1040 states that gunpowder was used in the new fire rockets, the record giving the correct formula for gunpowder, together with the details of its preparation. A Chinese edict of 1067 prohibited the export of sulphur and saltpetre to foreign lands, which is an indication of how valuable gunpowder was considered to be in China at the time.

Marco Polo, who obtained a high position in the Chinese bureaucracy under the Mongols, said that the Chinese had firearms by A.D. 1237. Chinese records of the time tell of several different kinds of gunpowder weapons. The first reference to a firearm projecting bullets occurs in 1259 when the Sung armies repelled the Tartars with firearms made of bamboo tubes. The Tartars in turn used gunpowder weapons against the Mongols. in 1231 they used a weapon called 'The Heaven Shaking Thunder', which appears to have been a grenade, an iron vessel filled with gunpowder and fitted with a fuse, that was hurled from a catapult. The Mongols captured a Chinese ammunition works in 1233, the general commanding, Souboutai, leading the Mongol invasions of Europe a few years later. During their invasion of Japan, 1274-81, the Mongols used iron cannon according to three different accounts, one of which adds that they used iron cannon balls. The oldest existing Chinese cannon that can be dated are A.D. 1354, 1357, and 1377, the oldest European cannon that can be so dated are A.D. 1380, 1395, and 1410.

Another Chinese development of the Sung period was the use of the magnetic compass for land and sea travel. In 1086 Shen Kua, a scholar-director of waterworks, wrote a work in which he described the various wonders he had seen in his time, such things as fossils, relief maps, and actual cases of transmutation of metals, and a magical means of finding direction. The transmutation he described was the conversion of iron into copper by means of a solution of copper sulphate, which was long regarded in the west as a real change of one metal into another. As regards the magnetic compass, he said that when magicians want to find direction, they rub a needle on a lodestone and then hang it up by means of a thin thread. The needle will usually point to the south, but sometimes, he adds, it will point to the north. By 1150 such compasses were used regularly for sea voyages and land travel, the declination of the compass from the true north and south being recognized by this time.

The Sung dynasty, like the Han, was rich in mathematicians, astronomers, calendar makers and surveyors. The official Chinese biography of such men, published in 1764, lists thirty-eight eminent mathematicians from the Han period, and twenty-nine from the Sung, whilst the highest number from the dynasties between was nine from the Sui. In 1247 the *Nine Sections of Mathematics* was published by Ch'in Kui Shao, in which place value and a sign of zero were introduced into Chinese numerals. He gave algebraic methods of solving trigonometrical problems, and dealt with higher numerical equations and indeterminate equations. Chon Huo, 1011-75, president of the bureau of astronomy, solved the problem of summing a series of squares over a given number of terms, and Chu Shi Kie, c. 1280, gave the first description of Pascal's triangle of binomial coefficients.

In astronomy little that was new appeared in Sung times, for Chinese astronomy was mainly concerned with calendar problems, and these had been largely solved during the Han period. However, the Confucians of the Sung dynasty developed a nature philosophy of their own, taking over elements of the earlier Taoist speculations and incorporating them into the official 'Celestial Sphere' cosmology. The most notable Neo-Confucian was Chu Hsi, 1131-1200. He held that, in the beginning, the universe was a primordial chaos of matter in motion. This motion assumed the form of a vortex, whereby the heavy matter was separated from the light, the heavy matter going to the centre of the whirlpool to form the earth, and the light matter remaining above to form the heavens. The centre was the only motionless part of a whirlpool, hence the earth must be at the centre of the world. The vortex ordered the universe, and kept the earth in position. 'Should heaven stop only for one instant,' he said, 'earth must fall down.' Chu Hsi thought that the heavenly bodies were kept in motion by 'hard winds', each heavenly body having its own wind that formed a separate shell of the cosmic vortex. There were nine such shells of wind, the outermost moving so rapidly, and being so 'hard', that it constituted the periphery of the universe, though space stretched infinitely beyond. The planets were carried round by the rotary wind of the outer sphere, though they had proper motions of their own. The sun moved only one degree a day against the outer heaven, whilst the moon moved thirteen degrees a day.

This was because the sun was a prince, whilst the moon was only a minister, and so hurried more on his course.

Chu Hsi also recognized fossils to be the remains of organic beings. He wrote that,

'one frequently sees on high mountains conches and oyster shells, sometimes embedded in rocks. These rocks in pristine times were earth, and the shell fish and oysters lived in water. Subsequently everything was inverted. Things from the bottom came to the top, and the soft became hard. Careful considerations of these facts will lead to far-reaching conclusions.'

Such a passage represents Chinese science at its best, the combination of speculation with a certain acuteness of observation. Beyond this point the Chinese never went. They could not combine theory with experimentation, for the scholars regarded practical work as degrading. Chu Hsi himself related that,

'Sun Szu-mo was a noted doctor of literature of the Thang dynasty, but as he practised healing as a profession he was relegated to the class of artisans. What a pity!'

The Thang Annals record that,

'Mathematicians, surveyors, physicians, and magicians were charlatans. The sages did not regard them as educated.'

Thus the work of the scholars was largely speculative, whilst the men who accomplished the technical work of surveying, calendar-making, and astronomical observation, were largely empirical and untheoretical in their labours. The Chinese astronomers were not very interested in the Copernicus-Ptolemy-Tycho Brahe controversy when they were introduced to it by the Jesuits in the sixteenth and seventeenth centuries. The biographer of the Chinese mathematicians, astronomers, surveyors, and calendar makers, Yuan Yuan, wrote in the eighteenth century:

'Our ancients sought phenomena, and ignored theoretical explanation. Since the arrival of the Europeans, the question has always been concerning explanations, circular orbits, mean movements, eclipses and squares . . . (but) really it does not seem to me the least inconvenient to ignore the western theoretical explanations and simply consider the facts.'

Such a separation of theoretical and empirical enquiry has been a feature of most civilizations that have had a stratified agricultural character. Before the industrial revolution of the eighteenth century, science had been promoted the most vigorously in commercialized civilizations, such as those of ancient Greece, and Renaissance Europe. The Chinese never had such a civilization, the policies of their rulers being directed constantly against the independent merchant and producer. Each commodity was nationalized as soon as it became important—salt under the Chou, iron under the Ch'in, tea under the Thang—so that the control of it was taken out of private hands. Thus the craft and merchant guilds of China were puny compared with those of medieval Europe, and they did not evolve an independent tradition of their own. The Chinese, for example, never had an atomic philosophy; such a point of view seeming to appeal to the individualistic, mercantile type of mind, if we are to judge by the periods when the atomic philosophy was popular, before it was of scientific use, notably in ancient Greece, and Renaissance Europe. The most independent figure in Chinese society was the Taoist hermit in the wilderness, and he always remained within the bounds set by the concepts of his idealized primitivism.

Chapter 8

The Science of India

CIVILIZED SOCIETY arose in India, as it did in Mesopotamia, Egypt, and China, with a bronze-age culture in a river valley. As yet, however, not a great deal is known concerning the civilization of the Indus which flourished *c*. 3000 B.C. The peoples of the Indus valley had a pictographic script and a decimal numeral system. They used the same fast-spinning potter's wheel as the Sumerians, and alloyed copper with tin to make bronze, but they wove cotton rather than the flax or wool of the west or the silk of the east. About 2000 B.C., however, the civilization of the Indus became extinct.

The end of the ancient Indus civilization was probably due to the Sanskrit-speaking Aryans who invaded India from the north and set up their caste society over the conquered Dravidians. They had their scholars in the Brahmin priests, who handed on their ritual and learning orally in ancient Sanskrit long after Prakrit had come into everyday use. An alphabetic script was adopted some time before the beginning of the Christian era, and the ancient learning was then transcribed. Such ancient texts, the Vedas, contain references to the sun, moon, and some star constellations, but the planets were not recognized. Some specific cases of the Pythagorean theorem were noted in connection with the building of altars, and mention was made of the metals, gold, silver, lead, iron, and possibly tin.

Buddhism arose in the sixth century B.C., the new religion being adopted by Asoka, *c*. 260 B.C., the third emperor of the Mauyra dynasty which had arisen from the confusion left by the Greek invasions of India, 327-323 B.C. According to inscriptions carved on stone, Asoka founded the first hospitals and herbal gardens of India, placing them under Buddhist control in opposition to the Hindu Brahmins. Such inscriptions indicate further that a numeral system with some similarities to the modern Hindu-Arabic system was in use at the time. Later inscriptions show the probable development of our number system, place value and then a sign for zero appearing. A plate of the year A.D. 595 gives the date 346 in decimal place value notation, and the earliest undoubted occurrence of a zero has been found on a monument at Gwalior where, in A.D. 876, the number 270 was inscribed as we write it today. The first reference to the Hindu numerals outside of India occurs in a work by Severus Sebokht, a titular bishop who lived in the convent of Kenneshre on the Euphrates. Comparing Greek and Syrian knowledge, he wrote in A.D. 662:

'I will omit all discussion of the sciences of the Hindus; their valuable method of calculation; and their computing which surpasses all description. I wish only to say that this computation is done by means of nine signs.'

There are few authentic Indian records before the Muslim invasions, (A.D. 664 on), apart from coins, stone inscriptions, and land grants engraved on metal. The Hindus were acquainted with some of the science of the Greeks and perhaps that of the Babylonians, but in the absence of records it is not known how and when the knowledge came to India. It is possible that the transmission occurred some time between 150 B.C. and A.D. 140, as the Hindu astronomers knew of the work of Hipparchus but not that of Ptolemy, and that the route was the sea trade between the Roman Empire and Ujjain, the Indian trading centre with the west. Ujjain had been the vice-regal seat of Asoka during his father's reign at Patna, Ujjain and Patna being the royal cities of subsequent dynasties, and, together with Mysore in the south, the main centres of Hindu science. The first Hindu scientists of whom we have definite knowledge were the two Aryabhatas, c. A.D. 475-550, who worked at Patna; Varahamihira, c. 505, who had an astronomical observatory at Ujjain; and Brahmagupta, c. 628, who also worked at Ujjain. Later figures were Mahavira, c. 850, at Mysore, and Bhaskara, 1114-85, who came from the south but worked at Ujjain.

Varahamihira gave the first notable account of the Hindu astronomical works, the Siddhantas. He described five such Siddhantas written before his time, four of them based upon Greek astronomy and the other on ancient Vedic astrology. One of the four, the Romaka Siddhanta, indicates by its name that it came from the west (Rome), whilst Varahamihira quoted the Yavanas, or the peoples of the west, as the source of his astronomy. He, and the other Hindu astronomers, supposed that the earth was spherical, with the sun, moon, and planets, at distances from it which were proportional to their periods of revolution. Such a view was based upon the assumption that all of the heavenly bodies moved in circles round the earth with the same uniform speeds. Most of the Hindu astronomers held that each of the bodies of the solar system possessed a proper motion of its own, caused by a wind, whilst in addition there was a larger aerial vortex which carried all of the heavenly bodies round the earth once in twenty-four hours. One or both of the Aryabhatas dispensed with the larger vortex, by supposing that the earth performed a diurnal rotation upon its axis by virtue of a wind about one hundred miles above the surface of the earth, though such a view was not generally accepted. To account for the complexity of the motions of the planets, the Hindus used the Greek mathematical device of the epicycle, introducing ovoid epicycles to obtain more exact agreement. In dealing with the motions of the moon, however, the Hindu astronomers employed methods which show distinctive traces of Babylonian influence.

The mathematical work of the Hindus was more notable than their astronomy, and it continued the algebraic tradition of Babylonian mathematics rather than the geometrical tradition of the Greek. The Aryabhatas studied the summation of arithmetic series, and they attempted to solve quadratic, and linear indeterminate equations. They also introduced the use of the sines of angles instead of the chords used by the Greeks, beginning the study of trigonometry. Brahmagupta developed the application of explicitly general algebraic methods to astronomical problems. He gave general methods for solving indeterminate equations of the first degree, and for extracting one root of a quadratic equation, finding also a general formula for the area of any quadrilateral with two parallel sides.

Mahavira discussed the operations of addition, subtraction, multiplication, and division, including the use of the zero. He maintained that the division of any number by zero gave zero as the result. Bhaskara later was the first to point out that the result would be infinity. Two problems given by Mahavira are found in the Chinese *Arithmetic in Nine Sections*, (*c.* 200 B.C.) one of the problems occurring in all Hindu mathematical works from the sixth century A.D. Buddhism provided the connecting link, Indian missionaries going out to China from the second century A.D., and Chinese pilgrims visiting India from the fourth. The official Chinese history of the Sui dynasty, which was completed by A.D. 610, lists a number of Hindu mathematical, astronomical, and medical works that had been translated into Chinese.

The medicine and chemistry of the Hindus were less notable than their mathematics and astronomy. The oldest Hindu medical work is the Bower manuscript, which dates from about the fourth century B.C. The manuscript consists of a list of drugs, and the lore of their use, and these are copied by later works, notably the Charaka, a medical compendium that has been placed in the second century A.D., and the Susruta, a fifth century treatise on surgery. The later works are also dependent upon Greek sources, for the Charaka gives rules of syllogistic reasoning taken from Aristotle. The Charaka distinguished three vital processes in the human body. The first was due to the operations of air in the region below the navel, the second was due to bile which controlled the region between the navel and the heart, whilst the third consisted of the activity of phlegm above the heart. These vital processes engendered the seven principles, chyle, blood, flesh, fat, bone, marrow and semen, health depending upon the quantitative harmony of the seven principles, any disorder resulting in a disease.

The surgical work, the Susruta, is superior to the Charaka. It describes some 121 different surgical instruments, and gives an account of most of the surgical operations known before modern times. The connection between malaria and mosquitoes was noted in the Susruta, and also the voiding of sweet urine by diabetic patients. Mention is made in these works of six metals, gold, silver, copper, tin, lead and iron, and also of the caustic alkalis, which were distinguished from mild alkalis. A later medical work, the Vagbhata of the seventh century, contains the first Indian mention of mercury.

The practice of alchemy appears to have begun in the seventh century, according to the Chinese pilgrims of that period. It seems to have been associated with the revival of Brahminism, for the main alchemical works of the Hindus are the Tantras, works based on the old Vedas, which were written to disseminate the Brahmin religion in opposition to Buddhism. It has been claimed that the Hindu alchemists knew of the strong mineral acids, the claim being based on a work which is said to be of the eighth century, that speaks of the 'killing' of metals by a liquid, and a Tantra, which has been placed in the twelfth century, that describes the preparation of this liquid from green vitriol. There is also a Chinese record of A.D. 780 which states that,

'In India there is a substance called Pan-ch'a-cho Water, which is produced from minerals in the mountains. . . . (it) can dissolve herbs, wood, metals and iron: indeed if it is put into a person's hand it will destroy it.'

The alchemy of the Hindus, like that of the Chinese, seems to have been concerned primarily with the search for the elixir of immortality, though this involved gold making. As elsewhere, the primary substances for the preparation of gold, and the elixir of life, were thought to be mercury and sulphur though the Hindus conceived of mercury as the male principle, and sulphur as the female, in contradistinction to China and the west where sulphur was thought of as male, and mercury as female. Besides the two principles of mercury and sulphur, the Hindus had five elements, earth, water, air, fire, and ether, or space in itself, which seem to have been taken over from the Greeks. So too was the atomic philosophy, which by the fifth century was well established amongst the Brahmins and the Jain sect of the Buddhists.

The most important contributions of the Hindus to modern science were our modern numeral system, and the development of generalized algebraic operations. Their other sciences were borrowed and somewhat debased by the persistence of Vedic lore, even their mathematics being mixed in quality. The Persian, Albiruni, 973-1048, lived at Ghazna in Afghanistan from 1017 to 1030, during which time he studied the

Sanskrit literature and wrote a history of the Hindus and their sciences. Writing of their mathematical sciences, he said:

'I can only compare their astronomical and mathematical literature . . . to a mixture of pearl shells and sour dates, or of costly crystals and common pebbles. Both kinds of things are equal in their eyes, since they cannot rise themselves to the methods of strictly scientific deduction.'

Chapter 9

Science and Technology in the Muslim World

THE INHABITANTS of Arabia, who carved out for themselves an empire stretching from the Pyrenees in the west to the borders of China in the east between A.D. 634 and 750, were not entirely barbaric nomads before they began their conquests. Arab merchants took part in the sea-borne trade from Ujjain in India to Alexandria in Egypt, largely monopolizing the route from Aden to India. Muhammad the Prophet, whose doctrines provided the inspiration for the Arab conquests, was said to have been himself a merchant. Border Arabian tribes had entered the service of the Romans and the Byzantine Greeks as auxiliaries, and they had learned something of their masters' ways. Some Arabs had become converts to Christianity and were taken into the civil service of the Byzantine empire, notably in Syria. Thus, there were educated elements amongst the Arabs before the rise of Islam, a factor which facilitated the later Muslim assimilation of Greek science.

One such border tribe, the Umayyads, who had been auxiliaries of the Romans, took over Syria intact and set up the first Muslim Caliphate at Damascus in 661. The Umayyads were Hellenized from the start. They gathered men of science to Damascus, and founded an astronomical observatory there as early as 700. However, Hellenized Arabs tended to lack the energy deriving from religious fervour, and the Umayyads fell in 749 to the more religious Abbasids who set up their Caliphate at Baghdad. The Abbasids became Persianized rather than Hellenized, adopting the culture of the land they had conquered. Following the example of the Persians, who had set up a school of medicine and astronomy at Jundishapur in the fifth century, the second Abbasid Caliph, Al-Mansur, brought men of science to Baghdad. An Indian astronomer, Manka, was presented to Al-Mansur in 773, and the Hindu scientific works, the Siddhantas, Charaka, and Susrata, were translated. The third Caliph, Harun al-Rashid, ordered the collection of original Greek treatises, and the fourth, Al-Mamun, set up a 'House of Widsom', c. 828, for the translation of such works. Here the chief translator was the Nestorian, Hunayn ibn Ishaq, c. 809-77, who translated most of Galen's medical writings, and began the translation of Ptolemy's astronomy. His work was continued by some ninety pupils, the chief of whom were his son, Ishaq, died 910, who translated the works of Ptolemy and Euclid, and his nephew, Hubaysh, who translated the works of Hippocrates and Dioscorides.

Al-Mamun also founded an astronomical observatory at Baghdad in 829. Here observations were begun by Al-Far-ghani, died c. 850, and continued by the Sabian star worshippers, Al-Battani, c. 858-929, and Thabit ibn Qurra, c. 826-901. These men came from Harran in Mesopotamia, where the ancient Babylonian religion, with its astrology and its worship of the stars, lived on in the form of the pagan Sabian sect, tolerated by all the successive conquerors of Mesopotamia until the thirteenth century when the Sabians were annihilated by the Mongols. Al-Battani obtained values for the obliquity of the ecliptic and the precession of the equinoxes which were more accurate than those of Ptolemy, and he discovered that the sun's eccentricity was changing (in modern terms, that the earth's orbit is a varying ellipse). About this time, Al-Khwarizmi, died c. 835, introduced the Indian numerals and Indian methods of calculation to the Muslim world, though his algebra was inferior to that of the Hindus.

The first original Muslim writer on medical matters was the Persian, Al-Razi, 865-925, who studied and worked at Baghdad under one of Hunayn's disciples. Al-Razi, or Rhazes as he was known to the west, wrote over a hundred works, the best known being the *Comprehensive Book*, which embraced the whole of the Greek, Indian, and Middle East medicine then known. He may have owed something to Chinese sources as well, for Al-Nadim, writing in 988 shortly after the time of Rhazes, related that a Chinese scholar stayed with Rhazes about a year, learning to speak Arabic, and translating the works of Galen into Chinese. Galen does not appear to have influenced the medicine of China, but pulse lore, which played a great part in Chinese medicine, appears in the encyclopedic *Canon of Medicine,* written by the next great Muslim physician, Ibn Sina, 980-1037, of Bokhara. Neither Rhazes nor Ibn Sina (known in the west as Avicenna) improved upon the theories of Galen, though on the practical side they knew of a much larger number of drugs.

Alchemy arose in Islam during the ninth century with the person of Jabir ibn Hayyan, called 'The Mystic'. The writings that have come down to us as the works of Jabir, or Geber as he was called in Medieval Europe, appear to be, however, a tenth-century collection due to a mystic sect, calling themselves the 'Brethren of Purity'. Alchemy always has had a somewhat 'unofficial' character, displaying a connection with mystical religion on the one hand, and with the chemical craft tradition on the other. In Islam such connections were perhaps more marked than elsewhere. The orthodox Muslim religion was that of the official Sunni sects, whilst amongst the rank and file of Islam the mysticism of the Sufi was widespread. One of the more radical sects of the Sufi was the Qarmati, who held that all men were equal, and who endeavoured to give parity to their fellows through educational activities, such as the founding of schools and the writing of encyclopedias. They were particularly interested in artisans, developing, if not originating, the guilds of Islam. They set up schools of 'Pure Brethren' in the various cities of Islam, to work for the dissemination of their ideas. The 'Brethren of Purity', who wrote the main corpus of Islamic alchemy, seem to have been such a school at Basra, which had been the birthplace of Sufism in the ninth century. Their alchemical works were only part of an encyclopedia which they wrote, seventeen out of fifty-two treatises in the encyclopedia being devoted to scientific matters. The work was declared heretical, and was burnt by the orthodox Sunni of Baghdad, and finally the 'Brethren of Purity' were suppressed in the eleventh century.

The 'Brethren of Purity' were opposed to the deductive, geometrical kind of reasoning which the orthodox Muslim scholars had inherited from the Greeks. They exalted mystery above reason, and they held that mysteries could be explored empirically. The conception that man is a microcosm of the whole world, which throughout history has found favour with the alchemists and religious mystics, was taken up by the 'Brethren of Purity' and made the basis of their world system. They were indeed the first to work out in detail the consequences of the idea that man is a microcosm or an epitome of the whole universe, finding analogies and correspondences between all aspects of the anatomy and physiology of man and the structure and workings of the world which were then known. In the specific field of chemistry they divided natural substances into two main classes, 'bodies' and 'spirits', by analogy with the view that man was made up of a body and soul. 'Spirits' were volatile substances: 'bodies' substances which were non-volatile.

They formulated the doctrine that all things, and in particular the metals, were formed by the interaction of the principles of mercury and sulphur. The beginnings of such a doctrine were present both in the alchemy of China and that of Alexandria. The principle of sulphur was the active, male, and fiery essence, the Yang of the Chinese, and the Smoky Exhalation of the *Meteorology* ascribed to Aristotle. The principle of mercury was the receptive, female, liquid essence, the Yin of the Chinese, and the Moist Exhalation of the *Meteorology*. The Muslim alchemists adopted the Greek doctrine of the four elements, and they suggested that a metal could be transmuted by changing quantitatively its elementary constitution. On the practical side, the alchemists of Islam were notable in that they used the balance and studied chemical operations quantitatively. They were also acquainted with chemicals unknown to the Greeks, such as the mineral acids, and saltpetre which they called 'Snow of China'.

A technique that the Muslims took over from the Chinese and handed on to the western world was paper-making. At the battle of Samarkand in 704 the Muslims captured some Chinese paper-makers who handed on their art to their captors. The first paper mill in Islam was set up at Samarkand in 751, and the second at Baghdad in 793. Paper-making had spread to Egypt by 900, and Spain by 1100, when it came to northern Europe, the first paper mill north of the Pyrenees being that at Herault, set up in 1189.

Baghdad gradually declined in cultural importance, as the Seljuk Turks, beginning as mercenaries, took over more and more the control of the Eastern Caliphate. Some scholars remained on under the Turks, like the Persian poet and mathematician, Omar Khayyam, died 1123, who developed the mathematics of Al-Khwarizmi, discussing cubic equations, whilst the latter had only dealt with quadratics. Others moved east to Muslim India, such as the Persian, Al-Biruni, who settled at Ghazna, Afghanistan, and wrote his history of India. However, most Muslim scholars moved west to Cairo, notably in the reign of the Fatimid Caliph, Al-Hakim, 996-1020, who had set up a 'House of Science' in 995.

From Basra came Al-Hazen, 965-1038, who is noted for his work on optics. He opposed the theory of Euclid, Ptolemy, and other ancients, that the eye sends out rays of light to view objects. Al-Hazen held that rays of light came from the object seen, light spreading out spherically from any source. His experimental study of magnifying glasses brought him near to the modern theory of convex lenses and, on the subject of refraction in general, he showed Ptolemy's crude law, that the angle of incidence is proportional to the angle of refraction for a given interface, to be true only at small angles.

Another scientist working at Cairo in the time of Al-Hakim was the astronomer, Ibn Yunis, died 1009. He gathered together the records of observations made in the previous two hundred years and prepared from them the Hakemite astronomical tables, so named in honour of his patron. A little earlier, Al-Masudi, died 957, of Cairo, wrote an encyclopedic natural history which contains the first description of windmills. Such mills had a vertical axis of rotation and were fitted with sails similar to those of a ship.

Scientific activity continued on in Egypt under the Ayyubide Sultans, the Jewish philosopher, Maimonides, coming from Spain to take up the post of physician to Saladin, 1174-93, the founder of the dynasty. The main work of Maimonides, 1135-1204, was of a philosophical nature, though he was interested in medical questions, and was critical of Galen's theories. A later Cairo physician, Ibn al-Nafis, 1210-88, who came from Damascus to take charge of the Nasiri hospital in Cairo, was more positive in his criticism of Galen. He pointed out that the dividing wall of the heart, the septum, was solid, and quite devoid of the pores permitting the passage of blood, which Galen had postulated. Thus, he argued, the blood must flow from the right to left ventricle of the heart through the lungs. In this way Ibn al-Nafis arrived at the theory of the lesser circulation of the blood, but his discovery did not pass into the mainstream of science as his work did not come to light until the present century.

A third group of Muslim scientists arose in Spain, where a surviving member of the first Umayyad Caliphate of Damascus had set up the independent kingdom of Andalusia in 755, his descendants calling themselves Caliphs of Cordova from the tenth century. A library and a scientific academy were established at Cordova, in 970, and similar institutions were established later at Toledo. Of this period was Abulcasis, died c. 1013, a court physician of Cordova, who wrote a great medical handbook in thirty sections, the last one dealing with surgery which up to that time had been neglected by Muslim authors.

Astronomy arose in the western Caliphate with Al-Zarkali, 1029-87, of Cordova, who drew up the Toledian astronomical tables in 1080, and modified the Ptolemaic scheme of the heavens by suggesting an elliptical deferent for the epicycle of the planet Mercury. The Ptolemaic system was much criticized by the Spanish Muslims, as they desired a physically real system of the world, and were greatly influenced by an Aristotelian current of thought which found a philosophical expression in the works of Averroes, 1125-98. The movement began with Avempace, died 1139, of Saragossa, and was continued by Abubacer, died 1185, of Granada, and Alpetrugius, died c. 1200. They rejected Ptolemy's device of the epicycle on the grounds that the planets must revolve about a physically real central body, not a geometrical point. Hence they endeavoured to work out a system of the heavens that was physically feasible, based upon the Eudoxian system of homocentric spheres which had been adopted by Aristotle. In this task they were not successful for even Eudoxus had not been able to explain the well-known approach and recession of the planets, and now the movements of the heavenly bodies requiring explanation were much more numerous and complex.

The rise of science in the western Caliphate was only just prior to the Christian invasions of Spain. Muslim science still flourished in the cities taken over by the Christians, notably at Toledo which fell in 1085, and Spain, therefore, was the main route whereby ancient science came to the west. The Chinese art of paper-making came by this route also, though the Muslims did not hand on a great number of practical techniques. The Muslims do not appear to have known of gunpowder and firearms before the Mongol invasions, nor of Chinese printing methods until quite late. The first Muslim description of the printing methods used by the Chinese occurs in the works of the Persian scholar, Al-Banakati, c. 1200, whilst the first block-printing in Islam took place in Egypt some time between 900 and 1350.

The Mongols, who conquered the Eastern Caliphate at one end of Asia, and the Sung dynasty of the Chinese at the other, were more uniformly barbaric in their early days than the Arabs had been, though they rose to the level of civilization obtaining amongst the peoples that they had conquered. When the Mongol conquests were completed, intercourse between east and west was much easier and much more direct than it had been previously. Marco Polo, 1254-1324, could go to China and obtain a high post in the imperial salt bureau, whilst the Chinese, Mar Jaballaha, 1244-1317, came west to become Patriarch of the Nestorians in 1281.

The first Mongol attacks were made upon China under Genghis Khan in 1214. In 1233 General Souboutai captured a Chinese munitions factory at Pien Ching, and he spared the munition-workers there, so that he had gunpowder and grenades, if not firearms, on his European campaign which began in 1235. It is possible therefore that gunpowder, and perhaps firearms, came to Europe through the Mongols. The idea of printing, though probably not the technical details, may also have come to Europe in this way, for playing cards, which had been printed in China for some time, appeared in Europe shortly after the Mongol invasions, and were recognized to be of eastern origin. The wheelbarrow and the casting of iron also appear in Europe about this time, though in these cases an independent development is likely. In return distilled alcoholic liquors and eye-glasses reached China from Europe through the Mongols in the thirteenth century.

The Mongols took over the civilization of China intact, utilizing the existing scholar administration, though they placed foreigners like Marco Polo in the higher posts. They set up an observatory at Pekin and staffed it with Muslims from the west as well as with native Chinese. Some of the instruments used still exist, notably the great armillary spheres and mural quadrants. At Paris is preserved a treatise with Arabic and Chinese characters on its title page, the work being a set of lunar tables prepared by Abn Ahmad of Samarkand in 1362 for his Mongol patron in Pekin.

Further west, Hulago Khan, the grandson of Gengis Khan, sacked Baghdad in 1258 and finally ended the eastern Abbasid Caliphate. Hulago Khan founded an observatory at Maragha in Azerbaidjan, south of Tabriz, and placed it under the direction of his vizier, Nasir ed-din, 1201-74, who was himself an astronomer. Here a library of some 400,-000 volumes was gathered together, whilst astronomers were brought from places as far distant as China and Spain. From China came Fu Meng-chi, and from Andalusia, Al-Maghribi, who wrote a monograph on the calendars of the Chinese and the Uigurs. Finally, after twelve years of observation, the Ilkohanic tables were brought out by Nasir ed-din and his astronomers. At Tabriz itself, paper money was printed in 1294, bearing both Chinese and Arabic characters, and a few years later, the Persian physician, Rashid al-Din, 1247-1318, gave a detailed account of printing methods used by the Chinese. The last burst of Tartar science came in 1420 when an observatory was set up at Samarkand by Ulugh Beigh, 1394-1449, the grandson of Tamerlane. Here the positions of the stars studied by Hipparchus were mapped afresh, these observations being the most accurate of all such made before the time of Tycho Brahe in the sixteenth century.

Chapter 10

Technology and the Craft Tradition in Medieval Europe

THE 'DARK ages', extending from about the time of the fall of Rome, 455, to the first intellectual reawakening of the west under Pope Sylvester II, 999-1003, are traditionally regarded as a somewhat barren period in the history of European civilization. Such was the case in the field of natural philosophy, but during those centuries appeared a number of fundamental technical innovations which provided the basis of a way of life materially superior to that of classical antiquity for the majority of men. The Teutonic barbarians who invaded the disintegrating Roman empire brought with them numerous small things to which we are accustomed; the wearing of trousers instead of the toga, the use of butter instead of olive oil, improved methods of felt-making, the ski, and the making of barrels and tubs. More important were the introduction by the barbarians of the cultivation of rye, oats, spelt, and hops, the use of the stirrup for riding horses, and, above all, the heavy wheeled plough that provided the means for the development of the three-field system on which the life of the medieval manor was based.

The older plough, used in antiquity, had remained substantially unchanged since the beginning of the bronze age, save that the ploughshare was generally made of iron instead of wood from about 1000 B.C. Having no wheels, the older plough was maintained at the correct height for cutting the soil by the ploughman, an operation which required considerable physical effort, and which gave furrows that were not very straight and not very deep. The soil was merely scratched, and so it had to be ploughed twice, the second ploughing being in a direction at right angles to the first. On the light, dry lands of the Mediterranean region where a two-course system was practised, one year cropped followed by one year fallow, the old plough was reasonably effective, but not on the heavier, wetter, and also richer soils of the north, where the new plough came into its own.

The new plough, which was used by the barbarians before 100 B.C., had wheels to control the depth of ploughing, thus conserving the energies of the ploughman. Equipped with a coulter to cut the soil and a mouldboard to turn it over, the new plough gave deep, regular furrows, rendering superfluous the cross-ploughing practised in the south. Hence the new plough led to the cultivation of long strips of land in the northern three-course system, as opposed to the square blocks of the two-course system practised below the Alps and the Loire. The chief novelty of the three-course system, which was first mentioned in A.D. 765, was the sowing of crops in the spring in addition to the traditional winter sowings. In a typical cycle, the first year would be devoted to a winter sowing of wheat or rye, the second to a spring sowing of oats, barley, beans, or peas, while the land would lie fallow in the third. Thus, given the same productivity per unit area of land, the northern three-course system would produce a third as much again as the southern two-course system.

The new plough was heavier than the old, and it required a considerable traction force, so that the peasants of a village pooled their oxen for the ploughing, before the use of horses became common during the tenth and eleventh centuries. Horses were rarely used for ploughing in antiquity as the harnessings then known wasted about half of the power which the animal could exert. The ancient harnessing of horses was based upon the neck-yoke of the ox, and so the horse could not exert a pull with its shoulders where it is most effective. The yoke was placed over the back of the horse's neck and was held in place by a strap running round the front. Thus, as soon as the horse exerted any pull, it tended to choke itself. Again the ancients did not know how to harness horses in file, in order to multiply their tractive power, nor did they know of horse shoes, with the result that their horses were often injured on stony ground.

By the ninth and tenth centuries the horse was utilized more efficiently in northern Europe. The horse collar was introduced, perhaps from Asia as the Teutonic and Slav word for horse collar, the English *hames,* is of Central Asiatic origin. The tandem harness and the iron horse shoe also came in, with the result that the effective pulling power of the horse was now some three to four times greater than it had been in antiquity. The horse was then applied to agriculture, but only, in general, on the northern three-field systems, where surplus grain was available for feeding it. Oxen were still used in the main on the less productive two-field systems, as the ox could be fed on hay, the ox being economical as regards food while the horse saved time and labour.

Another device which spared human effort was the water-wheel, which was widely applied to the grinding of corn during the dark ages. A vertical water-wheel was described by the Roman, Vitruvius, about 16 B.C., and it is probable that horizontal water-wheels were in existence even earlier. However, human or animal driven querns were generally used for grinding corn in antiquity, mills becoming common only during the dark ages. By the middle ages most villages in Europe had their own mill. The Domesday Book of 1086 lists some 5,000 mills in England at that time, indicating that there was about one mill to every 400 of the population. During the next century windmills appeared. The first mentioned was a Normandy mill of 1180, which possessed a horizontal driving shaft and vertical sails, so that it was probably an invention independent of the tenth-century oriental windmills which had vertical driving shafts.

These various innovations had the consequences that the majority of men were now relieved of some of the crude physical labour, which had been required of them in antiquity, and that a surplus of food was produced above the subsistence needs of the manorial estates. Such surpluses allowed of the growth of the towns with their crafts and commerce, and provided the wealth for the notable ventures which took place between the eleventh and the thirteenth centuries, the Crusades, the building of the cathedrals, and the foundation of the universities. The craftsmen and the students of the big cities were fed and supported with the food surpluses coming in from regions which were quite restricted by the poor communications of the period. So too were the cathedral builders, an American writer estimating that the eighty cathedrals built by the French between

1170 and 1270 cost the modern equivalent of a thousand million dollars in labour and materials.

Another consequence of the technical innovations introduced during the dark ages was that the centres of civilization shifted from the Mediterranean to northern Europe, where the improvements were more effectively utilized. The trading of food surpluses and craft products developed markedly between the eleventh and thirteenth centuries, particularly in northern Europe, and the Baltic and the North Seas came to rival the Mediterranean in regard to the quantity of commerce carried over them. Such developments found expression in the foundation of the Hanseatic League of trading towns in A.D. 1241. The main Hansa towns were Lübeck, Cologne, Breslau, and Danzig, though the League had concessions in places as far apart as Novgorod and London.

The extension of trade was accompanied by new navigational discoveries, such as the stern-post rudder and the bowsprit, which are said to have appeared first in Europe upon the ships of the Hanseatic League during the thirteenth century. The steering of ships in ancient times was carried out by means of an oar over the side at the stern, as in a canoe steered by a paddle. On large ships such a method of steering was not very effective, a factor which limited the performance of large sailing vessels in which a fine control of the steering was necessary in order to sail into the wind. The stern-post rudder overcame the limitation, while the bowsprit allowed the lower forward corner of the main-sail to be hauled beyond the bows, enabling the ship to sail more closely into the wind. The fore-and-aft rig, which made tacking into the wind possible, was itself a product of the dark ages. The earliest form of this rig, the lateen sail, was first depicted in a wall painting found in a pre-Muslim church of South Palestine, and later in a Byzantine miniature of the ninth century.

Such navigational discoveries again helped to save manual labour, that of the galley slave, and they greatly extended the scope of sea transport. Larger ships could be constructed and effectively steered, permitting ocean-going voyages to be made as opposed to the coast-wise sailing of earlier times. The magnetic compass appeared in Europe during the thirteenth century, and it proved to be essential for determining direction on ocean voyages under cloudy skies, when neither the heavenly bodies nor the land could be seen. Here the north came first once more as visibility was good in the Mediterranean. By the fifteenth century, the deviation of the compass from true north was allowed for by the compass makers of Flanders, but not by those of Genoa.

In the crafts upon which commerce depended there were parallel technical advances. In the textile trade, the spinning wheel was developed during the thirteenth century, and the loom was improved at dates which are uncertain. About the same time water power was applied to fulling, a process of beating cloth in water to cause shrinkage, and so to increase the density and durability of the cloth. The beating was carried out at first by hand, but in the second half of the twelfth century trip hammers worked by water-wheels were introduced for the purpose of fulling. Shortly afterwards water-driven trip hammers were used to crush the woad of the dyers and the tan bark of the leather workers. Subsequently a variety of trades were based upon water power. In the thirteenth century water power was applied to saw-mills and the bellows of the smithy's forge, in the fourteenth to forge hammers and to grindstones, and in the fifteenth to pumps for draining mines. Perhaps the most important single application of water power was to the bellows of iron furnaces, giving a blast which elevated the smelting temperature and melted the iron, so that it could be cast. Cast iron appeared first during the thirteenth in Europe, though blast furnaces did not become common until the fifteenth.

Other technical innovations were brought to Europe from China by the Muslims and probably the Mongols. First came paper-making, which was well established in Muslim Spain by 1150. Within a few years, 1189, the first paper mill in Christendom was set up at Herault in France. By 1276 paper-making had reached Montefano in Italy, and by 1391 Nuremberg in Germany, whilst in 1494 the first paper mill was set up in England. In regard to printing, it is probable that the Mongols brought accounts, and perhaps samples, of Chinese printing to Europe, where the technical details were worked out afresh. Wood blocks for the elaborate capital letters used in medieval manuscripts were cut at Engelberg monastery in

1174, some time before the Mongol invasions and the Muslim accounts of Chinese printing. The first record of block-printing in Europe was that carried out at Ravenna in 1289. The change-over to movable and metal type followed quickly, examples occurring at Limoges in 1381, Antwerp in 1417, and Haarlem in 1435. Finally Gutenberg at Mainz perfected the early modern method of printing between 1436 and 1450.

Gunpowder first appeared in Europe during the thirteenth century, the first mention of it occurring in a letter written by Roger Bacon in 1249, a few years after the Mongol invasions. Cannon were first mentioned in 1325, and first depicted in 1327. The 1327 picture shows the early cannon to be a vase-shaped device, shooting a bolt with an arrow head. It is possible that the first cannon were based on the grenade used by the Chinese, an iron pot filled with gunpowder and hurled from a catapult, the pot now being filled only partially with gunpowder, and firing a bolt from its orifice. Later European cannon consisted of iron bars bound together with hoops to form a cylinder, but soon they were cast, first in bronze and then in iron.

Printing and firearms at the end of the middle ages had effects similar to the inventions of the alphabet and iron making at the end of the bronze age. Printing, like the alphabet before it, served to increase the literacy of mankind, and made the accumulated records of human civilization more available. It made possible the rise of the vernacular and craft literatures, the artisan for the first time in history recording the experience and values of his tradition. Printing also aided the Protestant Reformation by making the Bible more readily available, so that men could seek religious truth in their own experience of the Scriptures as the Reformers suggested.

Gunpowder and firearms ended the days of the armoured knight and his fortified castle, just as iron weapons had eliminated the bronze-age knights with their chariots and bronze rapiers. However, firearms at first did not level down the fighting potentialities of men, for cannon destroyed the Swiss pike phalanx, which up to the sixteenth century had provided an effective answer to the mounted knight. Moreover, military power was concentrated in the hands of the prince who commanded the control of gunpowder-making and cannon-casting. Thus the development of firearms gave an impetus to the rise of the absolute monarchies of the sixteenth and seventeenth centuries.

The middle ages saw not only the development of new techniques, but also a considerable refinement of skills, and a differentiation of the crafts. Mechanical clocks appeared in the thirteenth century, the construction of thirty-nine clocks being recorded between 1232 and 1370. The first mechanical clocks were large, heavy and crude, being used only in large public buildings, monasteries or cathedrals. But craftsmanship rapidly improved, and by the sixteenth century pocket watches were being constructed at Nuremberg. Similarly in buildings, the ratio of the area of floor space to the cross-sectional area of supporting walls increased from four to eight during the middle ages, indicating a growing economy in materials and skill in construction. With such refinements in technique came a differentiation of the crafts. Engineers and instrument makers separated off from the mill-wrights and blacksmiths, sculptors and artists from the stone masons and the decorators. The more skilled of the specialist craftsmen became literate, and recorded the experience of their art, whilst later such men assimilated some of the learning of the scholarly tradition, and contributed towards the development of modern science.

From as far back as 1250 we have the notebooks of the master mason, Villard de Honecourt, who travelled widely, going to Hungary to rebuild churches after the Mongol invasions, and recording various things he had seen. His drawings of living beings are realistic and based on observation, he notes specifically that his lion is drawn from the life, though it was a heraldic-looking creature. The scholarly drawings of animals and plants, on the other hand, were copied from earlier manuscripts, so that there was a rapid degeneration of realism, biological illustrations becoming formal and conventionalized. A plant which appeared as a strawberry in a French herbal about 550, had come to look rather like a blackberry in a Rhenish herbal of 1050 after numerous copyings without reference to the plant itself.

One of the earliest works that expresses the values of the craft tradition is a handbook on painting, written in the vernacular by the Florentine painter, Cennini, about 1400. Here the beginnings of an experimental attitude emerges, for in describing the making of pigments, and the technique of painting, Cennini states that he will 'make note of what he has tried out with his own hand'. The sense of guild tradition is marked in Cennini's handbook. He wrote the work, he tells us, in reverence of his teachers, and for the benefit of apprentices to the craft.

The themes suggested by Cennini were developed by later craft writers. The Gothic master builder, Mathias Roriczer, wrote a treatise on building in 1486, in which he listed some geometrical constructions which he had discovered himself. His aim in writing the work was more general than that of Cennini; it was not just to improve his own craft, but 'to better wheresoever something is to be bettered, and to amend and explain the arts'. Such values were still alien to the scholarly tradition, for there empirical enquiry was at a discount, and the ancient systems of natural philosophy were regarded as the unsurpassable heights of human wisdom.

Finally with the artist-engineers of the Renaissance we get the assimilation of the learning of the scholars by the more skilled and talented elements of the craft tradition. The artists, Botticelli, 1444-1510, Dürer, 1471-1528, Michael Angelo, 1475-1564, and Leonardo da Vinci, 1452-1519, all studied human anatomy by the practice of dissection. Botticelli and Dürer studied optics, Dürer laying down the canons of proportion, whilst Dürer and Leonardo observed the heavenly bodies, the activities of Leonardo ranging further over most branches of science and technology then known.

It is possible to exaggerate the degree to which the crafts had differentiated by this time, for most of the Renaissance artists still covered several fields. Leonardo da Vinci was not only a painter and a sculptor, but also an inventor, and a civil and military engineer. The interests of the Renaissance artists were diverse, and thus their activities had a comprehensive character, covering many fields. Perhaps for this reason their work was amorphous and somewhat inconclusive, for they did not make any notable scientific discoveries.

However, they did develop the empirical side of scientific method. Albrecht Dürer published a work on geometry in 1525 in which he stated that, as most German painters did not know much geometry, he had written a work on the subject, so that a painter reading it, 'will not only get a good start, but will reach better understanding by daily practice; he will seek further and find much more than I now indicate'.

Leonardo da Vinci had an even fuller appreciation of the empirical side of the scientific method. In a note on method Leonardo wrote:

'In dealing with a scientific problem, I first arrange several experiments, since my purpose is to determine the problem in accordance with experience, and then to show why the bodies are compelled so to act. That is the method which must be followed in all researches upon the phenomena of Nature. . . .

'We must consult experience in the variety of cases and circumstances until we can draw from them a general rule that it contained in them. And for what purposes are these rules good? They lead us to further investigations of Nature and to creations of art. They prevent us from deceiving ourselves, or others, by promising results to ourselves which are not to be obtained.'

It was some time before the men of the scholarly tradition developed the experimental side of the scientific method and arrived at a similar conception of the place of empirical procedures in science. However, they originated the new scientific ideas which the Renaissance craftsman had not been able to formulate, and they gave the mathematical method its place in science.

Chapter 11

The Scholarly Tradition During the Middle Ages

THE REVIVAL of learning in the middle ages came with the other notable developments which took place between the eleventh and the thirteenth centuries, the expansion of the crafts and commerce, the building of the cathedrals, and the foundation of the universities. The revival of Greek knowledge might have come earlier than it did, for Toledo remained a Christian archbishopric throughout the Muslim occupation. Indeed, it is possible that earlier attempts to transmit Greek science to the west were discouraged, as a ninth-century Roman ecclesiastic expressed his concern at the fact that Spanish Christians were studying Arabic works. It seems that opportunities for the translation of works from the Arabic existed without the stimulus, just as there were opportunities for the direct translation of works from the Greek which were not fully utilized during the Latin occupation of Constantinople, 1204-61.

The western Crusade against the Muslims in Spain resulted in the fall of Toledo in 1085, and it was from that time that the Arabic versions of the Greek scientific works were translated, the most active period of translation being 1125-1280. Spain was the most important centre of contact between the Muslim and Christian worlds, for here were found bilingual Mozarabs, Christians who had been assimilated by the Muslims, and Mudejars, Muslims assimilated by the Christians, whilst there were also a large number of Jews, some of whom were trilingual. Archbishop Raymond set up a school of translation at Toledo soon after its fall, and scholars came from all over Europe to learn of Muslim science there. The most important translator was Gerard of Cremona, 1114-87, who came to Toledo from Italy to look for the astronomical works of Ptolemy, the Almagest as they were called by the Muslims. Gerard translated the Almagest in 1175, and by the time of his death he had translated a further eighty works, covering all fields of Muslim learning.

A secondary contact with Muslim science occurred in Sicily, which fell to the Christians in 1091 after one hundred and thirty years of Muslim rule. Here the inhabitants spoke not only vernacular forms of Latin and Arabic, but also Greek, and some, notably the Jews, were acquainted with the written scripts of all three. It was here, under the patronage of the emperor, Frederick II of Sicily, that Michael Scot, died 1235, translated the biological works of Aristotle, and much of the Muslim alchemy. There were also direct commercial ties between North Africa and Sicily, and some Arabic works came to Europe by this route, the most notable being the mathematical writings brought over by Leonardo of Pisa and the medical works transmitted by Constantine the African.

Leonardo of Pisa's *Liber Abaci*, written in 1202, illustrates the varied sources of Muslim mathematics, and incidentally of our own. He begins first of all with the reading and writing of the new Hindu-Arabic numerals, the '*Novem figure indorum*' as he calls them. Then follow the operations of arithmetic, and questions concerning the prices of goods, barter, partnerships, and alligations. Next he considered various problems, one of which copies almost verbatim a problem set out in two cuneiform texts, one of which dates from early Babylonian times, and the other from the Hellenistic period. Later comes the Rule of False Position, or the '*regulis elchatayn*', as Leonardo calls it, a transliteration of the Arabic '*hisab al-Khataayn*', which means the rule from China. The rule of false position was of course known to the ancient Egyptians, but it would appear that the European knowledge of it came through the Muslims from China.

The Christian translators did not bring all the knowledge gathered or discovered by the Muslims to western Europe, for they showed a preference for the Roman and Greek authors with whom they were acquainted through the direct transmission of knowledge from the Roman world. The discovery of the lesser circulation of the blood by Ibn al-Nafis, for example, remained unknown until the present century, and indeed many Arabic and Persian works remain untranslated to the present day. Thus the new learning for the most part amplified the old, and it was readily assimilated into the university curricula. The most important single innovation was that the works of Aristotle now became known in full, and thus the Aristotelianism of the Scholastic philosophers replaced the earlier Platonic orientation of the Church Fathers.

The universities had arisen from guild-like associations of masters and students gathered at the cathedral schools. During the eleventh century, the words 'university' and 'guild' were used alike to describe craft associations, but by the thirteenth century the term 'university' had come to mean specifically a student association. There were three main kinds of university. Firstly there were ecclesiastical foundations, in which students and masters formed a closed corporation under a chancellor, as in the case of Paris, Oxford and Cambridge. Secondly there were civic universities which were governed by a rector elected by the students, such as Bologna and Padua. Thirdly, there were state universities founded with Papal recognition by a monarch, such as Naples, formed by Frederick II of Sicily and Salamanca, set up by Ferdinand III of Castile.

Following the foundation of the universities and the translation of ancient science from the Arabic, there was a brief burst of experimentation in Europe during the thirteenth century, which was continued to some degree by the alchemists up to modern times. The most notable figure in this movement was Roger Bacon, c. 1214-94, a Franciscan of Oxford university. He was critical of the scholars who based their opinions upon fallible authorities, or the weight of custom, and who hid their ignorance by wordy arguments. The true student, he said, should know 'natural science by experiment, and medicaments, and alchemy and all things in the heavens or beneath them, and he would be ashamed if any layman, or old woman, or rustic, or soldier, should know anything about the soil he was ignorant of'. Bacon himself made experiments in optics, following the works of Al-Hazen. He studied the magnifying effects of plano-convex lenses, and suggested that a telescope could be made with them. Through the experimental study of science, Bacon held that man would be able to construct self-propelled boats and land carriages, as also submarines and flying machines. However, such a point of view was not popular, for Bacon was reprimanded, and placed under the surveillance of his order. Bonaventura, the General of the Franciscan order, had said that, 'The tree of science cheats many of the tree of life, or exposes them to the severest pains of purgatory.'

Another notable experimenter was the nobleman, Pierre de Maricourt, who was probably a friend of Bacon's. About 1269 he wrote a small work describing his magnetic experiments. In it he said that the student of magnetism must be 'industrious in manual work' to correct the errors of reason. He made a sphere out of lodestone, and studied its magnetic properties with short pieces of iron wire, thus discovering the magnetic meridians which he marked by chalk lines. He knew that unlike magnetic poles attract and that like poles repel, and also that when a magnet is broken into two, each half becomes a magnet. However, he believed that magnets pointed to the pole star, and not to the pole of the earth, and that spherical magnets spontaneously rotated.

Practical work was also carried out in anatomy by Mondino de Luzzi, c. 1275–90. He wrote a book on dissection, describing his own work on the dissection of two female bodies, which was used as the principal text-book of the medical schools on the subject during the middle ages. Dissection was practised throughout the middle ages, particularly for post-mortems, and sometimes for the teaching of medical students. However, after the time of Mondino, the dissection was carried out by an illiterate barber-surgeon under the supervision of a doctor who did not himself operate. Furthermore dissection was somewhat discouraged by the dictum that the 'Church abhors the shedding of blood', which was promulgated by the Council of Tours in 1163 in connection with the practice of dismembering and boiling down dead Crusaders for ease of transport on the journey home.

Alchemy was another practical activity which was revived in Europe during the thirteenth century. With the spread of alchemy came new chemicals, such as the mineral acids, first mentioned by the French Franciscan, Vital du Four, about 1295, and alcohol, prepared by the distillation of wines and beers, which was first described by the so-termed Magister Salernus, who died in 1167. Alcohol was called the 'water of life' and it was regarded as second only to the alchemist's elixir of life. The properties of alcohol appear to have been studied extensively by the monks and friars, perhaps to excess in some cases, as the possession of apparatus for distilling alcohol was forbidden to members of the Dominican chapter at Rimini in 1288. However, the monastic orders continued their studies, developing a number of celebrated liqueurs.

Alchemy in general was forbidden in a bull issued by Pope John XXII in 1317, a prohibition which indicates that the practice of alchemy must have been fairly widespread. In the theories of the medieval alchemists there was not a great deal that was new. The metals, they believed, were generated by the union of the male principle of sulphur and the female principle of mercury, and the base metals could be ennobled by a process of death and resuscitation. Inorganic substances in general were living beings, made up of a body and a soul, or matter and spirit. The constituents of substances could be separated by heating, when the spirit came off as a vapour, which could be condensed into a liquid in some cases. The characteristics and properties of a substance were determined by its spirit, and so a liquid obtained by distillation contained the concentrated essentials of the substance from which it came. Such liquids were therefore highly active and potent agents, giving new life to old bodies, and conferring noble properties upon base matter. Thus, in theory, a transmutation could be effected by transferring the spirit of a noble metal to the matter of base metals. However, of the various metals only mercury would distil and give an isolatable 'spirit'. In accordance with the theory of the alchemists, mercury vapour silvered the surfaces of base metals, and so mercury was regarded as the spirit of silver, the progenitor of the metals, and indeed the origin of all things.

Such a system in which mercury was central was put forward by Raymond Lull, c. 1232-1315, an alchemist and a mystic who was regarded as something of a heretic by his fellow Christians, and who was martyred by the Muslims whilst attempting to convert them to Christianity. In the beginning, Lull held, God created mercury. The mercury then circulated as in a reflux distillation and differentiated out into all other things. The finest parts of the primeval mercury separated out first and formed the bodies of the angelic beings, whilst the less fine parts formed the heavenly bodies and the celestial spheres. The coarse parts formed the four elements and the quintessence, which gave rise to all terrestrial things. From the four elements were formed the bodies of terrestrial substances, while from the quintessence were formed their spirits. The quintessence, for Lull, was not confined to the celestial sphere as Aristotle had supposed. It was a spiritous air, the pneuma, pervading the whole universe as an immediate and universal manifestation of God. Lull therefore was much opposed to the Muslim philosopher Averroes who, like his master Aristotle, had placed God, or the Prime Mover, outside of the universe beyond the sphere of the fixed stars.

Lull, like other alchemists, held that the quintessences or spirits of terrestrial things could be isolated and concentrated by distillation. Alcohol, he thought, was an important but an impure spirit. If it were subject to reflux distillation, Lull held that it would separate out into two layers, an upper one that was sky-blue and a lower one that was turbid, just as the primeval circulation of mercury had separated out into heaven and earth. The upper layer would then be the pure spirit of alcohol. Lull and his followers believed that one spirit would attract another, and hence they extracted substances, particularly plants, with alcohol to isolate their quintessences, their tastes, perfumes, and medicinal virtues. Such alcoholic extracts they used for pharmaceutical purposes, the followers of Lull tending more and more to the medical side of alchemy. At the same time they became increasingly critical of orthodox Galenical medicine, forming a movement that culminated with the medical Iatrochemistry of Paracelsus in the sixteenth century.

The mainstream of medieval learning in the universities passed by the alchemists, perhaps because they were associated with mystical religion on the one hand, and with practical, manual activities on the other. After the ephemeral experimentation of the thirteenth century, the scholarly tradition developed by rational discussion rather than by empirical enquiry, remaining largely isolated from the craft tradition throughout the middle ages. Perhaps for such reasons, little that was really novel originated from the scholars of the middle ages, though some propositions suggested by ancient scientists in opposition to Aristotle were developed at length.

The philosophy of Aristotle was integrated with Catholic theology by Albertus Magnus, c. 1206-80, and more particularly by Thomas Aquinas, 1225-74. These men did not go beyond the cosmology of Aristotle. The universe, they held, was a sphere filled throughout its volume with matter, a vacuum being impossible because all activities required a direct or indirect physical contact between the actuating force and the body moved. St. Thomas's first proof of the existence of God was that the motions of the heavenly spheres required a Prime Mover, namely God. The activity of God was not manifest directly in the celestial spheres. The motions of the heavenly bodies were mediated by the hierarchies of angelic beings postulated by Dionysius in the fifth century.

Such a scheme was not accepted without reserve in some of the medieval universities. An important critical movement began at Oxford with William of Ockham, c. 1295-1349, who denied the validity of St. Thomas's first proof of the existence of God. A body in motion, he argued, does not necessarily require the continuous physical contact of a mover, as, for example, in the case of a magnet which can move a piece of iron without touching it. The was a case of action at a distance, which presumably could take place through a void. Hence as space did not have to be filled with matter to transmit physical effects, a vacuum was possible. Ockham revived the impetus theory of John Philoponos, which had come down to medieval Europe through the Muslims. Philoponos had suggested that forces confer impetus upon bodies they set in motion, the impetus being a quality which gradually dies away so that such bodies ultimately come to rest. Ockham agreed with Philoponos that an arrow could fly through a vacuum, in opposition to Aristotle. He agreed further with Philoponos that in the beginning God might have conferred an impetus upon the heavenly bodies which did not decay with time, thus removing the necessity of postulating a variety of angelic movers to the heavenly bodies. 'It is vain to do with more what can be done with fewer,' Ockham declared, thus formulating his 'razor' principle. Discussions on impetus were continued at Oxford, notably by Walter Burley, 1275-1357, Richard Suiceth, c. 1345, and Walter Heytesbury, fl. 1330-71, though the theory quickly lost ground, and by the fifteenth century the Oxonians for the most part were Aristotelian in their physics.

The impetus theory was developed further at Paris, first of all by Jean Buridan, who became rector of the university in 1327. Buridan put forward two important arguments against the Aristotelian thesis that bodies in motion were propelled by displaced air rushing in at their rear to prevent a vacuum being formed. Firstly, he said, a spinning top rotates without changing its position, therefore it cannot be moved by displaced air. Secondly, a javelin with a flat rear end does not move faster than a javelin pointed at both ends, which would be expected if air is the propellent. In both cases, Buridan held, impetus was the sustaining force of motion. The amount of impetus a body received from a force he thought to be proportional to the density and volume of the body and its initial speed. Following Ockham and Philoponos, Buridan denied that angelic beings propelled the heavenly bodies round their courses, supposing that an impetus conferred upon the heavenly bodies in the beginning performed the task. Such an impetus would never decay, for there was no air resistance in the heavens. Buridan declared:

'One cannot read in the Bible that there exist intelligences charged with giving the celestial orbs their proper motions: it is then permitted to show that there is no necessity of supposing the existence of such intelligences. One might say in effect that God gave each celestial orb an impetus which has kept it going since.'

Buridan was followed by Albert of Saxony who became rector of Paris university in 1353. Albert distinguished between uniform motion, when a body moved with constant speed, difform motion, when the speed of a body varied regularly from point to point, and irregular motion, which satisfied neither of the previous definitions. He examined the view that the speed of a falling body depended upon the duration of the fall, and the view that it depended upon the extent of the fall. Both views he rejected, as both led to the conclusion that a body would reach an infinite speed, either after an infinite time or after covering an infinite distance. Instead he maintained that bodies falling freely reach a limiting speed, as the impetus conferred by gravity increased more slowly with the velocity than the air resistance.

The greatest figure of the Paris school of impetus theorists was Nicolas Oresme, who became master of the College of Navarre in 1362 and bishop of Lisieux in 1377. Oresme introduced a method of representing velocities graphically. He plotted the distance a moving body covered along a horizontal line, and the speed at given points by vertical lines drawn at right angles to the horizontal axis. By joining the upper ends of the vertical lines a geometrical figure was produced, such that a rectangle represented a uniform motion, a triangle a difform motion, and a curve an irregular or 'difformly difform' motion.

Oresme also revived the idea that the earth rotated daily on its axis. He averred that, 'No experience whatsoever could prove that the heavens rotate daily and not the earth.' He arrived at such a view from the supposition that perfection and nobility lay in repose. If the earth possessed a diurnal spin, the speed of each heavenly body would be proportional to its degree of imperfection. The earth, the basest body in the universe, would rotate once a day, 'the moon in a month, the sun in a year, Mars in about two years, and the same with the others', leaving the perfect sphere of the fixed stars with only a slight movement to account for the precession of the equinoxes. Such considerations he thought 'profitable for the defence of our faith'.

William of Ockham, Jean Buridan, and Albert of Saxony had all been sympathetic to the doctrine of the diurnal rotation of the earth, but Oresme was the first to adopt the thesis explicitly, and associate it with the impetus theory of mechanics. The spin of the earth, he argued, like the rotation of the heavenly bodies, would continue indefinitely under the primary impetus conferred at the Creation, because there was no resistance to stop it. The impetus theorists also favoured the notion that the universe might be infinite in extent, and that there might be other worlds like our own.

Such conceptions became explicit with Nicolas of Cusa, 1401-64, who was made bishop of Brixen in the Tyrol in 1450. The earth, Cusa thought, rotated daily on its axis, due to an impetus conferred upon it at the beginning of time. The heavens were not more perfect than the earth, for, he held, the universe was made of the same four elements throughout. Moreover, the heavenly bodies were similar to the earth in that they were inhabited by creatures similar to those on earth. The universe as such was infinite, for, he said,

'whether a man is on the earth, or the sun, or some other star, it will always seem to him that the position he occupies is the motionless centre, and that all other things are in motion.'

The impetus theorists constituted only a minority movement in the medieval universities. Their views were not widely accepted at any time, the orthodox theory being the Christianized-Aristotelian system developed by Dionysius and St. Thomas. Here there was only one finite universe, with a base motionless earth at the centre, surrounded by celestial spheres of ever increasing perfection, which were kept in motion by the continuous activity of angelic beings. It was this system and its refinements that the early modern scientists had to combat and displace, not the scheme of the impetus school, in which the earth rotated daily in the midst of an infinite universe, and from which intelligences and unseen hands causing motion had vanished.

The impetus school of thought in fact had degenerated somewhat in the fifteenth century, though impetus theory was still taught at the beginning of the sixteenth. The decay of the theory began with Marsile of Ingham, who was at Paris in 1379, later becoming rector of Heidelberg. He suggested that impetus was like heat. It was weakest in that part of a body furthest removed from the source of motion, just as a rod was coolest at the end furthest removed from a source of heat. When a body left its mover, the impetus evened itself out throughout its bulk, just as heat evened itself out in a rod removed from a fire, and finally the impetus decayed, just as the rod lost its heat.

The impetus theory was taught in Paris at the beginning of the sixteenth century by the Scots, John Majoris, and George Lockert, both at the College of Montaigue. Amongst their students were the Humanists, Erasmus, c. 1466-1536, and Jean Luiz Vives, the latter regarding the impetus theory no longer as an enlightened view, but as a curb on human knowledge. When Vives took up a chair at Louvain, he wrote that at Paris,

'one sees raised a vast edifice of assertions and contradictory propositions touching uniform motion, non-uniform motion, motion uniformly varied and non-uniformly varied. There are innumerable people who discuss, without getting anywhere, that which cannot present itself in nature.'

For his own part, he preferred the accumulated knowledge of the craft tradition, which was at once real and useful.

'How much wealth of wisdom', he wrote, 'is brought to mankind by those who commit to writing what they have gathered on the subjects of each art from the most experienced therein. . . . By such observation in every walk of life, practical wisdom is increased to an almost incredible degree.'

The discussion of impetus problems spread to Padua in the fifteenth century, but the Italian discussion did not go beyond the results achieved by the Parisians of the previous century. In 1404 the university of Padua was acquired by Venice, which was then the leading anti-clerical and anti-papal state, and thus permitted its scholars to hold unorthodox views. Padua was primarily a medical university, where natural philosophy was taught to illustrate the scientific method of the time. There were therefore discussions on method at Padua in the fifteenth and sixteenth centuries. Efficient causes came to be considered as the primary concern of the natural philosopher, and the necessity of making observations was stressed. However, purposive acts and final causes were still considered to be legitimate explanatory principles in natural philosophy, and the role of experimentation in science remained largely undeveloped by the scholars of Padua until the time of Galileo.

The scholars and the craftsmen contributed in different ways to the birth of modern science. There were two prime elements in the scientific revolution of early modern times; firstly, the rise of a new method of enquiry, the scientific method, and secondly, an intellectual transformation, the growth of a new way of looking at the world. The craftsmen contributed towards the experimental method of modern science, while the men of the scholarly tradition at first contributed more to the intellectual revolution, using traditional methods, as we shall find in the case of Copernicus. However, both elements of the scientific revolution were ultimately dependent upon the convergence and the interpenetration of the craft and scholarly traditions, a process which we have seen at work in the cases of the scholar, Vives, who concerned himself with the practical arts, and the craftsman, Leonardo, who was interested in the impetus theory. Thus, the machine analogy, which later formed part of the new way of looking at things, was drawn from the crafts, whilst the mathematics of the scholars were brought into the procedure of the scientific method.

PART THREE

THE SCIENTIFIC REVOLUTION
OF THE SIXTEENTH AND
SEVENTEENTH CENTURIES

Chapter 12

The Copernican System of the World

OBSERVATIONAL ASTRONOMY revived in the fifteenth century in connection with navigation, and the reform of the old Julian calendar, which was getting rather out of step with the solar year. The movement began with George Purbach, 1423-61, of Vienna university, and more particularly, his pupil, Johannes Muller, 1436-76, who went to Italy to study the original Greek versions of Ptolemy's astronomy. Muller settled in Nuremberg, and made observations with his friend and patron, Bernhard Walther, 1430-1504, a wealthy merchant who had a private observatory. Walther also had his own printing presses, and they prepared nautical almanacs which were of great service to the Spanish and Portuguese navigators. Muller was the first to correct astronomical observations for atmospheric refraction, and the first to use the mechanical clock in astronomy. Later he went to Rome in order to reform the calendar, but died before he could do so. His observations were continued by Walther, and his friend the artist, Albrecht Dürer, so that a considerable body of accurate modern observations were available when Nicolas Copernicus, 1473-1543, began his work.

Copernicus was the son of a prosperous merchant and municipal official of the old Hansa town of Thorn on the Vistula, but his father died when he was ten years old, and he was adopted by his uncle, Lucas Watzelrode, who became bishop of Ermland in 1489. During the years 1496-1506 he studied in Italy, returning to take up a canonry at Frauenburg on the Baltic when his uncle died in 1512. The activities of Copernicus during this thirty years at Frauenburg were many-sided, touching upon medicine, finance, politics, and ecclesiastical affairs, but he seems to have been primarily concerned with the new system of the world which he thought of when quite a young man, perhaps when he was in Italy.

His new system of the world placed the sun in the centre of the universe, and ascribed three motions to the earth, a daily spin on its axis, an annual orbit round the sun, and a gyration of the earth's axis of spin to account for the precession of the equinoxes. Copernicus wrote a small work called the *Commentariolus*, giving an account of his theory, which circulated in handwritten copies amongst his friends from about 1530. The theory became more generally known, and attracted George Rheticus, 1514-76, a mathematician of Wittenberg, who studied with Copernicus for two years and published the first printed account of the Copernican theory in 1540. Finally Copernicus himself published his main work, *On the Revolutions of the Celestial Orbs* in 1543.

The work was printed at Nuremberg, first under the supervision of Rheticus, and then of Andreas Osiander, a Lutheran pastor. Osiander added a preliminary note to the work of Copernicus stating that the new theory was not necessarily true, and that it could be regarded merely as a convenient mathematical method of accounting for the apparent motions of the heavenly bodies, and predicting their future positions. Copernicus himself did not share this view; he regarded his system of the world as real, for he discussed questions that were not of a mathematical character, such as the physical objections to the theory of the motion of the earth, which need not have been considered were the theory regarded as hypothetical.

The arguments used by Copernicus to substantiate his theory were mainly of a mathematical nature. He regarded a scientific theory as a group of ideas deduced from certain assumptions or propositions. True assumptions or propositions, he held, must do two things. Firstly, they must 'save the appearances', that is, account for the observed motions of the heavenly bodies. Secondly, they must not contradict the Pythagorean preconceptions that the motions of the heavenly bodies were circular and uniform. In the view of Copernicus an assumption that disagreed with observation possessed no more grave a defect than one which departed from the preconception that the motions of the heavenly bodies were circular and uniform.

Copernicus considered that Ptolemy's system was, 'not sufficiently absolute, not sufficiently pleasing to the mind', because Ptolemy had departed from the strict letter of the Pythagorean preconceptions. To explain the motions of some of the heavenly bodies, Ptolemy had assumed that they move in circles with angular speeds which were not uniform relative to the centres of their circles, they were uniform relative only to points outside of such centres. Copernicus thought such a device a serious shortcoming of the whole Ptolemaic scheme. However, this was not his main point. The most important criticism Copernicus made of the ancient astronomers was that, given their 'axioms of physics', and the necessity of 'saving the appearances', they had either failed to explain what was observed in the heavens, or had unnecessarily complicated their systems of the universe. Speaking of his predecessors, Copernicus wrote:

'Therefore in the process of demonstration, which is called method, they are found either to have omitted something essential, or to have admitted something extraneous and wholly irrelevant.'

Copernicus concentrated on the latter point. He saw that the ancients, from his point of view, had added the three motions of the earth to each of the heavenly bodies in order to arrive at a scheme in which the earth was stationary in the centre of the universe. Thus three circles, or systems of circles, had been added to each of the heavenly bodies in the Greek geometrical systems of the universe in order to account for the apparent movements of the heavens from the point of view of a stationary earth. Copernicus considered such circles an unnecessary complication of the Greek schemes, and he got rid of them by assuming that the earth spun on its axis daily and moved round the sun in an annual orbit. In this way Copernicus reduced the number of circles required to explain the apparent movements of the heavens from the eighty or so used in the elaborate versions of the Ptolemaic system to forty-eight. This system appeared in his *Revolutions of the Celestial Orbs*, which gave a more detailed account of the planetary motions than did the *Commentariolus*, where he had used only thirty-four circles. His disciple, Rheticus, remarked with regard to the suggested motion of the earth round the sun:

'Since we see that this one motion of the earth satisfies an almost infinite number of appearances, should we not attribute to God, the Creator of nature, that skill which we observe in the common makers of clocks? For they carefully avoid inserting in the mechanism any superfluous wheel, or any whose function could be served better by another with a slight change of position.'

In effect Copernicus provided the simplest answer to the Greek problem of how to explain the apparent movements of the heavens in terms of motions that were circular and uniform. There was nothing novel in his method; it had been used by astronomers from the time of Pythagoras on. Using the Greek preconceptions, he had overthrown the Greek systems of the world. However, there was one preconception that he did not use, the one in fact that his system confounded, namely, the idea that the heavens were divine and the earth imperfect.

In the Copernican system the earth revolved round the sun, like the other planets. It enjoyed the same uniform and circular motions as the other heavenly bodies, motions that had been characteristic only of perfect and incorruptible things in the old schemes. Moreover, Copernicus emphasized the similarity of the earth to the heavenly bodies by suggesting that they all possessed gravity. This gravity did not act across space. It existed only within aggregates of matter, like the earth and the heavenly bodies, providing their binding force, and bringing them into the perfect form of a sphere. His argument was purposive and teleological:

'I think gravity is nothing else than a natural appetency, given to the parts by the Divine Providence of the Maker of the universe, in order that they may establish their unity and wholeness by combining in the form of a sphere. It is probable that this affection also belongs to the sun, moon, and the planets, in order that they may, by its efficacy, remain in their roundness.'

The Copernican system was simpler and more elegant than the Ptolemaic scheme. On the old system the heavenly bodies had both east-west motions, and rotations in the opposite direction. Now the earth, and all of the planets, moved round the sun in the same direction with speeds that decreased with distance from the sun, whilst the sun at the centre, and the fixed stars at the periphery of the universe were stationary. Now it was seen why it was that the planets appeared to approach and recede from the earth. They were at one time on the same side of the sun as the earth, and at another time on the opposite side. By an ingenious combination of epicycles, Copernicus accounted for the fact that the apparent diameter of the moon does not vary widely, whilst the epicycles assumed by Ptolemy required that the apparent diameter of the moon should vary by a factor of four.

Astronomical computations were rendered easier by the Copernican scheme, owing to the smaller number of circles involved in the calculations, but the prediction of planet places and so on, was not more accurate than with the Ptolemaic system; both involved error of about one per cent. Moreover, there were serious physical objections to the Copernican system. One that was perhaps not very serious at the time was the fact that the centre of the universe did not lie exactly in the sun. Copernicus placed it at the centre of the earth's orbit, which was slightly displaced from the sun, to account for the inequality in the lengths of the seasons. Some philosophers had demanded that real physical bodies should serve as centres of rotation in the universe, but it was widely accepted that geometrical points would suffice for this purpose, as in the device of the epicycle itself. Moreover, the contemporary Aristotelians held that gravity acted towards a geometrical point, the centre of the universe, which did not necessarily coincide with the centre of the earth in their scheme.

More serious was the objection that, if the earth rotated, the air would tend to be left behind, resulting in a constant east wind. Copernicus gave two replies to this objection. The first was a medieval kind of explanation that the air rotates with the earth because it contains earthy particles, which are of the same nature as the earth, and thus compel the air to move with the earth. His second explanation is more modern; the air rotates 'without resistance since the air is contiguous to the constantly rotating earth'. A similar objection was that a stone thrown upwards into the air would get left behind by the rotation of the earth, and fall to the west of its point of projection. To this Copernicus gave the medieval kind of reply alone,

'Since the objects which are depressed by their weight are mainly earthen, there is no doubt that the parts retain the same nature as their whole', and thus rotate with the earth.

A further objection was that, if the earth rotated, it would fly to pieces by centrifugal force. Copernicus replied that if the earth did not rotate, then the much vaster sphere of the fixed stars must rotate with a very great velocity, and thus it would be much more susceptible to fragmentation through centrifugal force. This argument was not really conclusive at the time, as the heavens were thought to be composed of the perfect and weightless fifth element, the quintessence, which was not influenced by terrestrial actions, such as centrifugal force. However, the original Aristotelian idea of the quintessence had become somewhat coarsened during the middle ages, and the celestial spheres were thought of as rigid, glassy, and crystalline, which gave weight to the argument of Copernicus. He also found another way out of the difficulty by suggesting that centrifugal force was only found in unnatural, artificial motions, not in natural motions, such as those of the earth and the heavenly bodies. He argued that,

'Things governed by nature produce effects contrary to those governed by violence. Things upon which force and impetus are conferred must dissolve, and they cannot subsist for a long time; but what is done by nature is rightly ordered and is preserved in its best composition. Ptolemy is therefore wrong in fearing lest the earth and all terrestrial things might be dispersed in a rotation brought about by the efficacy of nature.'

It is seen that Copernicus accepted neither the Aristotelian nor the impetus theory of motion, for he considered both the action of movers and of impetus to be unnatural and artificial. He held that spin and uniform movement in a circle were spontaneous and natural attributes of the perfect spherical geometrical form, the form in which the earth and the heavenly bodies are found. Thus Copernicus did not employ a hierarchy of angels to propel the heavenly bodies round their courses, the angels of the higher spheres commanding those of the lower, as in the accepted Dionysian modification of the Aristotelian-Ptolemaic scheme. The heavenly bodies had spontaneous, natural motions of their own. Rheticus tells us that, 'in the hypothesis of my teacher, which accepts, as had been explained, the starry sphere as boundary, the sphere of each planet advances uniformly with the motion assigned to it by nature, and completes its period without being forced into any inequality by the power of a higher sphere. In addition, the larger spheres revolve more slowly, and, as is proper, those that are nearer to the sun, which may be said to be the source of motion and light, revolve more swiftly.'

Thus with Copernicus an entirely new set of cosmic values emerged. The Prime Mover at the periphery of the universe was no longer important. The sun at the centre of the universe was the governor of the heavens. Copernicus himself wrote:

'In the centre of everything rules the sun; for who in this most beautiful temple could place this luminary at another or better place whence it can light up the whole at once? . . . In fact, the sun sitting on a royal throne guides the family of stars surrounding him . . . the earth conceives by the sun, and through him becomes pregnant with annual fruits. In this arrangement we thus find an admirable harmony of the world, and a constant harmonious connection between the motion and the size of the orbits as could not be found otherwise.'

An intermediate figure between the Aristotelians who stressed the power of the Prime Mover at the periphery of the universe, and Copernicus who exalted the sun at the centre, is Nicolas of Cusa, who in arguing for the infinity of space had held that the world is such that its 'centre coincides with its circumference'. The reform of Copernicus was, however, much more thoroughgoing. Power in the universe was no longer delegated down a hierarchy of angels from the Prime Mover at the circumference of the universe to the base earth at the centre, it was wielded absolutely by the sun who ruled over bodies of more or less equal status, the earth along with the planets, which equally possessed gravity and circularity of motion.

It may be said that Copernicus was interested in promoting such new values, for if he had wanted merely a simpler system of the world it is not unlikely that the scheme later adopted by

Tycho Brahe, 1546-1601, would have occurred to him. Here the planets moved in orbits round the sun, whilst the sun and the planets as a whole moved round the stationary earth at the centre of the universe. Such a system was mathematically equivalent to the Copernican scheme, and moreover it did not raise the physical problems of the moving earth which the latter entailed. But for the most part it preserved the old cosmic values, and it may have been for this reason that Copernicus preferred his new heliocentric scheme.

It is curious that Copernicus was so novel in his values and conceptions, yet conservative in his method. Throughout his life, he held to the Greek preconception that the motions of the heavenly bodies must be circular and uniform, and thus, whilst his system was much simpler than that advocated by Ptolemy, it was complex compared with the later system worked out by Johann Kepler, 1571-1630. Kepler explained the apparent motions of the heavenly bodies with seven ellipses as opposed to the thirty-four circles used by Copernicus. As Kepler said, Copernicus was not aware of the riches that were within his grasp. Copernicus knew that a combination of circles gave an ellipse, but he never used this figure to describe the orbits of the heavenly bodies. Moreover, in his early days he had a high respect for the observational work of antiquity. He wrote a strong letter against the astronomer, Werner, who had suggested that the more recent observations of Purbach and Muller were more accurate than those of Ptolemy. In point of fact they were some three times more accurate.

The most important observational work of early modern times was carried out by Tycho Brahe, whose observations were some fifty times as accurate as those of Muller, reaching the limits of vision with the naked eye. Tycho Brahe was a Danish nobleman. Frederick II of Denmark gave him a pension, and the island of Hveen in the Copenhagen Sound for his astronomical work. Here he built a castle, workshops, a private printing press, and an observatory, where he worked with his numerous assistants from 1576 to 1597, collecting a large body of precise observations. He remarked that it was not possible to make observations without the guidance of a theoretical system of the world, and so he adopted the modified geocentric system that has been mentioned. However, his prime interest was in observation, and in this he excelled.

When Frederick II died, his patronage was not extended by his successor, and so Tycho Brahe went to Prague in 1599, where he was given a pension by the emperor Rudolph II. In the following year he was joined by the young German astronomer, Johann Kepler, who was primarily a mathematician in the Copernican tradition. Kepler was the son of a Würtemburg army officer and an innkeeper's daughter. He studied at Tübingen where he was converted to the Copernican theory by Michael Mastlin, the professor of astronomy there. His period of co-operation with Tycho Brahe at Prague was short, for the latter died in 1601, bequeathing his collection of observations to Kepler. Kepler remained at Prague, preparing tables of the planets' motions for his patron, and in connection with this task, pursuing his own researches on the nature of the planetary orbits. His *Rudolphine Tables*, which appeared in 1627, were much more accurate than those current at the time, namely Reinhold's *Prussian Tables* of 1551, based on the Copernican theory, and the *Alphonsine Tables* of the thirteenth century, based on the Ptolemaic system. These earlier tables were of about the same degree of accuracy. The *Rudolphine Tables* were more accurate than their predecessors because they were based on the precise observations of Tycho Brahe, and a new geometry of the planetary orbits that Kepler had derived from those observations.

Kepler's earliest work on cosmology, *The Mystery of the Universe,* which appeared in 1596, was of a somewhat mystical character. He sought for mathematical harmonies between the orbits of the planets in the Copernican system, finding that the five regular solids could be fitted between the spheres of the planetary orbits. When he came into possession of Tycho Brahe's observations his work became more conclusive, though for long he was obsessed by the notion that the motions of the heavenly bodies must be circular and uniform. However, he found that such an idea in either the Copernican, Ptolemaic, or Tychonian system failed to give predictions of the same degree of accuracy as Tycho's measurements. Hence he abandoned the idea, and trying out other geometrical figures, he found in 1609 that the ellipse fitted perfectly, giving predictions of the required degree of accuracy. The motions of the planets were now no longer circular nor uniform, for his two laws of planetary motion, published in 1609, stated firstly that each planet describes an ellipse with the sun in one focus, and secondly, that the line drawn from the sun to the planet sweeps out equal areas in equal times. Nine years later he discovered his third law, namely, that the squares of the times the planets require to complete their orbits are proportional to the cubes of their respective mean distances from the sun.

In his *Epitome of the Copernican Astronomy*, written 1618-21, Kepler gives an account of astronomical method which differs greatly from that of Copernicus. There are, said Kepler, five parts to astronomy. Firstly, the observation of the heavens; secondly, hypotheses to explain the apparent movements observed; thirdly, the physics or metaphysics of cosmology; fourthly the computing of past or future positions of the heavenly bodies; and fifthly, a mechanical part dealing with the making and the use of instruments. Kepler held that the third part, the metaphysics of cosmology, like the Greek preconception that the planetary motions must be uniform and circular, was not essential to the astronomer. If his hypotheses fitted into a metaphysical system, so much the better, but if they did not, then it was the metaphysics that had to go. The only restriction on hypotheses, Kepler said, was that they should be reasonable, and the main purpose of a hypothesis was 'the demonstration of the phenomenon, and its utility in common life'.

Kepler took over, and indeed developed, the cosmic values of Copernicus. He too thought that the sun was the governor of the universe, the World Soul in the sun directing the planets round their orbits. The sun, Kepler thought, was the one heavenly body

'which alone we should judge to be worthy of the most high God, if He should be pleased with a material domicile, and choose a place in which to dwell with the blessed angels.'

The whole universe, he held, was the image and analogy of the Trinity. The Father was the centre, the Son, the encircling spheres, whilst the Spirit was the complex of relationships within the universe.

With Kepler the spatial configuration of the solar system was finally cleared up, and the way was made open for the interpretation of the pattern of the heavens in terms of a dynamic equilibrium of mechanical forces. This was the great achievement of early modern science. The Greeks had been mainly concerned with the static pattern of the universe, dealing with motions only so long as they recurred uniformly and traced out geometrical figures. Because of their preconceptions, the patterns of the Greek systems of the world remained complex. By dropping one preconception, namely the qualitative distinction between the heavens and the earth, Copernicus obtained a much simpler system, and by dropping most of the other Greek preconceptions, Kepler obtained the simplest system of all. In so doing they opened the way for the interpretation of celestial motions in terms of terrestrial mechanics, a development that would have been inconceivable to the dominant Greek schools of thought had they possessed a science of dynamics, which they did not.

Chapter 13

Gilbert, Bacon, and the Experimental Method

DURING THE sixteenth century the barrier between the craft and scholarly traditions, which up to that time had separated the mechanical from the liberal arts, began to break down. Guild secrecy faded out, craftsmen recording the lore of their tradition and assimilating some scholarly knowledge, whilst some scholars became interested in the experience and the methods of craftsmen. A notable product of this movement was a work *On Pyrotechnics*, published in 1540 by an Italian metal worker, Biringuccio, who became head of the Papal foundry and munitions works. His book describes the smelting of metals, the casting of cannon and bells, the making of coins, and of gunpowder. A book covering similar ground, and methods of mining in addition, was produced in 1556 by the scholar George Bauer, a doctor of the mining regions in the Harz mountains.

Later new technical inventions and scientific discoveries are recorded in the craft literature. One such was the discovery of the dip of the magnetic needle by a retired mariner and compass-maker of London, Robert Norman, who published his discovery in a pamphlet called *The Newe Attractive* in 1581.

He found that a magnetic needle suspended at its centre not only pointed to the north, but also assumed an angle to the vertical, known as the angle of dip. He also weighed iron filings before and after magnetization to see whether magnetism was ponderable, finding that it was not so. Furthermore he floated a magnet on water, supported by cork, and found that the magnet only turned to the north-south direction; it did not move to the north or south, from which he concluded that magnetism was only an orientating force, not a motive force. These things were discovered, he said, by 'experience, reason, and demonstration, which are the groundes of artes'. He discusses various magnetic questions in relation to navigation, notably the diversity in the deviation of the magnetic compass from true north in various places. Such a deviation does not vary regularly from place to place, he said, as some mariners believe who, 'notwithstanding their travells mostley have more followed their Bookes than experience in this matter'. As to a theory of magnetism, he confesses that he cannot make any suggestions—'I will not offer to dispute with Logitians in so many pointes, as here they might seem to over-reach me in natural causes.'

Thus the craft tradition of the sixteenth century could produce good experimentalists, like Norman, but not theoreticians. However, the scholars of the age could, and those interested in the craft literature provided the theory that the craftsmen lacked. In the field of magnetism the most notable scholar of the period was William Gilbert of Colchester, court physician to Queen Elizabeth, who produced his book, *Concerning the Magnet*, in 1600. Gilbert took over and extended the experimental work of Robert Norman, and the thirteenth-century writer, Pierre de Maricourt. Following Maricourt, he made spherical lodestones, which he called 'little earths', and mapped out their magnetic meridians with a compass needle and chalk. He showed that the phenomenon of dip, discovered by Norman, was exhibited by his spherical lodestones, a compass needle inclining itself to the vertical on their surfaces. He showed further that a spherical lodestone with an irregular surface possessed irregular magnetic meridians, from which he supposed that the deviation of the compass from true north of the earth's surface was due to the presence of land masses. Not a great deal of his experimental work was really original, and for the most part it was of a qualitative nature. Exceptions here were his discovery that lodestones capped with iron have enhanced magnetic powers, and his study of the relation between the size of lodestones and their attractive powers for a given piece of iron, which he found to be a direct proportionality.

On the basis of the facts of magnetism known to him, Gilbert constructed a considerable body of theory. From his experiments with spherical lodestones, he presumed that the earth was a giant magnet constructed of lodestone throughout with only a superficial covering of water, rocks and soil. Magnetism in a lodestone he thought akin to a soul in a body, causing movement and change. Thus he was attracted by Pierre de Maricourt's theory that spherical lodestones spontaneously rotate, though he added, 'Until now we have not succeeded in seeing this.' He believed in the diurnal spin of the earth on its axis; the great magnet of the earth, he said, 'turns herself about by magnetic and primary virtue'. The magnetic power of the earth reached up to the heavens, he thought, and held the world together. Gravity, for Gilbert, was nothing other than magnetism.

Gilbert dedicated his book to a new tradition, that of men 'who look for knowledge not in books but in things themselves'. He rejected the old scholarly tradition, which, he said, was composed of 'people who blindly trust authorities, literary idiots, grammarians, sophists, pettifoggers, and perverse mediocrities'. Gilbert associated himself with the craftsmen, and the scholars interested in the craft tradition. George Bauer, the physician of the Harz mountains, Gilbert terms a man 'outstanding in science'. Maricourt, he thought, was 'learned considering the period', whilst Robert Norman he praises as 'an expert mariner, and ingenious artificer', who 'invented and made public, magnetic instruments and expedient methods of observation, necessary to navigators and long distance travellers'.

The work of Gilbert and Norman exemplifies the beginning of a union between craft lore and scholarly knowledge, and between the empirical study and the theoretical interpretation of nature. Norman did not fully transcend the old craft tradition, for he could not develop explanations for his discoveries. In the same way Gilbert could not avoid the influence of the old scholarly tradition which he rejected. His theories were of a speculative nature, even though they were based on experiment. Moreover, as Francis Bacon pointed out later, Gilbert did not use his hypotheses as a guide to further experimental work; he framed his theories after his experimental work had been accomplished, and did not proceed to think up further experiments that would confirm his explanation.

By the beginning of the seventeenth century the development of modern science was under way, though its procedure was somewhat halting, and its novel characteristics were by no means fully recognized. The craft and scholarly traditions had been gradually merging throughout the sixteenth century to give a new method of enquiry, but few people realized what such a development portended, and fewer still were aware of the nature of the new method, and the potentialities of its application. Francis Bacon, 1561-1626, Lord Chancellor of England under James I, was amongst the first to become conscious of the historical significance of science and the role it could play in the life of mankind. What he saw seemed to be good, and he sought to give impulse and direction to the new scientific movement, by analysing and defining the general methodology of the sciences, and indicating the mode of its application.

Bacon was primarily a philosopher, not a scientist. He set out to explore the possibilities of the experimental method, to be a Columbus in philosophy as he put it, and to interest other persons so that they could carry such possibilities into effect. His first work on the subject was *The Advancement of Learning,* published in 1605, which was an early and popular exposition of his views. His main work was the *Great Instauration of Learning,* which was published in part in 1620, and in fact was never finished. Bacon intended that it should be divided into six parts. Firstly a general introduction, for which, he says, the *Advancement of Learning* will serve. The second part, which is the most complete, consists of an analysis of the scientific method, or the *New Instrument* as he called it. Part three was to be an encyclopedia of craft lore and experimental facts, and the fourth part, which is missing, would show how the new method was to be applied to such facts. Part five would be concerned with previous and existing scientific theories, whilst the sixth part would deal with the new natural philosophy itself, the final synthesis of hypotheses culled from the encyclopedia of facts, and of existing scientific theory.

Very little of this vast design was accomplished by Bacon. His *Great Instauration* consists of little more than his analysis of scientific method, but this in itself was most influential in England during the seventeenth century, and in France during the eighteenth. As regards method, Bacon endeavoured to unite the procedures of the scholarly and the craft traditions, to bring about as he put it,

'the true and lawful marriage of the empirical and rational faculties, the unkind and ill-starred separation of which has thrown into confusion all the affairs of the human family.'

In assessing the two traditions as they stood in his day, Bacon contrasted the cumulative growth of the crafts with the more erratic course of philosophy.

'With their first authors', he wrote, 'the mechanical arts are crude, clumsy, and cumbersome, but they go on to acquire new strength and capacities. Philosophy is most vigorous with its early authors, and exhibits a subsequent decline. The best explanation of these opposite fortunes is that, in the mechanical arts, the talents of many individuals combine to produce a single result, but in philosophy one individual talent destroys many. The many surrender themselves to the leadership of one . . . and become incapable of adding anything new. For when philosophy is severed from its roots in experience, whence it first sprouted and grew, it becomes a dead thing.'

Thus, for Bacon, the contemporary learned tradition was sterile, because it had lost contact with experience, but at the same time, the experience of the craft tradition was not fully effective scientifically because it was largely unrecorded. Hence, he wrote, when 'experience has learned to read and write, better things may be hoped'. These 'better things' were new scientific principles and new technical inventions. Just as Gilbert took over the thirteenth-century experiments of Pierre de Maricourt, Bacon took over the ideals of his thirteenth-century namesake, Roger Bacon, who had seen the future great with technical inventions stemming from the application of experimental method. Francis Bacon too had a similar vision: the union of the theoretical interpretation and the practical control of nature, will produce, he wrote, 'a line and race of inventions that may in some degree subdue and overcome the necessities and miseries of humanity'.

However, Bacon was by no means a utilitarian in the narrower sense: the scientific understanding, and the technical control of nature went hand in hand, both were produced by the application of scientific method. Bacon was much impressed by the development of printing, gunpowder, and the magnetic compass, inventions he used as examples of the superior knowledge of modern man over the ancient Greeks. He noted that all of these things were based on new principles. Printing was not a means of writing quickly, nor firearms an improvement of the ancient catapult, they embodied principles different in kind from those employed previously in the crafts to which they were applied. Moreover, such principles were often of considerable scientific interest, as shown by the work of Maricourt, Norman, and Gilbert, which stemmed from the magnetic compass.

Thus, argued Bacon, the primary requirement of the new method of advancing the sciences and the arts was the searching out of new principles, processes, and facts. Such facts and principles could be derived from craft lore, and from experimental science. When they became understood then they would lead to fresh applications in both the arts and the sciences. Many principles, he thought, lay hidden and unnoticed in everyday craft processes, which were thus a valuable source of scientific knowledge. Such processes were of particular interest in that they were of an active, experimental character, involving the change and transformation of natural substances. Here nature revealed her hidden workings and forced them upon the attention of man, whilst in the passive contemplation of nature, as in the observation of animals and plants, the human mind too readily picked out the facts that supported its preconceived notions.

'There is one kind of natural history which is made for its own sake,' he wrote, 'another which is gathered for the information of the intellect, in order to the building up of philosophy and these two kinds of history differ on other points indeed, but especially in this, that the former of them contains the various Natural Species, the latter, the Mechanical Arts. For just as in civil affairs the abilities of each man are better drawn forth in the positions of disturbance than elsewhere, similarly the hidden things of nature more betray themselves when the Arts provoke them than when they wander on in their own course.'

Not all craft processes were equal in this respect. It was the 'arts which expose, alter, and prepare natural bodies and Materials', which reveal the hidden workings of nature, rather than those 'which consist principally in subtle motions of the hands or instruments'.

Bacon drew up a list of some one hundred and thirty topics and processes which he considered deserving of investigation, and he requested that James I should order the collection of information relating to such subjects, but to no effect. The collection of a large body of facts was the primary requisite of his method, indeed he thought that with an encyclopedia of information about six times as great as Pliny's *Natural History* he would be able to explain all natural phenomena. With such a body of data, he held that any subject could be investigated by classifying together the facts relating to that topic. First of all a list of 'positive instances' of the phenomenon in question should be drawn up, that is, instances of where the phenomenon was present. Thus in studying the nature of heat, positive instances would be the sun's rays, flames and so on. Secondly, a list of 'negative instances' would be required, cases where the phenomenon was absent. Heat, for instance, was not present in the moon's rays, air, water and so forth. Thirdly, 'degrees of comparison' should be noted, as, for example, the variation of animal heat with exercise or frictional heat with the vigour of the motion that produced it. From such lists, scientific knowledge could be obtained by trying out various hypotheses, excluding the improbable ones, and testing further the more likely ones. For this purpose, other 'instances' might be used to discriminate between rival hypotheses, namely 'solitary instances' where the phenomenon in question was isolated from confusing associations with other phenomena, and 'glaring instances', where the phenomenon manifested itself in its most intense form.

In this way Bacon tried out various hypotheses as to the nature of heat, and came to the conclusion that the essence of heat was motion, for wherever heat was found there occurred some form of motion. He did not put forward this idea in the obvious sense that friction always produces heat: it was the 'motion of the smaller particles of bodies' taking place beneath the surface of phenomena that produced the sensible effect of heat. Bacon held that behind the visible world of nature there were structures and processes that were hidden from us by the nature of our sense organs. Such structures and processes he called the 'Latent Configurations', and the 'Latent Processes' of nature, and it was the business of the scientist to find out what these were. Bacon himself felt that the 'Latent Configuration' of nature was of an atomic character, whilst the 'Latent Process' of heat was a motion of such atoms or particles.

The method of obtaining hypotheses from tables of facts could be applied, he thought, to hypotheses themselves to get axioms of wider generality. At each stage of the process the hypotheses, axioms, or theories were to be tested experimentally, and applied to human use if suitable. Thus a pyramid of scientific theory would be built up inductively, solidly based on an encyclopedia of factual information, with applications stemming off at every stage. Not all levels of the pyramid were equally fruitful in this respect, for, Bacon held, the middle generalizations are the most useful.

'The lowest axioms differ but little from bare experience,' he wrote, 'and those highest and most general axioms are conceptional and abstract, and have no solidity. But the middle ones are true, solid, living axioms, whereon depend man's affairs and fortunes.'

Such a generalization may be said to contain a considerable element of truth. Gilbert's generalized idea that magnetism holds the world together could not be applied to any great extent in either the arts or sciences, but his 'middle axiom' that land masses cause the distortion of the magnetic meridians, even though fallacious, was adopted both in science and navigation, and stimulated further enquiry.

Bacon's view of scientific method was essentially experimental, qualitative, and inductive. He distrusted mathematics and the art of deductive logic that went with it. He was not unaware that mathematics were a useful tool in science, but he thought they were already well developed, 'like logic, and hitherto these have not been the handmaids of science, but have exercised dominion over it'. He was averse to the method which was being developed by Galileo, of isolating phenomena from their natural context, studying only the aspects of such phenomena that were measurable, and then erecting a vast body of mathematical theory upon the results. Bacon wanted to consider all the facts that might be relevant to the matter on hand—the physical nature of the heavenly bodies in astronomy, which Copernicus had not thought important, or the role of air resistance in gravitational fall, which Galileo ignored.

Considering all the facts in astronomy, Bacon came to the view which was not unreasonable at the time, 'that both they who think the earth rotates and they who hold the Primum Mobile and the old construction, are about equally and indifferently supported by the phenomena'. To decide between the Copernican and the Ptolemaic systems, he thought that a great deal more work had to be done, particularly on physical questions, such as the nature of the 'spontaneous rotation' that Copernicus had attributed to the heavenly bodies. In this connection he opposed the view of Aristotle that the physics of the heavens and the earth are different in kind. Speaking of Aristotle's philosophy, he wrote:

'This philosophy, if it be carefully examined, will be found to advance certain points of view which are deliberately designed to cripple enterprise. Such points of view are the opinion that the heat of the sun is a different thing from the heat of a fire, or that men can only juxtapose things while nature alone can make them act upon one another.'

Bacon also rejected the view of the Greeks that the motions of the heavenly bodies are circular and uniform; this was merely, 'A thing feigned and assumed for ease and advantage of calculation.'

Thus Bacon rejected the methodological axioms of the Greeks—the superiority of the heavenly bodies, the circularity of their motions—though he accepted some of the content of their views, namely the central position of the earth in the universe. He was original, for the most part, only in the new method he advocated, and even this did not receive an immediate application. Progress in science during the seventeenth century was achieved mainly by the mathematical-deductive method, developed by Galileo, and elaborated by Descartes, and it was only in the nineteenth century with the development of evolutionary geology and biology that the qualitative-inductive method of Bacon came into its own. It was then that vast stores of facts, mainly of a qualitative nature, were collected together from all over the globe, and inductive reasoning applied to elaborate the theories of geology and biology.

In applied science, Bacon was interested mainly in craft and industrial processes; indeed he has been termed 'The philosopher of industrial science', and he was not so much concerned with the commerce and navigation that flourished in his own day. Here again his programme was not carried into effect until the nineteenth century, though his schemes for the advancement of the crafts attracted much attention during the seventeenth. Bacon's method was a development and clarification more of the values and procedures of the craft tradition than of those of the scholars. In the same way, Descartes' method expressed more the point of view of the scholar than of the craftsman, as we shall see. Thus neither of the men of the seventeenth century who set out to analyse and codify the new methodology of the sciences succeeded in fully integrating the two traditions, to unify 'the empirical and rational faculties'. Hence the shadow of the old barrier between craftsman and scholar still remained, and indeed still remains in the distinctions of status drawn between the experimental and mathematical scientists, and the pure and applied scientists.

Chapter 14

Galileo and the Science of Mechanics

THE SCIENCE of astronomy was associated with the priesthood and the scholarly tradition from remote antiquity up to modern times. There was no craft tradition in astronomy of any standing until the great geographical discoveries, when astronomy was brought into navigation for the purposes of determining latitude and longitude at sea out of sight of land. Accordingly we find that the early modern astronomers used the mathematical methods of the old scholarly tradition, being original in the theories they produced, but conservative in the methodology. Copernicus, and Kepler in his early days, did not regard mathematics just as an intellectual tool, a method of developing a scientific theory without prejudice to the content of that theory. Their mathematics were of a metaphysical character, embodying the preconceptions of Pythagoras and Plato. The heavenly bodies were necessarily spherical in form, and their motions were necessarily circular. Observations had to be accommodated to these preconceptions, for mathematical shapes, forms, and harmonies determined the structure of the universe, and were a reality prior to the perceptions of the sense organs. Descartes wrote in 1628:

'When I bethought to myself how it could be that the early pioneers of philosophy in bygone ages refused to admit to the study of wisdom anyone who was not versed in mathematics . . . I was confirmed in my suspicion that they had knowledge of a species of mathematics very different from that which passes current in our time.'

By the seventeenth century mathematics had become part of the logic of the scientific method; they were a neutral tool of enquiry rather than an *a priori* determinant of the nature of things, and Descartes noted the profound change that had taken place in the status of mathematics. The change did not take place primarily in astronomy, but in the science of mechanics. In this subject there had long been a tradition both of craft practice and scholarly discussion, and it was in mechanics that the experimental-mathematical method of science arose. The science of mechanics and the mathematical experimental method developed during the sixteenth century in northern Italy, which was then perhaps the most technically advanced region of Europe, notable for its architects and engineers. In contrast, England, which was less advanced technically, produced the science of magnetism and the qualitative inductive method, whilst the Germans, using old methods, developed the science of astronomy.

After the stagnation of the impetus school of thought in the universities, mechanics was developed during the sixteenth century mainly by engineers, though scholars continued the impetus discussions, and the subject finally assumed a modern form with the university teacher, Galileo. The engineers quickly went ahead of the impetus theorists in method, practising experiments rather than confining themselves to discussion. Moreover, their experiments were quantitative. They measured and correlated the variables they studied, to obtain empirical physical laws. The artist and engineer, Leonardo da Vinci, 1452-1519, studied various building problems experimentally. Using small-scale models, he investigated how the weight that vertical pillars and horizontal beams could support varied with their thickness and their height or length. His experiments led him to the results that the carrying power of a pillar, of a given material and height, varied as the cube of its diameter, and that the carrying power of a beam was directly proportional to its thickness and inversely proportional to its length.

Such experiments indicate that Leonardo appreciated the importance of quantitative experimentation in scientific method; nor was he unaware of the value of mathematics. 'There is no certainty in science where one of the mathematical sciences cannot be applied', he wrote. Mechanics, he thought, was the noblest of the sciences, 'seeing that by means of this, all animated bodies that have movement perform all their actions'. In his theoretical views, Leonardo did not get beyond the results of the impetus theorists, though he expanded the scope of mechanics beyond physical questions to animate nature. The bones and joints of animals he considered to be lever systems operated by the forces of their muscles.

A mechanical question that grew in importance with the development of firearms was the nature of the motions of projectiles. The Greeks had been able to deal only with combinations of forces or motions that were in the same straight line, or parallel as in the case of levers. The motions of projectiles always remained rather a problem as they were due to a force of projection and the force of gravity which were rarely linear or parallel. The Aristotelians of the middle ages were of the view that a projectile moved at first upwards along an inclined straight line until the projective force was exhausted, and then it fell vertically downwards under the force of gravity. Thus they did not combine the projective force with the force of gravity; they considered them to be consecutive. The impetus theorists thought that gravity might begin to act a little before the impetus of a projectile was exhausted, so that the uppermost point of its trajectory would not be a sharp angle, but rounded off. Leonardo adopted this view. There were three parts to the trajectory of a projectile, he thought; firstly, a straight line motion under impetus, secondly, a curved position where gravity and impetus are mixed, and thirdly, a vertical fall under the force of gravity.

The work of Leonardo was followed by that of Tartaglia, 1500-57, a self-taught engineer, surveyor, and book-keeper, who wrote on mathematics and mechanics. In 1546 he published a book dealing with military tactics, munitions, and ballistics, in which he laid it down that the impetus of projection and the force of gravity act together on a projectile throughout the course of its flight. Thus the path of a projectile is curved throughout its course, for, 'there is always some part of gravity drawing the shot out of its line of motion'. Tartaglia also found an empirical rule connecting the range of a cannon with its angle of inclination. The range is a maximum, he said, when the cannon is inclined at an angle of forty-five degrees; and as the angle is increased or decreased, the range falls off, at first slowly, and then more rapidly. Tartaglia was much concerned with the promotion of mathematics, and mechanics. He made the first Italian translation of Euclid's geometry, and published the first edition of Archimedes' mechanics in 1543. For him, mathematics were solely concerned with the method of science; they did not represent metaphysical truths to which scientific theories had to be adjusted.

'The purpose of the geometrical student', Tartaglia wrote, 'is always to make things that he can construct in material to the best of his ability.'

Thus mathematics were of use in science only in so far as they were applicable to concrete physical matters.

Contemporary with Tartaglia was Jerome Cardan, 1501-76, a wealthy scholar who lectured at the Platonic school of Milan. In contrast to Tartaglia, Cardan held that geometric forms and arithmetic harmonies determined the character of natural things, a knowledge of mathematics giving man occult powers over nature. In the field of mechanics he was not an experimentalist, or at least he did not test his generalizations experimentally, for he held that the force required to hold a body at rest on an inclined plane was proportional to the angle of that plane, a statement which experiment would have disproved for angles of appreciable magnitude.

A little later, Benedetti, 1530-90, at the university of Padua, continued the discussion of the impetus theory. His book *On Mechanics* (1585) was mainly a critique of the Aristotelian theory. Benedetti rejected Aristotle's idea that the speed of a body increased the nearer it approached to the centre of the universe, holding that the speed of such a body falling freely increased the further it receded from its starting point. Benedetti thought that a stone dropped down a shaft going right through the earth would not stop at the centre, as Aristotle had supposed, because the impetus which it had acquired would carry it on so that it would oscillate to and fro about the centre until its impetus decayed. However, Benedetti believed that bodies of the same shape and size would fall with speeds proportional to their densities, the heavier body falling the faster.

A notable student of mechanics at this time, outside Italy, was Simon Stevin, 1548-1620, of Bruges. Like Tartaglia, he began his career as a book-keeper and military engineer, but Stevin had more success, becoming technical adviser to Prince

Maurice of Nassau, and ending his days as Quartermaster General of Holland. Stevin was at first self-taught, but he acquired more education than Tartaglia, entering the university of Louvain when he was thirty-five years old. In 1586 Stevin published a work on mechanics containing several important results. He made an experiment disproving the Aristotelian view that heavy bodies fall more quickly than light ones, an experiment that has been ascribed incorrectly to Galileo.

'The experiment against Aristotle is this,' Stevin wrote. 'Let us take . . . two leaden balls, one ten times greater in weight than the other, which allow to fall together from the height of thirty feet upon a board or something from which a sound is clearly given out, and it shall appear that the lightest does not take ten times longer to fall than the heavier, but that they fall so equally upon the board that both noises appear as a single sensation of sound.'

Stevin also obtained an intuitive understanding of the parallelogram of forces, a method of finding the resultant action of a combination of two forces that are not in the same straight line nor parallel. The method was first explicitly formulated by Newton and Varignon in 1687: it consists in the representation of the two forces, in magnitude and direction, by two straight lines originating from a common point: the resultant is then given by the diagonal of the parallelogram formed by drawing two other lines parallel to the first two. Ancient mechanics, it will be remembered, had never been able to deal with combinations of forces that were neither linear nor parallel.

In connection with shipbuilding, Stevin advanced the science of hydrostatics, adding to Archimedes' principle of buoyancy, the proposition that any floating body assumes such a position that its centre of gravity lies in the same vertical line as the centre of gravity of the displaced fluid. He also was a keen exponent of the decimal system, advocating its use in the representation of fractions, for weights, measures and coinage. In method he was a confirmed experimentalist, and an applied scientist. Experiments, he said, 'are the solid basis on which the arts must be built'. Furthermore he advocated the co-operation of many persons in a joint scientific project, for he affirmed, 'the error or negligence of the one is compensated by the accuracy of the other', in this way.

In mechanics, as in magnetism, we find that craftsmen and engineers could develop the scientific method and new experiments, but not new bodies of theory. In both magnetism and mechanics it was the scholars who were interested in the craft tradition, and who were opposed to the old scholarly tradition, that originated the new theories. Ancient mechanics were overthrown and modern mechanics were founded by such a man, Galileo Galilei, 1564-1642, of the universities of Padua and Pisa. Galileo was born at Pisa, where he studied and taught for a short time at the university there. In 1592 he moved to the more liberal and enlightened university of Padua, where he remained eighteen years, carrying out the more important of his researches on mechanics. In 1610 he moved to Florence as the 'Philosopher and First Mathematician to the Grand Duke of Tuscany', and there he carried out his investigations in astronomy with the telescope. Finally he studied mechanics again when his astronomical work was condemned.

Galileo's two great works are the *Dialogue concerning the Two Chief Systems of the World, the Ptolemaic and the Copernican,* published in 1632, and his *Discourses on Two New Sciences,* published in 1638. Both of these works were written in the form of a dialogue between two of his friends and supporters, Sagredo and Salviati, and a supporter of the Aristotelian point of view, Simplicius. In this way Galileo endeavoured to give his works a wide appeal and to discredit effectively the Aristotelian mechanics and cosmology. Galileo's work in mechanics, like that of Leonardo, Tartaglia, and Stevin, was stimulated by the problems experienced in engineering. In a letter written in 1632 to Marsili, Galileo stated that it was specifically the problem of the flight of projectiles that led him to study the gravitational fall of bodies. Moreover, his book on mechanics, the *Discourses on Two New Sciences,* opens upon a scene set in the Venetian Arsenal, where Salviati remarks:

'The constant activity which you Venetians display in your famous arsenal suggests to the studious mind a large field for investigation, especially that part of the work which involves mechanics; for in this department all types of machines and instruments are constantly being constructed by many artisans, among whom there must be some who, partly by inherited experience, and partly by their own observations, have become highly expert in explanation.'

But even though these craftsmen knew a great deal, said Galileo, they were not really scientific, for they were not acquainted with mathematics and thus they could not develop their results theoretically. Galileo was much concerned with the role of mathematics in scientific method, and in particular with the problem of the degree to which physical objects correspond to geometrical figures. In the astronomical dialogue, the Aristotelian, Simplicius, pointed out that geometrical spheres touch a plane at one point, whilst physical spheres touch a plane at several points, over a whole area in fact, so that there would appear to be a lack of correspondence between mathematics and nature. Salviati replied that, whilst this was so, it was possible to imagine an imperfect geometrical sphere which touched a plane at several points. Thus mathematics could be adjusted to physical objects and could be used to interpret nature, the correspondence between the two being judged by 'well chosen experiments'. Any discrepancy was the fault of the scientist—'The error therefore lyeth neither in the abstract nor in geometry, nor in physicks, but in the calculator, that knoweth not how to adjust his accompts.'

The first set of mechanical problems to which Galileo addressed himself were those involving the scale effect; the problem of how it was that large machines often broke down and collapsed, though they were built in exactly the same geometrical proportions as smaller machines which were durable and adequately served their purposes. The properties of geometrical figures did not depend upon their size, the value of π was the same for all circles, yet large ships would collapse in the stocks whilst smaller vessels built in the same proportions could be launched with safety. Again there appeared to be a lack of correspondence between mathematics and nature, but the problem could be solved, Galileo affirmed, if the amount of matter a body contained were treated as a mathematical quantity, if matter were regarded 'as if it belonged to simple and pure mathematics'. Thus if the dimensions of a machine were doubled, its weight would go up eight times, but the strength of its individual parts increased in a smaller ratio, so that they could not sustain the greater weight. Galileo, like Leonardo, showed that the weight a horizontal beam would support actually decreased in proportion to its length, so that it had to be much thicker in order to sustain even the same weight. Similarly a large building required proportionately much thicker pillars to carry the weight of the roof than a smaller structure. Again like Leonardo, Galileo carried over his mechanics to the animate world, pointing out that the legs of elephants had to be proportionately much thicker than those of insects to carry their weight. He found that hollow cylinders were stronger than solid cylinders containing the same amount of material, and he suggested that this accounted for the fact that the bones of animals were hollow, and roughly cylindrical, as this form gave them the maximum strength for the minimum weight.

Galileo was of the view that suitably chosen mathematical demonstration could be applied to the investigation of any problem involving qualities which were measurable, beyond the spatial measures of length, areas, and volumes which had been the traditional subject-matter of geometry. Investigating the scale effect, he studied amounts of matter, later called masses, in this way; then he enquired into dynamical problems involving the measures of time and velocity in a similar fashion. Here the central problem for Galileo was that of the fall of bodies under gravitational force. First of all he argued his way out of the Aristotelian view that heavy objects fell faster than light ones. What would happen, he asked, if a heavy and a light body were tied together and then dropped from a height? From the Aristotelian point of view it could be maintained that the time occupied by their fall would be either the mean of the times the two bodies would have taken to fall separately, or the time taken by a body with their combined weights to fall from the same height. 'The incompatibility of these results', Galileo wrote, 'showed Aristotle to be wrong.' To find out what actually did happen in the gravitational fall of bodies Galileo made the experiment of measuring the time taken by smooth metallic spheres to roll down given lengths of a graduated inclined plane. The free fall of an object under gravity was too rapid to be observed directly, and so Galileo 'diluted gravity', using the device of the inclined plane, in order that his metallic spheres should move downwards under gravity with measurable speeds. Thus he found that all bodies, irrespective of their weights, fell through the same distance in the same time, the distance being proportional to the square of the time taken in the descent, or what amounted to the same thing, the velocities of falling bodies increased uniformly with time.

According to the Aristotelian physics, the constant action of a force kept a body moving with uniform speed. Galileo's results showed however that bodies did not move in uniform velocity under the constant influence of the force of gravity; on the contrary, over each time interval they received an extra increment of velocity. The speed a body had at any point continued on and was augmented by gravitational force. If the force of gravity could be switched off, the body would still move with the velocity it happened to have at that point. Such a phenomenon was observed when Galileo's metallic spheres reached the bottom of the inclined plane: they continued to move across a smooth horizontal table with uniform speed. From these considerations followed the principle of inertia which laid it down that a body remains in the same state of rest or of uniform motion so long as it is not acted upon by a force.

Thence Galileo proceeded to show the value of mathematical demonstration in science by developing the theory of the path traced out by the flight of a projectile. He considered the movement of a sphere rolling across a table with uniform speed until it came to the edge, when it traced a curved path to the floor. At any point on this path the sphere would have two velocities; one horizontal, remaining constant owing to the principle of inertia, the other vertical, increasing with time because of gravity. In the horizontal direction the sphere would sweep out equal distances in equal times, but vertically, the distances it covered would be proportional to the square of the time. Such relations determine the form of the path described, cannon would then be a full parabola, giving a maximum range namely a semi-parabola. The path of a projectile shot from a when the gun was elevated at an angle of forty-five degrees.

Thus what Tartaglia had observed as a fact, Galileo deduced theoretically from the results of his inclined plane experiments. Galileo wrote in this connection:

'The knowledge of a single fact acquired through a discovery of its causes, prepares the mind to understand and ascertain other facts, without need of recourse to experiments, precisely as in the present case, where by argumentation alone the author proves with certainty that the maximum range occurs when the elevation is forty-five degrees.'

Such a development was of the greatest importance for science. Hitherto new phenomena had been found only by chance or accident, and rival hypotheses, like the impetus and Aristotelian mechanics, could live side by side for many generations because of the lack of criteria to decide between them, other than logic. Now Galileo showed how it was possible to demonstrate 'what perhaps has never been observed' from phenomena already known; the demonstration providing an explanation for those phenomena, and the experimental discovery of the predicted facts verifying that explanation. For Tartaglia, a gun elevation of forty-five degrees giving the maximum range was a brute fact; for Galileo it was the resultant of the properties of the two velocities possessed by projectiles, his explanation being verified by the physical occurrence of the predicted fact. Similarly Galileo knew as a fact that the swings of a pendulum took the same time, no matter what the amplitude of swing, and later Christian Huygens of Holland demonstrated mathematically that this was a necessary consequence of the uniformity of the force of gravity.

With Galileo the mathematical-experimental method of science came to maturity. He took geometry away from its subject-matter of lengths, areas, and volumes, and applied it to other measurable properties, namely time, motion, and amount of matter, in order to discover the connections between them and to deduce the consequences of those connections. In order to apply mathematics to physical phenomena in this way, the field of investigation had to be narrowed down to the observation of qualities that were measurable. Mathematics could not be applied to non-measurable qualities, and thus they had to be ignored. Galileo had to disregard some of the less relevant measurable phenomena too, so that he could simplify his study and get to the fundamentals of his problem. He knew very well that air resistance, which was in principle measurable, modified the gravitational fall of bodies, but he ignored the matter. Galileo made his experimental conditions as perfect and as 'mathematical' as possible, using a polished inclined plane, and a smooth metal sphere. Only in this way could he get information that transcended the conditions of the particular experiment, information that described the fundamental behaviour of all bodies undergoing gravitational fall. Thence mathematical demonstration could be applied, giving a structure of abstract theory and predicted consequences which could be tested by further experiments.

There was a limit beyond which the mathematical-experimental method could not go. It could not deal with non-measurable phenomena, such as the qualitative properties that mark off one living creature from another. Here Bacon's qualitative-inductive method came into its own, though not for some time. During the seventeenth century the mathematical-deductive method received the wider application: indeed it was made into a philosophy. The non-measurable properties of matter, which the mathematical scientist ignored, came to be regarded as unreal. A clear distinction was drawn between the measurable primary qualities of matter, and the secondary qualities which were non-measurable. The measurable primary qualities, mass, motion, and magnitude, were regarded as real, objective properties of matter, whilst the non-measurable secondary qualities, colours, smells, tastes, were regarded as subjective products of the sense organs, possessing no reality as such in the external world.

Another development that accompanied the rise of the mathematical-experimental method was the elaboration of measuring instruments, so that mathematics could obtain a foothold in phenomena. Galileo used extensively those traditional measuring instruments, the ruler, balance, and water clock, and he developed others. He made the first thermometer to measure temperature, and used the pendulum to measure time, first in medicine to estimate pulse rates, and then more generally in the design he left behind him for the first complete pendulum clock. Galileo also developed the telescope and used it extensively for making astronomical observations, though curiously enough most of his observations of the heavens were of a qualitative character.

In 1609 Galileo heard that 'perspective glasses', which magnified distant objects, had been constructed by the Dutch spectacle makers of Middelburg, notably Hans Lippershey who patented the invention in 1608. Galileo investigated the optical properties of combinations of lenses, and he constructed several improved telescopes for himself. With these telescopes he examined the heavens and discovered a host of new facts. He found that the heavenly bodies were by no means as perfect and superior to the earth as the traditional Aristotelian view suggested. There were spots on the face of the sun, and the moon appeared to be much like the earth, possessing vast mountains the height of which he estimated from the length of the shadows they cast. He discovered that the Milky Way consisted of very many fixed stars, and with others, observed the nebula in the constellation of Andromeda. Galileo found further that the planet Venus had phases like the moon, changing from a thin crescent to a full orb, and that the planet Jupiter had four moons, presenting a picture, he thought, of the solar system in miniature according to the Copernican scheme.

Galileo had long been an adherent of the Copernican system of the world. Writing to his friend Johann Kepler in 1597, he said that he had been 'since many years already a follower of the Copernican theory', since it explains 'the reasons of many phenomena which are quite incomprehensible according to the views commonly accepted'. The Copernican system had not been widely accepted during the sixteenth century as it gave predictions of planetary positions which were no more accurate than those given by the Ptolemaic scheme, and it embodied suppositions that appeared to be unsound from the point of view of the traditional mechanics. Moreover, it was only a specialist departure from the integrated Aristotelian philosophy of nature, it was not yet part of a coherent view of the world as a whole.

However, developments in astronomy tended to favour the Copernican theory, and the discoveries of Galileo gave a considerable impetus to this trend. In 1572 a bright new star had appeared, probably a supernova, and lasted throughout the following year, disappearing in 1574. Furthermore, the path of a comet that came into view in 1577 was observed and measured by Tycho Brahe, Michael Mastlin, and others, who showed that it moved round the sun, through the solar system. Aristotle had held that the appearance of comets was a terrestrial phenomenon taking place below the orbit of the moon, and that the heavens were perfect and unchanging, subject to neither generation nor decay. Both of these tenets were disproved by the astronomical phenomena observed in the 1570's and now Galileo added the evidence of spots on the sun and mountains on the moon to illustrate the imperfection of the heavens. Moreover, it had been pointed out quite early on that if the Copernican theory were true, then Venus ought to show phases like the moon. With the naked eye Venus appeared always to be an orb, but Galileo with the telescope showed that the expected phases did occur. Again, it had been argued that there could only be one centre of rotation in the universe, and since the moon rotated round the earth, the other heavenly bodies must move in the same way. Galileo now showed that whatever view be taken concerning the arrangement of the solar system, there was certainly more than one centre of rotation in the world, as four moons circled round Jupiter.

Galileo published most of his astronomical discoveries in the second decade of the seventeenth century, and they were most effective in securing support for the Copernican theory. Now that evidence for the new astronomy was forthcoming, the opposition to it stiffened, for it could no longer be regarded as an unimportant minority opinion. Neighbouring ecclesiastics denounced Galileo's views as heretical, whilst the scholastic philosophers of Pisa declared his opinions to be false and contrary to the authority of Aristotle. They suggested that the sun spots were only clouds moving around the sun, or that they were due to imperfections in the telescope, and there could be no moons moving round Jupiter for there was no mention of them in the works of the ancients. In 1615 Galileo was summoned before the Inquisition at Rome, and there he was made to abjure the Copernican theory. The propositions that the earth rotated on its axis, and that it moved round the sun, were officially declared to be false and heretical, and in 1616 the work of Copernicus was placed on the Index of prohibited books, not to be removed until 1835.

Galileo did not undergo a change of opinion, for sixteen years later he published, with the permission of the Florentine Inquisitors, his *Dialogue concerning the Two Chief Systems of the World, the Ptolemaic and the Copernican*. This book opened straight away with an attack upon the Aristotelian doctrine that the celestial bodies were quite different and distinct from the earth in composition and properties. The appearance of new stars, comets, sun spots, and mountains on the moon, were all cited as evidence to the contrary. Galileo further rejected the notion current from antiquity that immutability and changelessness was noble and a sign of perfection. He advanced the conception which was later to become important that motion was not a transmutation, it led to neither generation nor decay, but it was, as he put it, 'a simple transposition of parts, without corrupting or ingendering anything anew'. Such a conception later became part of the mechanical philosophy, which held that the universe and its contents have remained, and will remain, much as they are, no new entities appearing nor disappearing, natural processes consisting merely of the mechanical movements of bodies and the interchange of their momenta.

In his arguments for the Copernican system, Galileo concentrated on meeting the commonsense mechanical objections that beset it. He repeated the answers that Copernicus had given to such objections, but these answers were now more cogent, for they were based on Galileo's new mechanics. According to the principle of inertia, the atmosphere would naturally rotate with the earth, it would not require a constant propelling force as Aristotle's mechanics suggested. Objects dropped from a height would not fall to the west, for they would share the movement of the earth. Galileo suggested similarly that a stone dropped from the mast of a moving ship would drop to the base of the mast, and not behind it, as the stone would share the general motion of the ship and its contents. Such an experiment was made by the Frenchman, Gassendi, in the 1640's and was found to give the result that Galileo had expected.

Thus Galileo attacked not only the astronomy, but also the

mechanics of the old cosmology. He presented a new way of looking at things which was coherent, in opposition to the Aristotelian view that also composed an integral whole. In Book III of his *Dialogues* he admitted that the arguments for both views were about equally decisive, but he thought he had a conclusive proof of the Copernican system in his theory of the tides, which he discussed in Book IV. The two motions of the earth, its daily rotation, and its annual movement round the sun, gave rise to jerks, he thought, which caused the waters of the oceans to wash to and fro, like water in a basin. He rejected the idea that the sun and moon caused the tides, because it implied that the heavenly bodies were superior to the earth and influenced terrestrial events, a doctrine to which he was much opposed. However, his theory demanded that there should be tides once a day, not twice as observed. Moreover, it contradicted his principle of inertia whereby bodies on the earth should share the motions of the earth.

Thus in the end, Galileo did not finally establish the Copernican theory, though he secured substantial support for it. His work was addressed to a wider audience than the professional astronomers and mathematicians. It was written first in the vernacular Italian, in the dialogue form of ordinary conversation, and it was simplified to convey his argument to the layman. He discussed only two world systems, the Ptolemaic and the Copernican, and left out their variants, such as the system of Tycho Brahe, and that of William Gilbert which was the same as the Tychonian system except that the earth rotated on its axis daily. Moreover, he ignored the system of his friend Kepler, which greatly improved upon the Copernican theory and provided overwhelming proof of the heliocentric hypothesis for professional astronomers and mathematicians, though perhaps not for laymen.

Galileo's great work on the systems of the world was published in 1632, some thirteen years after Kepler had made known the last of his three laws of planetary motion. But Galileo ignored his friend's work, and held to the last that the orbits of the planets were circles, not ellipses as Kepler had shown in 1609. According to Galileo's principle of inertia, if the surface of the earth were perfectly smooth, a sphere set in motion on that surface would continue to roll round the earth indefinitely. Hence he thought that uniform speed in a circle was the natural motion of all bodies which were not acted upon by a force. Such natural motions were enjoyed by celestial as well as terrestrial bodies, and since, in his opinion, the orbits of the planets were circular, no problem arose concerning the motions of the heavenly bodies; they were, as Copernicus had thought, entirely natural.

Thus Galileo did not have the modern conception of inertial motion as uniform speed in a straight line. If he had had, he might have shown that the force of gravity of the sun bent the natural rectilinear motion of the planets into an ellipse, for he had shown that the gravity of the earth bent the inertial motion of a projectile into a parabolic path. The two problems, that of the path of the cannon ball and that of the orbit of a planet were similar, and later Newton treated them as such. But Galileo failed at this point, for he was constrained by one ancient conception which he never overcame, namely the idea that the motions of the heavenly bodies were circular and uniform. His astronomical ideas were also limited by his neglect of Kepler's work, and perhaps by his adoption of a qualitative non-mathematical method in astronomy, a method to which he had been averse in the terrestrial sciences, for he had criticized William Gilbert's work on magnetism for its lack of mathematics.

Galileo and Kepler form a striking contrast. Kepler too heard of the new telescopes of the Dutch spectacle makers and investigated the optical theory of their construction. He designed a new type of telescope which was different in principle from that developed by Galileo, but he did not use the instrument for astronomical purposes; indeed, it appears that he did not even make it. Kepler was taken up with the ordering of the quantitative observations made by Tycho Brahe, not with the facts of qualitative telescopic observation, whilst the reverse was true of Galileo. Kepler was primarily concerned with making astronomy more precise and accurate technically, whilst Galileo was mainly interested in propagating the intellectual revolution inaugurated by Copernicus. Kepler was also interested in promoting the heliocentric theory, indeed he found the most permanent proofs for it, and in the same way Galileo did make some quantitative observations in astronomy of a technical nature. He prepared tables of the eclipses of Jupiter's moons, for the purpose of determining longitude at sea. But the aim of promoting the Copernican revolution was primary for Galileo. As we have seen, Galileo opens his work on astronomy with an attack upon the Aristotelians, not with a comment upon the interesting problems met with in navigation, whereas his work on mechanics opens with a discussion of the problems arising from a study of the machines in the Venetian arsenal.

The nature of Galileo's interests helps to explain why he largely abandoned the mathematical method in astronomy, and concentrated upon making qualitative telescopic observations. Any person could see the moons of Jupiter, the phases of Venus, and the mountains on the moon, with a telescope, but only skilled mathematicians could be convinced by Kepler's findings that the heliocentric theory was essentially correct. Thus Galileo was more effective historically in disseminating the Copernican system amongst the men of his time than Kepler; he brought simpler proofs to a wider public. The point was appreciated by his opposition, for in 1633 Galileo was again called to Rome to face the Inquisition, though his work had been sanctioned by the local Inquisitors of Florence. Again he was made to abjure the Copernican hypothesis, and this time he was condemned for his heresy, and was detained for the remaining nine years of his life in a villa near Florence. Here he wrote up his studies in mechanics, which were smuggled out of the country and published at Amsterdam in 1638 as his works were banned in Italy.

Chapter 15

Descartes: The Mathematical Method and the Mechanical Philosophy

DURING THE seventeenth century, the centres of science shifted from the late medieval centres of commercial prosperity and Renaissance culture, German and northern Italy, to the regions near to the Atlantic which had profited from the great geographical discoveries—France, Holland and southern England. Kepler and Galileo marked the peak of the German and Italian achievement in early modern science, and these lands were not to become prominent again in science until the turn of the eighteenth century. The change in the geographical locations of the centres of science during the seventeenth century was accompanied by a change in the type of man who became a scientist. Both Kepler and Galileo had been professional scientists. Both were patronized by princes, and held academic appointments at various times in their careers. The new type of scientist appearing in France, England, and Holland, was essentially an amateur. In England such men came from well-to-do landed or mercantile families, often the *nouveau riche* of the period, like Robert Boyle and Sir William Petty. Newton was here an exception, for he was a professional in that he held a chair at Cambridge, and was not wealthy. In France men coming from families connected with the administration were prominent in science, particularly lawyers connected with the legal administration of the local *parlements*, or law courts of France.

The shift of the centres of science is illustrated by the fact that both of the men who endeavoured to give a general analysis of the scientific method during the first half of the seventeenth century came from the new regions, Francis Bacon from England, and René Descartes from France. Descartes, 1596-1650, was a son of a counsellor in the *parlement* of Bretagne, and throughout his life remained a well-to-do amateur. In natural philosophy he set out to do two things. Firstly to examine and generalize the mathematical method which

165

had been developing in the science of mechanics. Secondly to build up by means of this method a general mechanical picture of the operations of nature. As the intellectual atmosphere of France proved somewhat unfavourable, Descartes moved to Holland in 1628, and here published his *Discourse on Method* in 1637. This work consisted of two parts; the first being an analysis of the mathematical-deductive method, and the second presenting an outline of his view of the physical world. The second part was later amplified in his *Principles of Philosophy*, published in 1644, and it was this part which was most influential during the seventeenth century.

Descartes had read Bacon's views on scientific method and sympathized with his aims, but he thought that Bacon had started his enquiries from the wrong end; he had started with the empirical facts of the natural world rather than the general principles which provided a basis for deductive enquiry. Descartes was impressed by the mathematical method developing within the physical sciences, and he saw that just as the student of mechanics limited the variety of things he could observe to those that were measurable, so too he had to cut down the variety of theories that could be suggested to those that could be mathematically developed. In the same way not all measurable qualities were of the same importance, some had to be disregarded to simplify the study, as Galileo had ignored air resistance in his study of gravitational fall. Similarly, Descartes thought, not all ideas susceptible to mathematical treatment were of equal importance; there were certain fundamental ideas 'given by intuition' which provided the surest starting point for deductions of a mathematical character. Such ideas were those of motion, extension, and God. The idea of God was the main foundation of his system, for God had made extension, and had put motion into the universe. Since motion had been given to the universe only once, at the Creation, the amount of motion in the world must be constant. By such an argument Descartes arrived at the principle of the conservation of momentum.

In this way, by arguing in a quasi-mathematical fashion from undoubted and certain principles, Descartes thought it possible to deduce all the salient features of the natural world. On the detail of the nature of things there was bound to be some uncertainty, for different consequences could be deduced from the same proposition, and here experiment was brought in to decide between rival views. However, Descartes did not stress this side of his method; he was primarily concerned with the deduction of the general scheme of things from first principles. Moreover, he regarded experiments as merely illustrating the ideas that had been deduced from the principles given by intuition: he did not regard experiments as determinative of the principles from which deductions could be made, as Galileo had done. Galileo had obtained his principle of inertia, and the law of the gravitational fall of bodies, from his experiment with the sphere rolling down an inclined plane, and it was from these experimentally given principles that he had deduced mathematically the path of projectiles, and so on.

The Dutch scientist, Christian Huygens, later remarked that Bacon had not appreciated the role of mathematics in scientific methods, whilst Descartes had disregarded the role of experimentation. In this sense the works of the two philosophers were complementary, Bacon perpetuating the empiricism of the craft tradition, and Descartes the speculative tendencies of the scholarly tradition. Thus neither of them fully integrated the two traditions, though both approached near to this end. Bacon was aware of the role that mathematics could play in science. 'The investigation on nature goes on best when physics are limited by mathematics,' he said, but he did not develop the point. Similarly Descartes distinguished between 'analysis', the practical way things are discovered, and 'synthesis', the theoretical way the same things can be deduced from first principles. Synthesis 'alone was used by the ancient geometers in their writings', he said, 'but it does not satisfy as does the other method, nor does it content the minds of those who desire to learn, for it does not show the way in which the thing was discovered'. Thus Archimedes discovered his principles of mechanics experimentally, but he presented them as deductions from axioms given by intuition, so we do not know what his experiments actually were. Galileo on the other hand gave an account of his experiments. Hence Descartes' method was more akin to that of Archimedes than that of Galileo, though he appreciated the advantages of the latter.

Descartes was not attracted by the ancient Pythagorean notion that mathematical considerations determined the structure of the universe; the idea that the perfect heavenly bodies must have the perfect form of a sphere, and that their motions must be circular and uniform. For Descartes, mechanical considerations determined the form and motion of the heavenly bodies, and indeed of all the operations of nature. Mathematics he considered only as a methodological device, and he was unsympathetic to the attitude of the pure mathematician. 'There is nothing more futile than to busy oneself with bare numbers and imaginary figures,' he wrote. Like Bacon, he considered utilitarian projects an important end of science. With his new method Descartes said, we shall come to know the nature of the elements,

'as distinctly as we understand the various trades of our artisans, and by application of this knowledge to any use to which it is adapted, we could make ourselves masters and possessors of nature.'

In developing his mathematical method, Descartes made notable advances in mathematical technique; in particular, he invented co-ordinate geometry. Galileo had relied upon geometrical demonstrations to prove his propositions in mechanics. He extended the scope of geometry to other measurable quantities, amounts of matter, velocities, and time. Here the application of geometry was rather forced and clumsy. Algebra was a far more flexible and generalized mathematical technique and it could well deal with problems involving masses and motions. But algebra was then rather a novelty; it was abstract and remote from the predominantly geometrical thinking of the mathematicians of the age. Even so there were important developments in the field of algebra during the sixteenth and seventeenth centuries. Tartaglia gave the first algebraic solution to a cubic equation, as opposed to the earlier geometric methods of solving such equations. François Viete, 1540-1603, who like Descartes' father was a member of the Bretagne *parlement*, improved the notation of algebra, and Viete's results were developed by the English mathematician, astronomer and surveyor, Thomas Harriot, 1560-1621. Following on Harriot, Descartes made geometry algebraic; he represented geometrical figures graphically by means of algebraic equations. Later Newton and Leibniz were to describe geometrical figures by algebraic equations representing the movement of a geometrical point, thus developing the calculus. These methods served the purpose of analysing the relations between masses and motions, and they were the more important of the mathematical techniques used by scientists up to the present century.

Descartes also developed the conception that mass and time were fundamental dimensions of the world, as important as the three dimensions of space. He wrote that

'it is not merely the case that length, breadth, and depth are dimensions, but weight also is a dimension in terms of which the heaviness of objects is estimated. So, too, velocity is a dimension of motion, and there are an infinite number of similar instances.'

However, Descartes lost the idea, and it was not until 1822 that Fourier developed the method of representing physical units, such as velocity, acceleration, and so on, by their fundamental dimensions of mass, time, and length, to obtain relations between them. Descartes reverted to the view that extension and motion were the fundamental quantities composing the world. Like Galileo, he tended to identify matter with volume, that is with extension in three dimensions. However, Descartes improved upon Galileo by suggesting that natural motions took the form of uniform speed in a straight line, not a circle as Galileo had supposed. Descartes was thus the first to enunciate the modern principle of inertia.

With the conception of extension and motion Descartes thought it possible to deduce with his method the main outlines of the working of the universe. 'Give me motion and extension', he wrote, 'and I will construct the world.' Descartes was of the view that the laws of nature were such as to develop any possible arrangement of the primordial chaos of matter into the kind of world we find ourselves in at present. Irrespective of its form at the beginning, the universe of necessity assumed its present-day configuration, and so too would any other possible world composed of matter and motion. As Descartes identified matter with volume, he denied that there could be such a thing as a vacuum, an empty hole in space, and that matter was composed of atoms with empty spaces between them. Matter permeated the whole of space, and in the beginning, therefore, the primordial matter could only undergo a rotatory motion. Thus a giant vortex was set up in which the primary blocks of matter were swept round and were gradually worn down by friction. Irrespective of their original shape, the primary blocks of matter were abraded into dust, the first matter, and small spheres, the second matter. The cosmic dust, or the first matter, was the fire element, forming the sun and the fixed stars. The second matter was the air or ether element, composing the material of inter-stellar space. There was also a third matter, namely original blocks of matter which had not been ground down to dust but had been merely rounded off. These large spherical blocks of matter were the earth element, forming the earth, planets and comets.

As the universe developed, secondary whirlpools were set up round each conglomeration of matter. There was a vortex surrounding the earth which carried the moon on its course, and another round Jupiter that maintained its four moons in their orbits, whilst the earth and all of the planets were caught in a wider vortex around the sun. In the cosmic vortices heavy matter was drawn towards their centres, whilst light matter was dispersed towards their outer edges. Hence heavy objects fell towards the earth and fire rose upwards.

Descartes pointed out that in his system all motions were really relative. Motion, he wrote, is nothing other than

'the transference of one part of matter, or one body, from the vicinity of those bodies that are in immediate contact with it, which we regard as in repose, into the vicinity of others.'

Hence the earth could be considered to be immobile at the centre of its own vortex. From this point of view the earth was at rest in the centre of the world as the traditional view required. However, Descartes' theory of the relativity of motion did not save his works from condemnation. They were placed on the Index of prohibited books at Rome and at Paris in 1663. Later, in 1740, they were removed from the Paris Index in order to provide an alternative to the Newtonian system of the world which was then becoming popular in France.

In its content Descartes' natural philosophy was diametrically opposed to the traditional world view based on the theories of Aristotle. In Descartes' system all material beings were machines ruled by the same mechanical laws, the human body no less than animals, plants, and inorganic nature. Thus he dispensed with the traditional conception that nature was hierarchically ordered, the idea that the beings composing the world formed a vast chain of creatures stretching down from the most perfect of all beings, the Deity, at the periphery of the universe, through the hierarchies of angelic intelligences in the heavens to the grades of men, animals, plants, and minerals on earth. For Descartes the physical and organic world was a homogeneous mechanical system composed of qualitatively similar entities, each following the quantitative mechanical laws revealed by the analysis of the mathematical method. The world was not, as the scholastic philosophers had believed, a heterogeneous but ordered diversity of entities, each finding its rank in the cosmic order through the purely qualitative analysis of a classification in terms of the kind of soul it possessed, vegetable, animal, or rational. Besides the mechanical world, Descartes supposed that there was also a spiritual world in which man alone of the material beings participated by virtue of his soul. Hence as the Cartesian philosophy gained ground, the traditional view, that the world was made up of a vertical scale of creatures, gradually disappeared and was replaced by the conception that the universe was composed, so to speak, of two horizontal planes, the one mechanical and the other spiritual, man alone sharing in both. From the time of Descartes such a dualism has been fundamental to European thought.

Before modern times the workings of the natural world were thought to be governed by custom, the principle of retribution, and acts of purpose, will, and design rather than by laws of nature and mechanical force. Descartes supposed that nature was governed in its entirety by laws and he identified the laws of nature with the principles of mechanics. 'The rules of nature', he wrote, 'are the rules of mechanics'. Descartes in fact was the first to use consistently the term and the conception of 'laws of nature' which, like the earlier usage of the notions of 'custom' and 'retribution', was an analogy based upon the practices of civil society. The ancient Greeks had rarely used the phrase 'the laws of nature'. The quantitative rules which they discovered were called 'principles', like the 'principle of levers' and Archimedes' 'principle of buoyancy'. Galileo called his quantitative rules 'principles', 'ratios', or 'proportions', though in the English version of his *Two New Sciences* these words have been translated as 'laws'. Galileo's 'principle of inertia' is the same thing as Newton's 'first law of motion', Newton using the term 'laws of nature' freely as it had become a commonplace by his time, though objected to by some. Robert Boyle thought the term 'an improper and figurative expression'. When an arrow is shot from a bow, he wrote, 'none will say that it moves by a law, but by an external impulse'.

Descartes supposed that God ruled the universe entirely by 'laws of nature' which had been decided upon at the beginning. Once he had created the universe, the Deity had not interfered at all with the self-running machine He had made. The amount of matter and the amount of motion in the world were constant and eternal, and so too were 'the laws which God has put into nature'. During the middle ages it had been thought that God participated in the day-to-day running of the universe, delegating power to the hierarchies of angelic beings who propelled the heavenly bodies round their courses and who observed and guided terrestrial events. Exceptional happenings were then of great interest, such as miracles or more evil portents, like the appearance of comets, which were thought to be due to Divine or diabolical interference with the customary movement of the cosmic process. The men of the seventeenth century on the other hand were interested in the ordinary run of events, looking for their 'lawful' mode of operation. Exceptional happenings, like the new star of 1572 and the comet of 1577, were now scientific problems rather than theological object lessons and they led to the abandonment of theoretical systems which could not account for their occurrence.

The historian[1] of the idea of 'laws of nature' has suggested that the term derived from two primary sources: firstly, from an analogy based upon the practice of civil government by statute law introduced by the absolute monarchs of the sixteenth and seventeenth centuries, and secondly, from the Jewish conception within Christianity of God as the Divine Legislator of the universe which had come down from the ancient despotisms of Babylonia. The term 'laws of nature' was used the most frequently in the ancient world by the Stoic school of philosophers which was influenced by the ideas of the Babylonians, notably by their astrology, and which was prominent during the period of the ancient absolutisms, the school rising in the time of Alexander the Great and flourishing under the Roman emperors. During the middle ages the term was not much used, for then civil society was ordered more by custom rather than by positive law, a monarch delegating his power to the various estates of the feudal order each with its traditional privileges and duties. The absolute monarchs of the sixteenth and seventeenth centuries were more powerful, ruling all of their subjects by means of the statute law which they introduced. Jean Bodin in the latter part of the sixteenth century advocated the development of civil government by statute law, a policy which was the most thoroughly implemented in France, the homeland of both Bodin and Descartes. 'It is not mere chance', wrote Zilsel, that the Cartesian idea of God as the legislator of the universe developed only forty years after Jean Bodin's theory of sovereignty.' Perhaps it was not also a matter of chance that some forty years before Bodin another Frenchman, John Calvin, in the field of theology, was working towards the conception of God as the Absolute Ruler of the universe, governing by laws decided

[1] E. Zilsel, 'The Genesis of the Concept of Physical Law', *Philosophical Review*, 1942, 51, 245.

It is possible that the idea of specifically *quantitative* 'laws of nature' came from another source, namely, the international associations of merchants, like the old Hansa League, who had laws of their own, transcending national boundaries, which dealt with numbers, weights and measures. Gerard Malynes, in his *Lex Mercatoria*, published in 1622, wrote: 'True law is a right reason of nature, agreeing therewith in all points, diffused and spread in all nations, consisting perpetually, whereby meum and tuum are distinguished and distributed by number, weight, and measure, which shall be made apparent.'

upon at the beginning. As we shall see, there were notable similarities between the cosmic values of the Calvinists and those of the man of the scientific revolution. Moreover, the Cartesian philosophy found favour with the Calvinists who were interested in science. During the seventeenth century the theories of Descartes were taught at the universities of Calvinist Holland and at Cambridge, the more Puritan of the two English universities, whilst in France they were supported notably by the Jansenists, the Calvinists within the Catholic creed. Descartes' philosophy was not welcomed by the Catholic theologians of the seventeenth century for it effectively rivalled the system of Aristotle which gave shape to their theology. Descartes provided constructive alternatives to both the method and the cosmology of Aristotle, though he was perhaps less successful than his ancient predecessor for he did not bring together all of the significant intellectual currents of the time in natural philosophy. Bacon had put forward another concept of scientific method whilst Galileo and Kepler, within the confines of their subjects, had formulated more adequately for science the mathematical method which he advocated. Moreover, Newton after him was destined to provide the final and most lasting world system of the seventeenth century, using the Galilean rather than the Cartesian method.

Chapter 16

The Scientific Revolution and the Protestant Reformation

IN 1873 a French botanist, Alphonse de Candolle, noted in his *Histoire des Sciences et des Savants* that of the ninety-two foreign members elected to the Paris Academy of Sciences during the two centuries from its foundation in 1666, some seventy-one had been Protestant in their religion, and sixteen had been Catholic, whilst the remaining five were indeterminate or Jewish in creed. Correlating these figures with the respective religious populations outside of France—107 million Catholics and 68 million Protestants—Candolle showed that more than six times as many Protestants as Catholics outside of France became eminent enough in science to be elected to the Paris Academy of Sciences. Such a correlation left out of account the scientists of France, and so Candolle examined the religious affiliations of foreign members of the Royal Society of London at two periods, 1829 and 1869, when there were more French scientists included than at other times. He found that at both periods there were about equal numbers of Catholic and Protestant foreign members of the Royal Society, yet, outside the United Kingdom, there were 139 million Catholics and 44 million Protestants, figures which substantiated his view that Protestants had tended to predominate over Catholics amongst the great scientists of modern Europe. Subsequent studies of the religious affiliations of modern scientists, which have been listed by R. K. Merton[1] in his study of the relation of Puritanism to the English science of the seventeenth century, have confirmed and amplified the general burthen of Candolle's findings.

Apart from the absence of an Inquisition in Protestant lands, the predominance of Protestants over Catholics amongst the great scientists of modern Europe may be ascribed to three main factors: firstly, to a congruence between the early

[1] R. K. Merton, 'Science and Technology in 17th Century England,' *Osiris*, 1938, IV, 360.

Protestant ethos and the scientific attitude, secondly, to the use of science for the attainment of religious ends, and thirdly, to an agreement between the cosmic values of Protestant theology and those of the theories of early modern science. The first factor was common to both of the two main Protestant movements, for in the early days of the Reformation both the Swiss and the German Reformers taught that man should reject the guidance and authority of the priests of the Catholic faith and should seek for spiritual truth in his own religious experience: he should interpret the Scriptures for himself. Similarly the early modern scientists turned away from the systems of the ancient philosophers and the medieval schoolmen to search for scientific truth in their own empirical experience: they interpreted nature for themselves. Such a consonance of aim between Protestantism and early modern science was well expressed by Thomas Sprat in his *History of the Royal Society,* published in 1667. He noted the 'agreement that is between the present design of the Royal Society and that of our Church in its beginning'.

'They both may lay equal claim to the word Reformation; the one having compassed it in Religion, the other purposing it in Philosophy. They both have taken a like course to bring this about; each of them passing by the corrupt copies, and referring themselves to the perfect Originals for their instruction; the one to the Scripture, the other to the huge Volume of Creatures. They are both accused unjustly by their enemies of the same crimes, of having forsaken the Ancient Traditions, and ventured on Novelties. They both suppose alike that their Ancestors might err; and yet retain a sufficient reverence for them. They both follow the great Precept of the Apostle, of trying all things. Such is the harmony between their interests and tempers.'

Sprat was in the Calvinist tradition, but it seems that early Lutheranism was also congruent with the scientific attitude, for the first technical developments of the new Copernican theory of the universe came from two scholars of the university of Wittenberg which was the centre of the German Reformation. Rheticus, the professor of mathematics at Wittenberg, went to study under Copernicus at Frauenburg and in 1540 he published the first printed account of the new heliocentric theory which he endeavoured to apply later to the motion of the planet Mars. His colleague, Reinhold, the professor of astronomy at Wittenberg, prepared the first astronomical tables based on the Copernican theory, the *Prussian Tables,* published in 1551.

The second factor having some bearing upon the prominence of Protestants amongst the great scientists of modern Europe, the utilization of science for religious ends, became important at the hands of seventeenth-century Calvinists, notably the English Puritans, who stressed the religious duty of performing 'good works' and who placed scientific activity amongst the good works considered beneficial to humanity. Neither Luther nor Calvin had laid much emphasis upon the religious importance of performing good works. For Luther inner faith sufficed for the achievement of man's salvation, whilst for Calvin a certain number of elect persons had been predestined to salvation, the works of man in neither case having much influence upon his future state. However, the followers of Calvin experienced an imperative need to know whether they were predestined to salvation or not, and the original doctrine was modified successively by the Scottish, Dutch, and English Calvinists so that by the mid-seventeenth century it had become generally accepted by the Puritans that the continuous performance of good works indicated that a man was saved. Amongst the good works sanctioned by Puritan ethics were scientific studies. John Cotton, a Puritan divine who emigrated to America, writing in 1654, went so far as to declare that the study of nature was a positive Christian duty.

'To study the nature and course and use of all God's works is a duty imposed by God upon all sorts of men; from the King that sitteth upon the Throne to the Artificer. . . .

'Reason 1. God's glory which is seen in the creatures. . . .

'Reason 2. Our own benefit; both of body for health, as in the knowledge of many medicinall things; and of soule for Instruction, which may be learned from the creatures; and of the estate for gain, when we know the worth and use of each thing. . . .

'Studying the nature of all things, which by observation and conference men might learn one of another, would enlarge our hearts to God and our skil to usefulnesse to ourselves and others. . . . Yea Schollers here are not to be excused who study onely some general causes and properties of the creatures, as the principles of naturall bodies, their motion, time, place, measure, etc, but neglect to apply their studies to the nature and use of all things under heaven.'

Of the two factors discussed hitherto with regard to the prevalence of Protestants amongst the scientists of modern Europe it was perhaps the second, which was marked amongst the seventeenth-century Calvinists, that had the greater weight. The anti-authoritarianism and empirical individualism common to the early Protestant and modern scientist gave at best a relation of congruity between their science and religion whilst the later Calvinist promotion of good works gave a positive impulse to scientific activity. In this connection it is of interest to note that after Galileo and Kepler the main centres of scientific activity passed from Catholic Italy and Lutheran Germany to lands which had come specifically under the influence of Calvinism: England with her Puritans and Reformed Anglican Church, Holland with her numerous Calvinist sects, and France with her Huguenots and her Calvinists within the creed, the Jansenists. After the foundation of the Royal Society of London in 1662 and the Paris Academy of Sciences in 1666, the English and the French remained preeminent in the field of science for the next century and a half, and whilst the Dutch lost ground during the eighteenth century the earlier homes of Calvinism, Switzerland and Scotland, became notable for their scientists in the same period. In Germany and Italy, however, it was not until the nineteenth century that scientists of the calibre of Kepler and Galileo appeared again.

The third factor which helps to account for the predominance of Protestants amongst the scientists of modern times, the agreement between the cosmic values of Protestant theology and of the new theories of modern science, is not an obvious one since both Luther and Calvin were opposed to the new Copernican astronomy on the grounds that it conflicted with the literal word of the Scriptures. However, the medieval view of the world had been composed of a theology and a natural philosophy which were closely integrated and its overthrow was accomplished simultaneously, though in a piecemeal fashion, on the one hand by the Protestant Reformers who attacked the theological elements, and on the other by the scientists who attacked the cosmological elements. Indeed it is possible to discern that the attacks of the Calvinists and of the scientists proceeded along lines which bore some similarity to each other and that both prepared the way for a new mechanical-theological world view, based on the work of Newton, which enjoyed considerable popularity during the eighteenth century.

The *leitmotif* of the medieval view of the world to which both the Protestant Reformers, particularly the Calvinists, and the early modern scientists took exception was the concept of hierarchy which was rooted in the idea that the universe was peopled with, or even composed of, a graded chain of beings stretching down from the Deity at the periphery of the world to the most imperfect entities on earth. It was supposed that the hierarchy of nature had a twofold character possessing both material and spiritual components. The four terrestrial elements, the natural motions of which were rectilinear, possessed the ascending order of perfection, earth, water, air and fire, whilst the heavens were composed of a more perfect fifth element which moved naturally with a circular motion. The spiritual hierarchy was more fine grained, many gradations existing within each of the main divisions of vegetable, animal, and rational souls. On the basis of such a scheme it was thought that nature was governed by a flow of control down the scale of creatures, so that a given being had dominion over those below it in the scale and served those above it. Plants and animals served man whilst man served God. The Deity delegated power to the angelic beings who moved the heavenly bodies and who observed and guided terrestrial events. It was here that an important integration of ancient natural philosophy and medieval theology had taken place for the angelic beings, which had been arranged into a hierarchy of nine orders by pseudo-Dionysius in the fifth century, had come to be accepted as the movers of the nine celestial spheres by the medieval schoolmen, the ordering of the heavens being considered to be such that the higher celestial spheres and their movers controlled the motions of the lower.

Pseudo-Dionysius, by means of his celestial hierarchy of angelic beings, had justified the existence of the ecclesiastical hierarchy of Church government on earth to which the Protestant Reformers, particularly Calvin, took the strongest exception.

'To the government thus constituted', Calvin wrote, 'some gave the name of Hierarchy—a name in my opinion improper, certainly one not used by Scripture. For the Holy Spirit designed to provide that no one should dream of primacy or domination in regard to the government of the Church.'

To the justifications offered by pseudo-Dionysius, Calvin replied that there was no 'ground for subtle philosophical comparisons between the celestial and earthly hierarchy', and he averred that man could not know whether a celestial hierarchy existed or not. Formulating his own view on these matters, Calvin minimized the role of the angelic beings in the government of the universe and assigned to the Deity a more direct and absolute control over the world. Speaking of the relation between God and the angelic beings, Calvin affirmed: 'Whenever He pleases, He passes them by and performs His own work by a single nod; so far are they from relieving Him of any difficulty.' Also, according to Calvin, the Deity had predetermined all events by means of decrees laid down in the beginning.

'We hold that God is the disposer and ruler of all things,— that from the remotest eternity according to his own wisdom, He decreed what He was to do, and now by His power executes what He has decreed. Hence we maintain that, by His Providence, not heaven and earth and inanimate creatures only, but also the counsels and wills of men are so governed as to move exactly in the course which He has destined.'

Thus the workings of the Calvinist world were orderly and fully predeterminate. John Preston, 1587-1628, the Puritan Master of Emmanuel College, Cambridge, laid it down in 1628 that 'God alters no law of Nature'. Miraculous happenings contravening the laws of nature were no longer to be expected, whilst the angelic beings lost their power and ultimately their place in the cosmic scheme. This was regretted by some Calvinist divines, but they admitted that it was the case. Richard Baxter, 1615-91, a Presbyterian theologian, remarked in a work published in the last year of his life:

'It is a doleful instance of the effect of a perverse kind of opposition to Popery, and a running from one extream to another, to note how little sense most Protestants show of the great benefits that we receive by Angels: How seldom we hear them in publick or private give thanks to God for their ministry and helps, and more seldom, pray for it?'

Baxter confessed that he had not come across many instances of the ministry of angelic beings: most of the stories he related were concerned with the activities of evil spirits. However, it seems that even the evil spiritual beings had disappeared from educated English opinion by this time, for the biographer, John Aubrey, 1626-97, tells us:

'When I was a child, and so before the civill warres, the fashion was for old women and maydes to tell fabulous stories night-times, and of sprights and walking of ghosts. This was derived down from the monkish balance, which upheld the holy Church; for the divines say, "Deny spirits and you are an atheist." When the warres came, and with them liberty of conscience and liberty of inquisition, the phantomes vanish.'

Some doubts concerning the existence of spiritual beings appear to have been raised before the Civil Wars, for the philosophical physician, Thomas Browne, 1605-82, writing about 1635, remarked that it was a riddle to him 'how so many learned heads should so far forget their metaphysics, and destroy the ladder and scale of creatures, as to question the existence of spirits'. The removal of the angelic beings from the government of the universe in Calvinist theology was indeed an attack upon the idea that the world was peopled by a graded scale of beings, or rather it was an attack upon the concept of hierarchy which was the kernel of the idea. The Deity no longer ruled the universe by delegating power to a hierarchy of spiritual beings, each possessing a degree of power which decreased as the scale was descended, but now He governed directly as an Absolute Ruler by means of decrees decided upon at the beginning. These decrees were nothing other than the laws of nature, the theological doctrine of predestination thus preparing the way for the philosophy of mechanical determinism.

Whilst the Calvinists were moving away from the hierarchical conception of the government of the universe towards an absolutist theory of cosmic rule in theology, the early modern scientists were effecting a not dissimilar transformation in natural philosophy. Copernicus rejected, implicitly at least, the gradation of the material elements for he assigned to the earth that circularity of motion which hitherto had been the prerogative of celestial matter. He further emphasized the similarity of the earth to the planets by investing the heavenly bodies with the property of gravity which previously had been considered to be peculiar to the lowliest of the terrestrial elements, earth and water. Again, according to his pupil Rheticus, he rejected the hierarchical view that the higher celestial spheres governed the motions of the lower.

'In the hypothesis of my teacher', wrote Rheticus, 'the sphere of each planet advances uniformly with the motion assigned to it by nature, and completes its period without being forced into any inequality by the power of a higher sphere.'

In the stead of the traditional view Copernicus advanced the conception that the sun had an absolute rule over the solar system: 'The Sun sits as upon a royal throne, ruling his children, the planets, which circle round him.' In most of the important new systems of the world put forward during the sixteenth and seventeenth centuries the sun assumed a position of particular importance in the cosmic order. Even in the conservative system of Tycho Brahe the planets moved in orbits round the sun, although the sun and the planets as a whole moved round a stationary earth. William Gilbert of Colchester adopted a third system, similar to the Tychonian save that the earth performed a diurnal spin on its axis, but his cosmic values were identical with those of Copernicus. The sun, Gilbert thought, was the noblest body in the universe 'as he causes the planets to advance on their courses' and acts as 'the chief inciter of action in nature'. The status of the earth was equal to that of the other planets for 'the earth's motion is performed with as little labour as the motions of the other heavenly bodies, neither is it inferior in dignity to some of these'. Gilbert in fact seems to have sensed some connection between theology and natural philosophy in rejecting the concept of hierarchy.

'The sun is not swept round by Mars' sphere (if sphere he have) and its motion, nor Mars by Jupiter's sphere, nor Jupiter by Saturn's,' Gilbert affirmed. 'The higher do not tyranise over the lower, for the heaven of both the philosopher and the divine must be gentle, happy, tranquil, and not subject to changes.'

With Kepler the connection became more explicit for he located the domicile of the theological Ruler of the world on the physical governing power of the solar system.

'The sun,' Kepler wrote, 'alone appears, by virtue of his dignity and power, suited for this motive duty (of moving the planets) and worthy to become the home of God himself.'

The physician-philosophers, Robert Fludd, 1574-1637, and John Baptist van Helmont, 1577-1644, similarly placed the dwelling place of the Deity upon the sun.

Such a change in cosmic values appears to have exerted some influence upon the metaphors and similes of majesty used at the period. It had been customary to compare a monarch in his realm to the primum mobile at the periphery of the world, but now the sun at the centre came in as the image and analogy of majesty. John Norden, 1548-1626, a surveyor and divine, in his *Christian Familar Comfort* of 1600 described Queen Elizabeth I as the prime mover of England.

'Her Majestie and Counsell is, as it were, *primum mobile,* whatsoever moveth must begin from thence, and by direction from thence must all the rest move as upon the axeltree, which carrieth about all the government of the commonwealth.'

Francis Bacon too adopted the same analogy in his *Essay on Sedition,* but when William Harvey published his theory of the circulation of the blood in 1628 he dedicated his book to Charles I as 'the sun of the world around him, the heart of the republic', and when Louis XIV came of age in 1660 he was hailed not as the primum mobile of France but as the sun-king, *le Roi Soleil.*

The new cosmic values of the early modern scientists also had an influence upon the theology of one of the Protestant sects, the Mortalists, so called because they held that the soul of man perished with his body and was resurrected with his body at the Second Coming of the Lord. The most notable adherent of the sect was the poet, John Milton, whilst the chief exponent of the Mortalist theology was Richard Overton, one of the leaders of the Leveller movement during the English Civil Wars. Overton held that God must reside in the most noble parts of the universe, and since the scientists considered the sun to be the most exalted of the heavenly bodies, the Deity must dwell in the sun, or rather the sun must be the body of God. Jesus Christ, Overton argued,

'ascended upward from the Earth into some part of the coelestiall bodies above, therefore without doubt he must be in the most excellent, glorious, and heavenly part thereof, which is the SUN, the most excellent piece of the whole Creation, . . . and according to the famous Copernicus and Tycho Braheus, it is the highest in station to the whole creation. . . . As for the Coelum Empyreum which the Astronomers have invented for his residence, I know no better ground they have for it than such as Dromodotus the Philosopher in his Pedantius had to prove there was devils: Sunt Antipodes: Ergo Daemones. Sunt Coeli: Ergo Coelum Empyreum.'

The central doctrine of the Mortalists, the view that the soul of man perished with his body, served to carry further the

attack upon the concept of hierarchy which had been initiated by the Calvinists and the early modern scientists. Calvin
and the astronomers had questioned the existence of the
celestial hierarchies, but the Mortalists now came to doubt the
reality of the terrestrial hierarchy of plants, animals, and
man. If man were wholly mortal, Overton argued, then he
did not differ essentially from the animals for beasts as well
as men would be resurrected in bodily form ultimately. The
Mortalists seem to have disappeared from England after the
Restoration, but it is probable that they have had a continuous history in America for they returned from thence to
England during the nineteenth century as the Christadelphian
sect. By this time, with the advance of astronomy, the domicile of the Deity had been transposed to the unknown centre
of the universe. Thomas Wright of Durham, 1711-86, a pioneer
of sidereal astronomy, appears to have been the first to suggest that the dwelling place of God lay at the unknown centre
of the universe. In 1750 he put forward the hypothesis that
the sun and the stars of the Milky Way moved round a common
centre, forming a giant sidereal system. At this centre, Wright
suggested, 'the Divine Presence or some corporeal agent full
of all virtues and perfections more immediately presides
over His Creation'.

The transposition of the dwelling place of the Deity from
the periphery to the centre of the world was not an important
element integrating Protestant theology with early modern
science—the Mortalists after all were only a minor sect. Of more
significance was the idea that the universe was infinite, each
star being the centre of a solar system of planets which themselves were peopled with creatures similar to those on earth.
The doctrine of the plurality of worlds even in its most restricted form whereby it was supposed that the planets of our
solar system, or the moon alone, were inhabited carried implications which ran counter to the traditional concept of
hierarchy, for it derived from, and gave added weight to,
the notion that the heavenly bodies were of the same qualitative nature as the earth. The French essayist, Montaigne,
1533-92, wrote that through arrogance man 'dareth imagin-
arily place himselfe above the circle of the Moone, and reduce
heaven under his feet. It is through the vanity of the same

imagination that he dare equall himselfe to God, that he ascribeth divine conditions to himselfe, that he selecteth and separateth himselfe from out the ranke of other creatures.' In its extended form, whereby it was supposed that there was an infinity of inhabited worlds in the universe, the doctrine tended to assign the same status to all finite beings. After he had abandoned science for religion, Pascal, 1623-62, wrote:

'In comparison with all these Infinites all finites are equal, and I see no reason for fixing our imagination on one more than on another. The only comparison which we make of ourselves to the finite is painful to us.'

Others who remained in the scientific movement found the doctrine elevating rather than depressing. Fontenelle, 1657-1757, who later became secretary of the Paris Academy of Sciences, wrote in 1686:

'When the heavens were a little blue arch, stuck with stars, methought the universe was too straight and close: I was almost stifled for want of air: but now it is enlarged in height and breadth, and a thousand vortices taken in. I begin to breathe with more freedom, and I think the Universe to be incomparably more magnificent than it was before.'

Most of the natural philosophers of the seventeenth century accepted the doctrine of the plurality of worlds. Tycho Brahe, Kepler, and Galileo thought that the planets of the solar system were inhabited, whilst Descartes believed that there was a plurality of inhabited solar systems in the universe.

The doctrine of the plurality of worlds helped to reconcile Calvinist theology with the theories of modern science during the seventeenth century. The most important obstacle to the reconciliation, the acceptance of the Scriptures as literally true, showed a notable decline in England with the growth of a coherent and organized scientific movement in the 1640's. The Puritan clergyman, John Wilkins, 1614-72, was a prominent figure in both of these developments, being the leader of the 'Philosophical College', a group of scientists who began to meet regularly about 1644. In 1638 Wilkins published his

Discovery of a New World, a work attempting to prove that there was another world of animate and rational creatures on the moon. Here he had no texts from Scripture against him, indeed some were in his favour, but even these he rejected. Wilkins did not defend the Copernican theory in this work but he endeavoured to establish the doctrines which had derived from that theory or which had become ancillary to it. Quoting the work of the astronomers who had shown that there were mountains and apparently seas on the moon and spots on the face of the sun, Wilkins affirmed, 'That the Heavens do not consist of any such Pure Matter which can Priviledge them from the like Change and Corruption as these Inferior Bodies are Liable unto.' Since the moon in particular resembled the earth, 'we may guess in the general that there are some inhabitants in that Plannet: for why else did Providence Furnish that place with all such Conveniences of Habitation?' A similar consequence, Wilkins noted, followed from the heliocentric theory of the world adopted by the disciples of Copernicus: 'Now if our Earth were one of the Plannets (as it is according to them) then why may not another of the Plannets be an Earth?'

Wilkins' next work, *A Discourse concerning a New Planet*, published in 1640, was a full defence of the Copernican theory. Much of the book, about half of it in fact, attempted to reconcile the Copernican theory with the Biblical texts which seemed to favour the diurnal motion of the heavenly bodies or oppose the theory of the motion of the earth. Here again he rejected the practice of interpreting the Scriptures literally, declaring that the Bible was not a philosophical treatise but a work intended for the capacity of the popular mind. In supporting the Copernican theory, Wilkins brought together his Calvinist theology and the natural philosophy of the day. Discussing the medieval theory of the motion of the heavenly bodies, he rejected the hierarchical conception of cosmic rule and adopted Calvin's view that the angelic beings were largely superfluous, God governing the universe directly. It was unreasonable of the schoolmen, Wilkins wrote,

'who make the faculty whereby the Angels move the Orbs to be the very same with their Understandings and Will: So that if an Angel do but meerly suspend the Act of willing their Motion, they must necessarily stand still; and on the contrary, his only willing them to move shall be enough to carry them about in their several Courses. Since it were then a needless thing for Providence to have appointed Angels unto this business, which might have been done as well by the only will of God.'

Wilkins' argument here exemplified his more general principle that natural processes were governed economically. It was 'agreeable to the Wisdom of Providence', he felt, that nature 'does never use any tedious difficult means to perform that which may as well be accomplished by shorter and easier means'. Such an idea was perhaps the root concept of the various 'minimum' principles evolved during the development of modern science: Fermat's principle that in reflection and refraction light travels by such a path that it takes the least possible time (1662): the principle of Leibniz that light travels by the path of least resistance (1682): and Maupertuis' general principle of least action (1744) which applied to a variety of physical phenomena.

Whilst Wilkins thought that the processes of change in the universe took place by routes of minimum effort, he believed that there was a maximum diversity of beings in the world.

'There may be many other Species of Creatures beside those that are already known in the World,' he wrote. ' 'Tis not Improbable that God might create some of all Kinds, that so he might more Compleatly Glorifie himself in the Works of his Power and Wisdom.'

Such a view was not novel for it was the basis of the traditional conception that the world was peopled by a graded chain of creatures. The late medieval theologian, Raymond de Sebonde, had argued in a similar way that there must exist a vast diversity and multiplicity of angelic beings.

'We must believe', Sebonde had written, 'that the angels are there in marvellous and inconceivable numbers, because the honour of a king consists in the great crowd of his vassals,

while his disgrace or shame consists in their paucity. . . .
But remember that one must not conceive of their multitude as
confused; on the contrary, among these spirits a lovely order
is exquisitely displayed.'

However Wilkins, unlike Sebonde, took the chain of creatures
to be only a scale of perfection, not a hierarchy of powers,
for it was to the latter conception that the Calvinists and the
early modern scientists objected. Giordano Bruno, 1548-1600,
with whom the doctrine of the plurality of worlds had come
into vogue in modern times, had shown how it was possible
to reconcile the traditional notion that all of the beings in
the world were graded in perfection with the new conception
that, apart from the Supreme Ruler of the world, all beings
had more or less equal powers. He supposed that the Supreme
Ruler had given each being an inner source of power so that
all creatures were autonomous and free from the traditional
relations of domination and servitude one to the other. Wilkins
inclined to such a view. In her operations, he observed, 'Nature does commonly make use of some inward Principle'. Another point of view, which was more generally adopted, derived
from the Cartesian philosophy according to which there was
only one kind of power governing physical events, that of
mechanical motion, which externally activated all beings. All
creatures were machines, but they varied one from the other
in perfection according to the complexity of their machinery
or the degree of their organization.

It was presumed that in this world where there was a maximum diversity of beings and where motion took place by
routes of minimum effort there could be no fundamental
change or improvement, for nature had already reached a
state of unchanging perfection.

'The most sagacious man', wrote Wilkins, 'is not able to
find out any blot or error in this great Volume of the World,
as if anything in it had been an imperfect essay at the first,
such as afterwards stood in need of mending: But all things
continue as they were from the very beginning of the Creation.'

The German philosopher, Leibniz, 1646-1716, was also of the
view that ours is the best of all possible worlds and that it is

'based on consideration of maximum and minimum, such that the greatest effect is obtained with the least, so to speak, expenditure'. Thus the solar system was a self-running machine, whilst the organic species were fixed for all time in the various diverse forms in which they had been first created. It was upon this point that the alliance between Protestant theology and modern science ultimately broke down, for the coming of the theories of evolution in the nineteenth century put an end to the view that the world and its creatures had preserved their present forms from all eternity. Science then seemed no longer compatible with Protestant theology, and religious opposition to the theories of evolution in the Protestant countries during the nineteenth century was strong.

However, the alliance lasted for a century and a half during which the physico-theological system of Newton was generally accepted in intellectual circles. The Newtonian theory encountered little religious opposition in England as Wilkins and the men of his generation had met the brunt of the Anglican resistance to the new astronomy and had made explicit the congruent features of the scientific revolution and the Calvinist Reformation. Both science and religion in seventeenth-century England adopted the Baconian aim of contributing to the 'relief of man's estate', the former through the applications of science and the latter through the performance of good works, the one being identified with the other. In the realm of ideas both modern science and Calvinist theology moved away from the hierarchical conception of cosmic rule, into which an element of the arbitrary had found its way, towards an absolutist theory of the government of the universe, in which events were subject to certain and irrevocable laws. Prominent amongst them were laws prescribing that action should be conserved or should be minimal in all motions in order to preserve the pattern of the world's unchanging perfection. The same end was served by the supposed existence of a maximum multiplicity of creatures forming a continuous chain of beings. The gradation of creatures into a scale of beings, for the most part, remained as a principle by which animals, plants, and other entities could be classified according to certain fixed characteristics, such as the complexity of their organization, but it was no longer a principle governing motion and action in the natural world. However, Lamarck and others were soon to suggest that the chain of animals and plants was not a static scale but a branched evolutionary series, thus bringing to its historical end a conception which had been common both to Catholic and Protestant theology and to ancient natural philosophy and early modern science.

Chapter 17

The Theory of Universal Gravitation

THE ANCIENT conception of gravity differed profoundly from that of modern times. From the seventeenth century on, scientists have thought of gravity essentially as a property of matter. One body attracts another with a force that depends upon the amounts of matter which those bodies contain, and upon the distance which separates them, the force being reciprocal. In antiquity and the middle ages gravity was considered to be more the property of a position, rather than of an aggregate of matter. Everything in the universe of Aristotle had its appointed station, to which it strove to return if displaced. Stones fell towards the earth because they were aspiring to reach their proper place at the centre of the universe, which happened to coincide with the centre of the earth, or very nearly so. But earthy and watery things would move towards the centre of the universe even if the earth were not there, just as airy and fiery things moved to their proper stations below the orbit of the moon, where there was nothing to 'attract' them except their appointed positions.

Such a conception of gravity gave rise to difficulties in the Copernican theory. Stones clearly fell towards the earth, but the earth could not be at the centre of the universe if it moved in an annual orbit round the sun. Copernicus therefore suggested that each body, the earth, sun, moon, and planets, had its own system of gravity so that a stone in space would fall towards the nearest heavenly body. Gravity, Copernicus thought, was the tendency of aggregates of matter to congregate together in the form of a sphere anywhere they might be placed, not just at the centre of the universe. But he still thought of a geometrical point as being the focus of gravity, though such points now had to be centres of spheres of matter. 'The element of the earth', he wrote, 'is the heaviest, and all heavy things are driven towards it, striving to its innermost centre.'

Copernicus did not think that the bodies of the solar system exerted an influence upon each other by virtue of their own private systems of gravity. The arrangement of the heavenly bodies and their motions were not determined by gravity or any other mechanical consideration. The motions of the heavenly bodies were entirely natural, whilst their arrangement was dictated by the mathematical harmonies obtaining between the speeds of the planets and the sizes of their orbits. Copernicus did not explicitly indicate what he considered the matter of interplanetary space to be, but Kepler after him stated that Copernicus believed the heavenly bodies to be embedded in solid crystalline shells which rotated one inside the other and carried the heavenly bodies on their courses.

Such a conception had to be abandoned, and an alternative explanation found for the arrangement and motions of the heavenly bodies, when Tycho Brahe and others in 1577 followed the orbit of a comet across the skies, and showed that it moved through the solar system, cutting across the supposed solid crystalline shells of the old Aristotelian cosmology.

'Now it is quite clear to me that there are no solid spheres in the heavens,' wrote Tycho Brahe, 'and those that have been devised by the authors to save the appearances, exist only in the imagination, for the purpose of permitting the mind to conceive the motion which the heavenly bodies trace in their course.'

But if the crystalline spheres did not exist, the question arose as to what actually did move the heavenly bodies and preserve their regular arrangement. It was possible that the heavenly bodies moved of their own accord independently of each other and had no regular order. Francis Bacon wrote that the

'first question concerning the Celestial Bodies is whether there be a system, that is whether the world or universe compose together one globe, with a centre, or whether the particular globes of earth and stars be scattered dispersedly, each on its own roots, without any system or common centre.'

However, the early modern scientists for the most part felt that the sun, moon, earth, and planets did compose a system with a common centre and that the system was united by a single principle upon which the diverse regular movements of the heavenly bodies were based. In 1600 William Gilbert of Colchester suggested that magnetism was the principle holding together the solar system. As we have seen, Gilbert presumed from his experiments with spherical lodestones that the earth itself was a giant lodestone with only a superficial covering of rocks and soil. As lodestones exerted a considerable force upon iron objects at a distance, he suggested that gravity was the magnetic force exercised by the giant lodestone of the earth upon surrounding objects, and that gravity extended throughout the solar system, acting as the integument of the universe.

In one of his experiments, Gilbert showed that the magnetic force exerted by a lodestone upon a given piece of iron increased with its size, the greater the mass of the lodestone, the greater its attraction for the piece of iron. Moreover the action was reciprocal, the iron attracted the lodestone, as much as the lodestone attracted the iron. Thus the properties of magnetic force, as investigated by Gilbert, provided the model for the modern conception of gravitational force. Concrete masses of matter rather than geometrical points were the foci of gravity, the force increasing with the amount of matter.

Like Tycho Brahe, Gilbert was of the view that the planets moved round the sun whilst the sun and planets as a whole moved round the earth at the centre of the world. However, he differed from Tycho, and agreed with Copernicus in that he supposed the fixed stars to be stationary and the earth to rotate on its axis daily. As the earth, sun, moon, and planets were all magnetic bodies, they orientated themselves in space, just as compasses orientated themselves on earth, so that all the bodies of the solar system moved in the same plane with their axes parallel; they 'took positions in the universe according to the law of the whole', as he put it. Thus the axis of the earth always pointed to the pole star because of the orientating effect of the earth's magnetism. Gilbert held that all of the bodies in the solar system mutually influenced each other's movements through the interaction of their magnetic forces: there was no Prime Mover controlling their movements from outside. 'Whatever in nature moves naturally, the same is impelled by its own forces, and by a consentient compact of other bodies,' he wrote, 'such is the circular propulsion of the planets' bodies, each the others career observing and inciting.'

Gilbert's theories were temporarily quite influential. They were taken over by Johann Kepler, who used them to explain why it was that the planets moved in elliptical orbits. Kepler also developed Gilbert's conception of gravity, supposing it to be 'a mutual affection between cognate bodies tending towards union or conjunction, similar in kind to magnetism'. Such a force of gravity between two bodies was dependent upon their masses. The earth was fifty-three times as large as the moon, so that—

'if the moon and earth were not held in their orbits by their animal force or some other equivalent, the earth would rise to the moon by one fifty-fourth part of their distance apart, and the moon would fall to the earth through the other fifty-three.'

We see incidentally that Kepler did not have the conception of inertia, it was 'animal force or some equivalent' which required to keep the bodies of the solar system moving.

To account for the movements of the planets, Kepler supposed that the sun sent out magnetic effluvia which rotated like the spokes of a wheel with the spin of the sun in the plane of rotation of the planets. These magnetic effluvia propelled the planets round in their courses by a tangential force. Thus the outermost planets moved more slowly than those nearer the sun, because they were heavier, and because the magnetic effluvium had become enfeebled with distance by the time it reached them. To explain the elliptical orbits of the planets, Kepler supposed that the sun was a giant magnet consisting of only one pole, whilst the planets were magnets with both a north and a south pole. The axes of the planets possessed a constant orientation in space, so that first the north pole of the planet, and then its south pole, were presented to the sun as the planet moved in its orbit. Hence the planet would be alternately attracted and repelled by the sun, and the circularity of its orbit would be distorted into an ellipse. Kepler thought that the spin of the earth on its axis was due mainly to its own magnetism: in a year three hundred and sixty of the earth's rotations were due to its own magnetism and the other five to that of the sun. Like Gilbert, Kepler accepted the cosmic values of Copernicus. The earth was much the same as the other planets, whilst the sun ruled the universe, possessing a special sort of magnetism which impelled the planets round their courses, and distorted their orbits from circles into ellipses.

As we have seen, Kepler subscribed to the old mechanical notion that a moving body required the constant application of an impelling force to keep it going, whilst his friend Galileo adhered to the old astronomical preconception that the motions of the planets were circular and uniform. They did not assimilate each other's work and so, whilst both of them could have brought together astronomy and mechanics, neither of them did. For Galileo the arrangement and motions of the heavenly bodies presented no problem. His principle of inertia laid it down that natural motions were uniform and circular and as he thought that the planets moved in circles with uniform speeds, he presumed that their motions were natural and self-sustaining. Galileo suggested that in the beginning God might have dropped the planets from a great height, so that they fell towards the sun, gradually increasing in velocity. When they had reached their present speeds, the planets were switched over into circular orbits, in which they continued to move indefinitely. It was an idea, Galileo said, that he had obtained from Plato, and from which he had removed 'the mask and poetic dress'.

Galileo in astronomy and Descartes after him in science generally were primarily concerned with propagating the new general ideas of the scientific revolution, and not with explaining the detail of the latest scientific discoveries. Descartes, like Galileo, believed that the planets moved in circular orbits with uniform speeds and not in elliptical orbits with varying speeds as Kepler had found. Moreover, Descartes rejected the idea that there was such a thing as a force of gravity operating between aggregates of matter across empty space. He criticized Galileo for determining the laws of the free fall of bodies, without first ascertaining whether the fall of a body could be free. According to Descartes, matter and extension were coterminous, so that space was full of matter and nothing could fall 'freely'. The fall of a stone to the earth was due to the suction effect of the vortex of matter which surrounded the earth. In the same way, the circular orbits of the planets were due to the suction effect of the vortex matter surrounding the sun, which distorted their natural straight-line motions under inertia into circles. The views of Descartes were influential, and at the time they served to divert attention from the problems of gravitational force. One of his followers, Christian Huygens, 1629-95, a Dutch gentleman-amateur scientist, made an experiment in 1669 which seemed to confirm Descartes' theory of the fall of bodies to the earth. He set up a whirlpool in a bowl of water, and he found that pebbles were drawn to the middle of the bowl at the centre of the vortex. Huygens therefore thought that gravity was nothing other than 'the action of the aether, which circulates round the centre of the Earth, striving to travel away from the centre, and to force those bodies which do not share its motion to take its place'. Whilst investigating the swing of a pendulum, Huygens had already discovered in 1659 that a centripetal force was required to keep a body in circular motion, and he ascertained the law governing such a force. But, in the case of the planets, he failed to see that this force was provided by gravitation, owing perhaps to his adherence to the views of Descartes at that time. In astronomy Descartes' system explained less than that of Kepler; it failed notably to account for the elliptical orbits of the planets, and it obscured Kepler's promising conception of gravity as a force acting across space between two bodies and dependent upon the amount of matter which those bodies contained.

Kepler's theories were revived in 1666 by Alphonse Borelli, 1608-78, who was a professor of mathematics at Pisa and a member of the Florentine Academy of Experiments which was composed of the pupils and disciples of Galileo. Borelli suggested that the elliptical orbit of a planet was the resultant of a balance between two opposing forces; firstly, the force of gravity attracting the planet to the sun, and secondly, a centrifugal force tending to move the planet away from the sun, similar to the force exerted on a stone when it is whirled in a sling. Borelli adhered to the impetus theory of mechanics, and so like Kepler he supposed that the planets were impelled round their courses by rays of force radiating from the sun and rotating with the sun, like the spokes of a wheel. He was of the view that bodies tended to move naturally in a straight line, not in a circle as Kepler and Galileo had believed, so that a force of gravity from the sun was necessary to constrain the planets to move in closed orbits. However, Borelli was unable to find exactly how great a force of gravity was required to bend the natural straight line motions of the planets into the elliptical orbits observed, and his theory of planetary motion remained a conjecture.

Borelli's ideas were the last flicker of theoretical physics in seventeenth-century Italy, whilst in Holland and France the theories of Descartes were dominant. So the problem of the mechanics of the heavens came full circle. Starting with Gilbert's largely qualitative speculations, the gravity problem returned to England again, the later Stuart school of scientists providing the definitive solution. The main figure of the school was Isaac Newton, 1642-1727, though the contributions of other members, notably Robert Hooke, 1635-1703, Christopher Wren, 1632-1723, and Edmund Halley, 1656-1742, were not without importance.

Granted the modern version of Galileo's principle of inertia which lays it down that the unimpeded motion of a body is a uniform speed in a straight line, the problem of explaining the motions of the heavenly bodies in mechanical terms resolved itself into two main subsidiary questions. Firstly, the derivation of the law governing the centripetal force required to bend such linear motions under inertia into circular or elliptical motions. Secondly, the demonstration that gravity could provide the centripetal force constraining the planets to move in closed orbits, which required the derivation of the law governing the variation of gravitational force with the distance between the gravitating bodies. Robert Hooke, the Curator of the Royal Society, endeavoured to study the second problem experimentally. He inclined to Gilbert's view that gravity was similar in kind to magnetism, and since Gilbert had shown experimentally that the magnetic force between two bodies varied with the distance between them, Hooke thought that the same might be demonstrated for gravitational force. In 1662 and 1666 he made experiments to test the hypothesis by comparing the weights of bodies measured at the bottom of deep wells or mine shafts, at the surface of the earth, and at heights above the earth but, as he put it, he found 'nothing of certainty'. In 1664 Hooke discussed with Wren the orbit of the comet which appeared in that year. Wren was of the view that the comet moved in a straight line under inertia, but Hooke pointed out that the path was curved when the comet was near to the sun and he suggested that the curvature was produced by the gravitational attraction of the sun. About the same time Hooke attempted to find a relationship governing the force required to keep a body moving in a circular path by studying the motions of circular pendulums, but he did not find the crucial law of centripetal force.

The same problems were investigated by Newton when he was away from Cambridge at his home, Woolsthorpe Manor, near Grantham, during the Great Plague of 1665-66. According to a statement he made thirty years later, Newton then discovered both the law of centripetal force and the relationship connecting the decrease of gravitational force between two bodies with the square of the distance separating them. It seems that he deduced the law of centripetal force and with it derived from Kepler's third law of planetary motion the inverse square law of gravitational force. The validity of these laws could be tested by calculating the gravitational force which the earth exerted on the moon from the inverse square law and the known value of the force of gravity at the surface of the earth as measured by experiments with falling bodies. From the speed of the moon and the size of its orbit it was possible to calculate the centripetal force required to constrain the moon in its orbit. If the calculated values of the centripetal and gravitational forces agreed quantitatively, then it was proved that the earth's gravity provided the centripetal force maintaining the moon in its orbit.

If Newton made these calculations when at Woolsthorpe Manor he did not make his results known. Several reasons have been suggested to account for Newton's failure to publish his work in 1666. Firstly it has been said that Newton's calculated values of the centripetal force and the gravitational force did not tally as the distance of the surface of the earth from its centre was not known with any accuracy at the time. Later it was suggested that Newton could not show at the time that the extended mass of the earth gave rise to the same gravitational field as it would if its entire mass were concentrated at the geometrical centre of the earth. More recently it has been suggested that Newton did not in fact discover the law of centripetal force and the inverse square law in 1666, since there are no documentary records to support the claim, and Newton's recollection thirty years later was faulty in a number of other respects and perhaps also here. Moreover, from 1666 to 1679 Newton developed a theory of gravity based on a supposed circulation of an etherial medium from the heavens to the earth and back again, the ether in its descent carrying ponderable bodies with it. In 1679 he suggested that the sun and planets might be ordered by 'some secret principle of un-sociableness in the ethers of their vortices', a statement which implies some leanings towards the Cartesian view.

By 1679 other scientists had arrived at the law of centripetal force and the inverse square law of gravitational force. Christian Huygens in Holland had made experiments upon the motion of pendulums and motion in a circle generally from which he derived in 1673 the law of centripetal force. With this law it was possible to deduce the inverse square law from Kepler's third law of planetary motion as we have seen. Huygens did not make the deduction, but Hooke, Wren, and Halley did, obtaining the inverse square law by 1679. In that year Hooke wrote to Newton asking if he could show that a planet would move in an elliptical orbit, given the law of centripetal force and the inverse square law. In his letter Hooke pointed out the difficulty that the sun and planets were extended bodies and yet they had to be treated theoretically as though their masses were concentrated at their centres. Newton did not reply to Hooke's question, but in 1684 Wren offered a prize for a solution to the problem Hooke had posed to Newton. By that time Hooke claimed that he had solved the problem though Wren, according to Halley, was not satisfied by the solution. In the following year the problem was posed to Newton again, this time by Halley, and later he sent Halley a proof that the planets would move in elliptical orbits under the influence of the sun's gravitational field. Halley pressed Newton to work out his proof in detail and to publish the results. Newton was now able to show that the gravitational field of an extended spherical body, such as the earth or the sun, would remain the same if the total mass were concentrated at the centre of the sphere, so that the heavenly bodies could be treated as heavy points. More accurate values of the earth's radius, the sun's distance from the earth, and other measures of the solar system were also now available, and with them Newton was able to show that the earth's force of gravity provided exactly the centripetal force required to keep the moon in its observed orbit. Similarly he demonstrated that the gravitational field of the sun accounted for the observed motions of the planets according to Kepler's laws, and that comets moved in approximately parabolical paths round the sun. Developing his system further, Newton suggested that each planet should be flattened at the poles and should bulge at the equator owing to its spin motion. Such a shape, an oblate spheroid, was observed telescopically in the case of the planet Jupiter and, in the case of the earth, was inferred from the fact that the earth's gravitational field was smaller at the equator than in the regions nearer to the poles. Since the earth was not exactly spherical, Newton showed that the gravitational attraction of the sun and the moon would not pass through the earth's centre and so would cause the earth's axis to perform a slow conical motion which accounted for the precession of the equinoxes. Finally, he ascribed the tides to the differential gravitational effects of the sun and the moon upon the oceans, demonstrating that very high tides occurred at new and full moon when the gravitational pulls of the sun and moon act together whilst low tides occurred at the quarters when the pulls tend to neutralize one another.

These demonstrations were incorporated in the *Principia Mathematica* which Newton had completed by late 1686. The Royal Society intended to publish the work but adequate funds were not available and Hooke, the Curator of the Society, laid claim to the prior discovery of the inverse square law of gravitational attraction, and so Edmund Halley published Newton's book at his own expense. Hooke put forward his claim again in 1693, this time before a meeting of the Royal Society. Immediately afterwards Newton became afflicted with a nervous illness and when he had recovered he announced for the first time that he had made the basic discoveries of the law of centripetal force and the inverse square law in 1666. Newton undoubtedly made the most important single contribution to the theory of universal gravitation though, as in the later case of the discovery of the calculus which culminated in the controversy with Leibniz over priorities, Newton was only one of a number of scientists who were working on the same problem and who independently and simultaneously contributed to the solution.

In studying the problem of the mechanics of the solar system Hooke and Newton worked by methods which differed somewhat one from the other. Hooke belonged to the generation of English scientists who came to maturity in the Commonwealth period, a generation that was the most strongly influenced by the empirical and utilitarian point of view of Bacon's philosophy of science. Newton belonged to a later generation which came to maturity in the Restoration period, and he adopted a more deductive methodology, akin to those of Galileo and Descartes. Hooke endeavoured to find out experimentally how the force of gravity between two bodies varied with the distance between them by measuring the weights of bodies at various heights above and below the surface of the earth, while Newton probably deduced the inverse square law of gravitational force from the law of centripetal force and Kepler's third law of planetary motion. In his discussion of the scientific method Newton adopted the view that the starting point for physico-mathematical deductions should be experimentally observed effects or laws and that the deductions should lead to the explanation or prediction of other observable effects. In the preface to the *Principia Mathematica* he wrote: 'All the difficulty of philosophy seems to consist in this: from the phenomena of motions to investigate the forces of nature, and then from these forces to demonstrate the other phenomena.'

Newton thus specified that it was the observed effects and laws of mechanical motion which should serve as the starting point for mathematical demonstrations in natural philosophy. Descartes had advocated the same view, suggesting that natural phenomena should be explained in mechanical terms since we were the most familiar with the workings of machines and other mechanical contrivances, and we must explain the unknown in terms of the known. Such a method of explaining the unknown in terms of the known was explicit in Newton's work. He made the connection which Galileo had failed to establish, comparing the orbit of a planet to the trajectory of a projectile.

'That the planets may be retained in certain orbits,' he wrote, 'we may easily understand if we consider the motions of projectiles: for a stone projected is by the pressure of its own weight forced out of the rectilinear path . . . and made to describe a curve line in the air, and through that crooked way is at last brought down to the ground: and the greater the velocity is with which it is projected, the farther it goes before it falls to the earth. We may therefore suppose the velocity to be so increased that it would describe an arc of (many) miles before it arrived at the earth till at last, exceeding the limits of the earth, it should pass quite by without touching it.'

Thus if the moon were analogous to a projectile, it should obey the same laws and its motion should be susceptible to the analysis Galileo had made of the motion of projectiles; that is, the moon's curved motion should be resolvable into two simple straight line motions, each governed by mechanical law. Newton's method in fact closely resembled that used by Galileo, both in its general aspects, such as the basing of mathematical demonstrations upon experimentally given principles, and in some of its details, like the procedure of resolving complex curved motions into simple rectilinear motions.

However, Newton differed from Galileo in that he sharply distinguished between principles given by experiment and principles given by intuition. Newton was averse to the Cartesian method of basing scientific demonstrations upon supposedly sure and indubitable ideas given to the unclouded mind by intuition. Such ideas for Newton were mere hypotheses, and he declared that he did not use hypotheses. Newton of course did formulate hypotheses of a speculative character and he based arguments and demonstrations upon them, but he held that such speculations were separate and quite distinct from his experimental philosophy. Thus in his later work, the *Opticks* (1704), Newton placed his experimental and speculative natural philosophies in separate sections, appending the latter in the form of a number of *Queries* to the end of the section dealing with his experimental work on light, so that the book consisted of two parts which, so to speak, were respectively Galilean and Cartesian in method.

Other scientists, such as Huygens and Leibniz, found it difficult to appreciate the distinction which Newton had drawn between experimental philosophy and speculation from hypotheses, and they suggested that Newton, with his conception of gravity as a force which acted at a distance between bodies, was bringing back the occult qualities and spiritual forces which had been so recently rejected by natural science. Newton replied that gravity was merely a name given to the supposed cause of the fall of bodies to the earth, the movement of the planets in closed orbits round the sun, and so on, and that experimental philosophy, having no relevant observations to go on, could not specify what this cause was. Experimental philosophers could only indicate how gravity operated, showing by how much a body increased its velocity in a given time during its fall to earth, and so on, as Galileo had done when he discovered that objects fell with a constant acceleration. To speculate upon the 'cause' of gravity, ascribing it to occult forces or to vortices in a cosmic ether, was of no value to experimental philosophy for all such explanations were of their nature hypothetical.

'To tell us that every species of things is endowed with an occult specific quality by which it acts and produces manifest effects is to tell us nothing,' Newton wrote. 'But to derive two or three general principles of motion from phenomena, and afterwards to tell us how the properties and actions of all corporeal things follow from these manifest principles, would be a very great step in philosophy, though the causes of those principles were not yet discovered: and therefore I scruple not to propose the principles of motion above mentioned, they being of very general extent, and leave their causes to be found out.

Newton of course did not confine his activities to experimental philosophy as he defined it. He suggested speculative hypotheses to explain why the observed laws of mechanics were obeyed, proposing notably a physical 'cause' for the force of gravity observed between corporeal bodies. By Newton's time the Aristotelian theory of gravitational phenomena, which suggested that the four elements constantly strove to reach their appointed stations in the universe, was no longer important.

More influential was Descartes' theory that interplanetary and interstellar space was packed with a matter which moved with a whirlpool motion round the earth and each of the heavenly bodies. These vortices had the property of drawing corporeal bodies towards their centres, so that heavy objects on the earth tended to fall downwards, for example, and the vortices round the larger heavenly bodies, such as the sun, swept round the smaller bodies, like the earth and the planets, so that they moved in closed orbits. In his later work Newton rejected the Cartesian theory on the grounds that it could not account for the precise form of Kepler's laws of planetary motion and was at variance with such astronomical phenomena as the movement of comets across the solar system. In the stead of the Cartesian view Newton proposed several alternative theories to account for gravitational phenomena, the most mature of which was that given in the *Queries* appended to his *Opticks*. Here he suggested that there was a stationary etherial medium pervading the whole of space, which was akin to the atmosphere surrounding the earth, but much finer. Such a medium was composed of very small particles which repelled one another and which were repelled by the particles of corporeal bodies. Owing to such a repulsion the etherial medium was rare within the interstices of massive objects, such as the heavenly bodies, and became denser in the space surrounding those bodies as the distance from them increased. Thus the etherial medium within a heavy object placed at a distance from the earth would be denser in the part facing away from the earth than in the part facing towards it, and owing to the repulsive effect of the etherial medium the object would fall towards the earth.

Newton calculated that if the average density of the etherial medium were about one 700,000th part that of air, and if it were proportionately more elastic, then the resistance of the medium to the motion of bodies through it would be very small, one 600,000,000th part of that of water in fact, so that there would be no sensible alteration the motions of the planets through frictional resistance even after 10,000 years. However, over very long periods of time the planets would be gradually slowed down and motion in general would be lost from the universe through frictional effects. Thus the amount of motion in the universe could not be constant as Descartes had supposed, and the world as a whole could not be a perfect self-running machine. Newton therefore suggested that the Deity constantly replenished the motion lost by friction from the universe, and corrected such disturbances as those caused by the planets and comets deranging one another's orbits. According to Newton, God was present everywhere at all times, so that no matter where a derangement occurred in the universe it was observed and corrected.

'The Deity', wrote Newton, 'endures for ever and is everywhere present, and by existing always and everywhere, He constitutes duration and space. . . . (He) governs all things and knows all things that are, and can be done. . . . Who, being in all places, is more able by His will to move the bodies within His boundless uniform sensorium, and thereby to form and reform the parts of the Universe, than we are by our will to move the parts of our body?'

Here Newton approached the mechanical-pantheistic point of view advocated by the Dutch philosopher, Spinoza, 1632-77. Theologically Newton was somewhat unorthodox for he was a Unitarian, rejecting the doctrine of the Trinity of the Godhead, like other notable figures of the early modern scientific tradition, such as Servetus, 1509-53, John Locke, 1632-1704, and Joseph Priestley, 1733-1804.

Newton always claimed that he did not use speculative hypotheses in his experimental philosophy, but here in his doctrine that the Deity constituted duration and space, and his theory that there was a stationary etherial medium pervading the whole of space, his hypotheses influenced and shaped his science. They led him to postulate that time, space, and motion were absolute quantities which could be determined in principle by reference to something that was fixed and immovable in the universe. For Newton the stationary and immovable entities in the universe were the Deity, whose existence constituted time and space, and the etherial medium which pervaded all space.

'The Father,' wrote Newton, 'is immovable, no place being capable of becoming emptier or fuller of Him than it is by the eternal necessity of nature. All other beings are movable from place to place.'

Such a view that time, space, and motion were absolute quantities persisted right down to the twentieth century, for in all the subsequent theories involving an etherial medium which filled all space there was one set of systems and observers in the universe who could measure in principle absolute velocities, namely, those that were at rest in the cosmic ether.

Chapter 18

Optics During the Seventeenth Century

THE GROWTH of experimental science during the sixteenth and seventeenth centuries gave an impulse to the study of the primary vehicle of all observations—light—and to the development of instruments extending the observing power of the human eye. The Renaissance artists had investigated optical questions in order to obtain naturalistic representations and to improve the perspective in their paintings. Then viewing instruments more powerful than the existing spectacles and magnifying glasses were sought for, a movement which culminated with the invention of the telescope and the compound microscope by the spectacle makers of Middelburg in Holland, Hans Lippershey and Zacharias Jansen, about the beginning of the seventeenth century. The scholars of the time—notably Galileo and Kepler—took up these craft discoveries and studied the theoretical principles which they embodied. During the seventeenth century attention was concentrated on the telescope, for it was of great use in astronomy and navigation, and its defects were less serious than those of the early microscope. However, the microscope was made and used by the scientists of the time. Galileo studied the anatomical structure of insects with the microscope about 1610, and his work was continued in England by Robert Hooke in the 1660's, whilst towards the end of the seventeenth century a flourishing school of microscopists developed in Holland.

Galileo did not add much either to the theory or to the instruments of optical science. In principle his telescope was the same as that of the Dutch spectacle makers, consisting of a convex and a concave lens, though he improved the performance of the instrument. In contrast Kepler designed several new telescopes, notably the astronomical telescope with two convex lenses, and he founded modern experimental optics, just as Galileo founded modern experimental mechanics, and Gilbert the modern science of magnetism. Kepler formulated intuitively the inverse square law of the diminution of light intensity with distance from the consideration that light radiated spherically from a given source. Studying the bending of a light beam at an interface between two transparent media, Kepler showed that Ptolemy's approximate law of refraction, which supposed a direct proportionality between the angles of incidence and refraction, was true only for angles less than about 30°. He thought that the refractive power of a medium was proportional to its density, but the English mathematician, Harriot, pointed out to him that oil is more refractive than water, though less dense.

The correct law for the refraction of light was discovered in 1621 by Willebrod Snell, 1591-1626, a professor of mathematics at Leiden, who found that the sines of the angles of incidence and refraction always bore the same ratio one to the other for a given interface between two media, the ratio being termed the refractive index for that interface. This law of refraction was first made known in 1637 by Descartes, who endeavoured to explain it and other optical phenomena, on the supposition that light consisted of small particles in rapid linear motion. He held that the reflection of light was just a rebound of the light particles from an elastic surface according to the laws of mechanics. Similarly, the refraction of light on passing from a dense to a light medium was analogous to a ball breaking through a thin cloth. The component of the ball's velocity at right angles to the cloth was reduced by the resistance of the cloth, but the velocity component parallel to the cloth remained unchanged. Hence the over-all velocity of the ball would be decreased and its path would be bent towards the cloth, just as light was bent towards the interface when it passed from a dense to a light medium. The analogy implied that light travelled faster in dense than in light media. Such a consequence we may well understand, said Descartes, if we remember that a ball rolls more easily along a hard, dense table than across a soft, light carpet.

Descartes had a second theory of light, according to which light was an action or pressure transmitted from an object to the eye through the closely packed matter of the intervening space. He suggested that light was like the pressure transmitted from an object to the hand of a blind man through his stick. Descartes believed that it was the pressure of light from the sun which maintained the vortex of the solar system rigid against the pressures of the vortices of the stars outside of it. Thus the centrifugal force of the cosmic vortices was nothing other than the pressure of light from their central regions. The different colours of light were produced by the different speeds of rotation of the matter in space, red was produced by the fastest motion, and blue by the slowest. The theory that light was an action transmitted by the ether of space was developed by the Cartesians, whilst the corpuscular theory of light was taken up by Newton and his followers.

Descartes' deduction that light must travel faster in a dense than in a light medium was questioned by Pierre de Fermat, 1608-65, a counsellor of the provincial *parlement* of Toulouse in France. Fermat based his argument upon a principle of economy, according to which it was supposed that the actions of nature were such as to take always the least possible time. He then showed that the laws of reflection and refraction of light followed as necessary consequences of the principle of least time if it were presumed that light travelled more slowly in dense than in light media. Such a result was opposed to Descartes' particle theory of light, and indeed to emission theories of light in general, whilst it agreed with theories which supposed light to be an action transmitted by an etherial medium, such as the wave theory of light suggested by Christian Huygens and developed by other Cartesians.

A wave theory of light was first suggested by Francesco Grimaldi, 1618-63, a Jesuit professor of mathematics at Bologna. Grimaldi found that light did not travel exactly in straight lines, for he discovered that shadows were a little larger than they should be on the supposition that the propagation of light was rectilinear. Moreover, he found that the edges of shadows were often coloured, and so he suggested that light was a fluid capable of wave-like motions, different frequencies being different colours, like the notes of sound vibrations. If the motions of the light-fluid were wave-like, then the edges of shadows should be blurred and coloured, he said, for water waves can easily go round an obstacle they encounter.

Grimaldi in 1665 supposed that his light-fluid moved with great speeds, undulating all the time. Christian Huygens in 1678 and 1690 approached nearer to the classical wave theory of light, suggesting that the luminiferous fluid or ether was stationary, light consisting of waves propagated longitudinally through this medium. Huygens' colleague in the Paris Academy of Sciences, Olaus Romer, 1644-1710, a Danish astronomer, had found between 1672 and 1676 that light travels with a definite velocity. He discovered that the moons of Jupiter had longer periods of rotation, and less frequent eclipses, when the earth was receding from them than when it was approaching towards them. Romer interpreted these phenomena to mean that light was propagated with a finite velocity, taking eleven minutes to cross the earth's orbit. Huygens brought together the supposition that light travels with a finite velocity and the idea that light is a form of wave motion. He supposed that there was a continuous luminiferous ether throughout all space, consisting of hard elastic particles which transmitted impulses without being displaced themselves. Thus each particle oscillated about a mean position and transmitted its motion to neighbouring particles, so that a disturbance from any source was propagated spherically throughout space with a finite velocity. Such an ether, he thought, filled all transparent bodies, so that light could pass through them. However, the light waves had to make detours round the particles of the body so that they were slowed down, thus accounting for the refraction of light.

In 1670 Erasmus Bartholin of Denmark found that a ray of light was split into two by a crystal of Iceland spar—one ray, the so-called ordinary ray, obeying the law of refraction, whilst the other, the extraordinary ray, did not. Huygens explained this phenomenon by supposing that the particles of Iceland spar were ellipsoidal in shape, so that the rays of light had to make longer detours when going in one direction than when proceeding in another. Thus a ray of light would be split into two halves, corresponding to the longer and the shorter paths, when it entered into such a crystal. However, he found that two such crystals gave four rays when arbitrarily orientated, and that the ordinary and extraordinary rays were interconverted when the crystals were placed at right angles. These phenomena he could not explain, nor could be account for the colours produced at the edges of shadows and the colours of thin films of oil, air, and glass.

Newton was led to the study of optics by the imperfections of the contemporary lens telescopes, which gave coloured and distorted images. Such defects were due to the light passing through curved pieces of glass, the lenses, and he thought that such defects could not be remedied in the lens telescope. Accordingly he designed and constructed in 1668 the first reflecting telescope, in which light was concentrated by means of a concave mirror. He then went on in 1672 to investigate the breaking up of white light into colours by pieces of glass with surfaces which were not parallel. With a prism he resolved white light into the spectral colours and, isolating each colour, he showed with a second prism that the colours had their own characteristic degrees of refrangibility, and that they could not be further resolved. Newton at first was attracted by the wave theory of light, but in his *Opticks*, published in 1704, he adopted the corpuscular theory, though he retained elements of the wave theory. He supposed that the particles of light, moving in straight lines, excited vibratory movements in the surrounding ether which could reinforce or impede the motions of the light particles. Newton considered that this would explain why light was partially reflected and partially refracted at the interface between two media. The reinforced particles possessed the necessary inpetus to get through the interface, whilst the impeded particles had not, and were reflected. He explained similarly the series of concentric light and dark rings which were produced when a double convex lens was placed in contact with the flat side of a plano-convex lens. As the distance between the two lenses varied, Newton thought that the light rays would be alternatively reinforced and impeded by their accompanying ether vibrations, depending upon whether the distance between the lenses was a whole number of wavelengths or not. Thus the light would alternately get through and be turned back from the lenses outwards from their point of contact, producing a series of light and dark rings. To explain the phenomenon of double refraction observed with crystals of Iceland spar, Newton supposed that the particles of light possessed 'sides', so that a ray of light was like a rod of rectangular cross section. The particles of Iceland spar, he thought, had a similar duality, so that a ray of light took two paths through the crystal.

Following Newton, there was little advance in optics for another century. Most scientists adopted the corpuscular theory of light, though the Swiss Cartesians, John Bernoulli, 1710-90, and Leonard Euler, 1707-83, adopted the wave theory. Pierre Maupertuis, 1698-1759, a French mathematician of the Berlin Academy of Sciences, saved Fermat's minimum principle for the corpuscular theory of light, by supposing that the *action* of a beam of light, the distance it travelled multiplied by its velocity, was a minimum, rather than the *time*, the distance it travelled divided by its velocity, as Fermat had thought. On this assumption, light would travel faster in denser media, as Newton held, and Descartes too in his corpuscular theory of light.

Newton's belief that lens telescopes would always give coloured images was challenged by David Gregory, 1661-1708, a professor of mathematics at Edinburgh and later at Oxford. The human eye, he said, is a lens system, yet it does not give the chromatic aberration found in lens telescopes. Hence it should be possible to construct an achromatic combination of lenses. His uncle, James Gregory, 1638-75, who preceded his nephew as professor of mathematics at Edinburgh, had been interested in the same problem, and had designed, though not constructed, a reflecting telescope in 1663, a few years before Newton. The method of making achromatic lens telescopes was first made known in 1758 by John Dolland, 1706-61, a London instrument maker of Huguenot descent, though a telescope producing images devoid of colour was first constructed by an amateur-gentleman scientist, Chester More Hall, in 1733.

Chapter 19

Medicine, and the Theory of the Circulation of the Blood

ASTRONOMY AND medicine were the two main subjects connected with natural science which were practised by the scholars of the middle ages and early modern times. Medicine was the more important of the two disciplines for, together with theology and law, it was one of the three subjects which could be studied up to the doctorate level in the medieval universities, while astronomy was only part of the more elementary bachelor of arts course. Professional physicians were comparatively numerous and, possessing the best scientific training of their time, they were prominent amongst the leaders of the scientific revolution. In England William Gilbert of Colchester, the student of magnetism, was physician to Elizabeth I, and William Harvey, who first put forward the full theory of the circulation of the blood, was physician to Charles I.

Another feature of the medical profession which was perhaps of importance for the development of modern science was the quite close contact between the craft and scholarly elements of the profession compared with the almost complete divorce between those elements in other subjects, such as that between the millwrights and the scholars who discussed theories of mechanics in the middle ages. As we have seen, there was no appreciable craft tradition in astronomy before modern times, but in medicine there were two such traditions—that of the barber-surgeons and that of the apothecaries. Such craftsmen had some contact with the medical scholars for university trained physicians prescribed medicines which were made up by the apothecaries, and upon occasion they superintended operations performed by the surgeons. Moreover, the training of the university educated physicians was a little more empirical than that of other scholars, as medical students were required to study anatomy and inspect dissected bodies. Even so the training of medical students as depicted in the first printed medical works appears to have been somewhat abstract and doctrinaire. A lecturer or reader read out passages from the books on anatomy by ancient authors, such as the Greek, Galen, whilst the professor in his chair commented thereon. A demonstrator pointed out the correspondence between the description given in the texts and the parts of the human body revealed by a dissection. The dissection itself was performed by a barber-surgeon who contributed nothing to the discussion, though he might be a skilled sectionist and know intimately the anatomical structure of the human body, for such discussions were carried out in Latin, a knowledge of which was probably unusual amongst the barber-surgeons.

Such a method of training was not conducive to the advance of anatomical knowledge as the men who dissected the human body had little knowledge of the ancient writings and those who read the medical authorities did not practise dissection. During the sixteenth century, however, the barrier between the scholarly and the craft traditions began to break down in medicine as elsewhere. The apothecaries and barber-surgeons became more literate and learned, while the medical scholars began to undertake practical dissections for themselves. From such men came notable advances in anatomical knowledge and surgical practice. The revival of the study of anatomy by practical dissection was most marked in northern Italy. Here at the turn of the fifteenth century the Renaissance artists, such as Leonardo da Vinci and his contemporaries, dissected numerous animal and human bodies, making careful and exact drawings of their findings. Then the professors of the great medical school at Padua began to dissect for themselves, continuing the tradition of the artists by appending excellent illustrations to their books on anatomy.

In the new medical developments of the sixteenth century something of the old division between the craft and scholarly elements of the medical profession remained, for anatomical knowledge was advanced mainly by the scholars, notably the professors of medicine at Padua, Andrea Vesalius, 1514-64, Realdus Columbus, 1516-59, and Hieronymus Fabricus, 1537-1619, while surgical technique was advanced notably by the barber-surgeons, such as the Frenchman, Ambrose Paré, 1510-90. Paré was only a rural barber's apprentice when he first went to Paris in 1529 but he became the foremost surgeon of the sixteenth century during the course of his work at the great Paris hospital, the Hôtel Dieu, and his service as army surgeon in the French Wars of Religion. His fame was such that, though a Huguenot, he was spared by royal edict from the massacre of the Paris Huguenots on St. Bartholomew's Eve in 1572. Paré made four main contributions to surgical technique. He showed firstly that gunshot wounds were best healed by soothing applications rather than by the traditional treatment with boiling oil, since they were not in themselves poisoned, contrary to current belief. Secondly, he found that bleeding after amputations was best arrested by tying up the severed arteries, not by burning them with hot irons as was generally practised at the time. Thirdly, he discovered that certain abnormal cases of childbirth were faciliated by turning the child round before birth. Fourthly, he invented ingenious artificial limbs which could perform mechanically a variety of operations. Furthermore Paré introduced the French barber-surgeons to the anatomical writings of the Paduan scholar, Vesalius, by epitomizing his Latin works into French.

The Fleming, Vesalius, at Padua published his major work, *Concerning the Fabric of the Human Body*, in 1543, the year in which Copernicus brought out his new theory of the motions of the heavenly bodies. The astronomer Copernicus arrived at views which differed from those of the ancient scientists by a process of logical demonstration, the method of the old scholarly tradition, but Vesalius, belonging to the medical profession with its more practical tradition, did the same by a process of empirical enquiry. The ancient medical writer, Galen, whose works occupied an authoritative place in medieval anatomy and physiology, had supposed that in the human body the blood flowed from the right to the left chamber of the heart through the dividing wall, the septum. In 1543 Vesalius pointed out that the septum of the heart was very thick and muscular, and in the second edition of his major work (1555) he denied that the blood could pass through it.

'Not long ago,' Vesalius wrote, 'I would not have dared to diverge a hair's breadth from Galen's opinion. But the septum is as thick, dense, and compact as the rest of the heart. I do not see therefore, how even the smallest particle can be transferred from the right to the left ventricle through it.'

Although Vesalius showed that Galen was incorrect with regard to the passage of the blood through the septum, he did not suggest any alternative explanation as to how the blood could pass from the right to the left ventricle of the heart or, in general, how it could get from the veins to the arteries. Before Vesalius went to Padua, he had worked in the medical faculty of the Sorbonne with Michael Servetus, 1511-53, who was the first of the moderns to put forward such an alternative explanation, suggesting that the blood circulates from the right to the left chamber of the heart through the lungs. Servetus was primarily a religious reformer, and in his work the scientific revolution and the Protestant Reformation found perhaps their most direct and intimate connection. His main work, *The Restoration of Christianity* (1553), was chiefly an exposition of Unitarian theology, but in a small section of some half a dozen pages he suggested the theory of the lesser circulation of the blood between the heart and lungs. Servetus put forward the theory not only on scientific, but also on religious grounds, his particular theology enabling him to overcome some of the difficulties which had stood in the way of the circulation theory. The main intellectual obstacles to the development of the idea of the circulation of the blood were Galen's theory that the human body was governed physiologically by three distinct and graded sets of organs, fluids and spirits, and the Aristotelian conception that only celestial matter could move naturally with a circular motion, natural terrestrial motions possessing a beginning and an end. Galen had been of the view that the human body possessed a triadic hierarchy of physiological functions. Firstly, there was the vegetative function of nourishment and growth which had its seat in the liver and which was mediated by the dark red venous blood and its natural spirit. Secondly, there was the animal function of motion and muscular activity, which had its seat in the heart, and which acted through the bright red arterial blood and its vital spirit. Thirdly, there was the nervous function, governing the irritability and sensitivity of the body, which had its seat in the brain and which was superintended by the nervous fluid and its animal spirit.

In the world view of later antiquity and the middle ages the arrangement of living entities and their functions into triadic hierarchies was quite general. All of the beings inhabiting the universe fell into one or other of three classes: material entities, like minerals, plants, and animals, spiritual entities, such as the angelic beings, and entities which were both material and spiritual, namely, human beings. Each one of the three groups of material entities subdivided into a further triad: animals were either birds, fishes, or terrestrial beasts, and each one of these sub-groups was subject to a further threefold division. The same triadic classification held for the ranks of men and for the hierarchies of the angelic beings above them, whilst at the head of the scale of beings in the universe stood the supreme Trinity of the Godhead.

The most important of the theological unorthodoxies of Servetus was his Unitarianism: he rejected the doctrine of the Trinity of the Godhead. He denied that the Son was co-eternal with the Father, and he was of the view that the Holy Ghost was nothing other than the breath of God, or the all-pervading *pneuma* of the world. Just as he denied the supreme Trinity, so Servetus denied the triadic hierarchy of natural, vital, and animal spirits in the human body, claiming that 'in all of these there is the energy of the one spirit and of the light of God'. In particular there were not two kinds of blood, differentiated by the natural and the vital spirits which they contained respectively, but only one blood, for there was only one kind of spirit in the blood, as 'the vital is that which is communicated by the joins from the arteries to the veins, in which it is called the natural.' Servetus suggested that the single spirit of the blood was the soul of man, or rather, as he put it, 'The soul itself is the blood.' One of the charges brought against Servetus, when he was caught by Calvin at Geneva and tried for heresy, was that he held the soul to be the blood, for it implied the unorthodox view that the soul perished with the body.

The traditional view that there were three different physiological fluids in the human body, two being different sorts of blood, has been a great obstacle to the development of the circulation

theory, for any large-scale movement of blood from the arteries to the veins and back again, which the circulation theory
required, would have involved the complete mixing of what
were regarded as quite distinct fluids, each with its own separate
function. Once the venous and arterial bloods were thought
of as identical, the way to the circulation theory was clear,
though Servetus himself put forward positively only the theory
of the lesser circulation of the blood from the right to the
left chamber of the heart through the lungs, as he was primarily concerned with the relations between the blood and the
atmosphere. He was of the view that 'the Divine breath is in
the air,' and that in the lungs the inspired air mixed with the
blood circulating through them. Thereby the blood was purified, whilst the soul participated in the Divine and was replenished: 'Just as by air God makes ruddy the blood, so does Christ
cause the Spirit to glow.' Servetus observed that the pulmonary
artery connecting the right chamber of the heart with the
lungs was very large, and carried a much greater quantity of
blood than was required for the mere nourishment of the
lungs, so that some other purpose was served by the movement of such an amount of blood. He suggested that this
purpose was the transformation of the dark red venous blood
to the bright red arterial blood in the lungs, where the blood
took up the inspired air and gave off impurities. Thence the
purified blood moved from the lungs through the pulmonary
vein to the left chamber of the heart, completing the lesser
circulation. As further evidence for his theory Servetus pointed
out that the lesser circulation of the blood was by-passed and
did not operate in the embryo which could not take up air.

In 1553 Servetus was burnt at the stake by Calvin for his
heresies, and with him were burnt most of the copies of his
newly-printed book. A few years later, in 1559, Realdus Columbus, professor of anatomy at Padua, put forward again the
theory of the lesser circulation of the blood. It is possible that
he had heard of Servetus' views, for Columbus too regarded
respiration as a process purifying and vitalizing the blood,
and not just as a process cooling the blood as was currently
believed. The work of Servetus was too heretical to quote at
the time: in fact it is not until we reach the comparatively
tolerant atmosphere of late seventeenth-century England that

we find the first mention of Servetus in connection with the theory of the circulation of the blood. The evidence which Columbus offered for the theory of the lesser circulation was purely anatomical and physiological. He argued that, as the septum of the heart was solid, the blood must pass from the right to the left chamber of the heart by the only alternative route, through the lungs. Furthermore, the pulmonary artery was large and carried much more blood than that required for the nourishment of the lungs, while the blood passing from the lungs through the pulmonary vein to the left chamber of the heart was bright red and vitalized. Thus the blood could not be vitalized in the heart with fumes, or air carried from the lungs through the pulmonary vein, as was generally believed, but the lungs themselves were the vitalizing organ, the process requiring the lesser circulation of the blood between the heart and lungs.

After Columbus little further empirical evidence for the theory of the circulation of the blood appeared for almost another half-century, though, as part of the intellectual revolution of the time, the theory was put forward purely as a speculation by several writers. As we have seen, the men of the scientific revolution and the Protestant Reformation moved away from the old hierarchical conception of the government of the universe towards an absolutist view, and this was true in the biological sciences as it was in the physical. The heart and the blood were given the same primacy of place in the human body as the sun assumed in the new world systems. 'The heart', Servetus had written, 'is the first to live, the source of heat in the middle of the body.' Such a conception also found favour with later theorists of the circulation of the blood, such as Columbus' pupil, Cesalpino, 1524-1603, who put forward the idea in a work published in 1571; Giordano Bruno, 1548-1600, who speculated upon the idea of the circulation of the blood in several of his writings, published from 1584 on; and notably William Harvey, 1578-1657, with whom the theory became finally established in 1628.

'The heart', wrote Harvey, 'is the beginning of life; the sun of the microcosm, even as the sun in his turn might well be designated the heart of the world; for it is the heart by whose virtue and pulse the blood is moved, perfected, made apt to nourish, and is preserved from corruption and coagulation; it is the household divinity which, discharging its function, nourishes, cherishes, quickens the whole body, and is indeed the foundation of life, the source of all action. . . . The heart, like the prince in a kingdom, in whose hands lie the chief and highest authority, rules over all; it is the original and foundation from which all power is derived, on which all power depends in the animal body.'

Here, curiously enough, Aristotle came into his own, for he had assigned the central control of the human body to the heart, in contrast to the later theory of Galen which assumed a more decentralized control through the triadic hierarchy of the brain, heart, and liver, and the three fluids and spirits associated with them. But the early modern development went beyond Aristotle, and the biologists ascribed to the blood that circularity of motion which had been considered unique to celestial matter, while the Copernicans in astronomy did the same for the substance of the terrestrial sphere as a whole. For such a purpose the biologists called upon the ancient conception that man was a microcosm of the whole world for, suitably interpreted, the notion implied that circulation should be as much a feature of the microcosm as of the macrocosm. Cesalpino, Bruno, and Harvey also drew upon the one example which Aristotle had given of natural circular motion within the terrestrial sphere, the example, Harvey indicated, which had suggested to him the idea of the circular motion of the blood.

'I began to think whether there might not be a motion as it were in a circle,' Harvey wrote, 'in the same way as Aristotle says that the air and the rain emulate the circular motion of the superior bodies: for the moist earth warmed by the sun evaporates; the vapours drawn upwards are condensed, and descending in the form of rain moistens the earth again; and by this arrangement are generations of living things produced; and in like manner too are tempests and meteors engendered by the circular motion, and by the approach and recession of the sun. And so in all likelihood does it come to pass in the body, through the motion of the blood.'

Harvey indeed seems to have searched constantly for examples of circular motion in terrestrial beings in order to give them parity of status with the supposedly superior heavenly bodies. In his later work *On Generation* (1651) he suggested that the succession of individuals constituting a species was a form of circular motion emulating the movements of the heavenly bodies.

'This is the round which makes the race of common fowl eternal; now pullet, now egg, the series is continued in perpetuity; from frail and perishing individuals an immortal species is engendered. By these, and means like to these, do we see many inferior or terrestrial things brought to emulate the perpetuity of superior or celestial things.'

The theory of the circulation of the blood and the doctrine of the supremacy of the heart in the human body were therefore particular applications of the new sixteenth-century conception that both the microcosm and the macrocosm were governed by an Absolute Ruler with all other entities in both the great and little worlds enjoying parity of status under the Supreme Power. However, the circulation theory was not based solely upon the intellectual revolution of the sixteenth and seventeenth centuries, for Harvey sought to verify his ideas empirically and he found solid evidence to support the hypothesis of the circulation of the blood. As a young man, Harvey studied medicine under Fabricus who was professor of medicine at Padua from 1565 to 1619.

In 1603 Fabricus made the important discovery that there were valves in the veins which permitted the blood to flow only in the direction towards the heart. Fabricus did not see the true significance of his discovery, for he held, with Galen, that the blood moved to and fro to some degree in the veins, the valves serving only to prevent the blood collecting at the extremities of the body and to retard the motion of the blood so that the nourishment it carried was absorbed efficiently by the tissues. His pupil, Harvey, was the first to note that the valves in the veins permitted the blood to flow only from the veins to the heart, while the valves in the heart allowed the blood to pass only into the arteries, so that there was a one way movement of the blood from the veins through the heart to the arteries. Thence, Harvey argued, the blood must return from the arteries to the veins, completing the circulation, for it was unlikely that blood would be continuously prepared at the ends of the veins and continuously destroyed at the ends of the arteries. Harvey made his argument much more cogent by calculating the amount of blood passing through the heart in the space of an hour, showing it to exceed the whole weight of an average man. Such an amount of blood could not be formed at the ends of the veins and destroyed at the ends of arteries during the space of an hour, and so the same blood must circulate continuously round the body. By dissections, Harvey traced out most of the course of the circulation, though he did not see the joins between the veins and the arteries, the capillaries, as he did not use the microscope. However, he showed that such joins must exist by applying a tourniquet to an arm, when the veins of the hand and forearm, the side of the limb away from the heart, became engorged with blood. The veins, lying near to the surface of the arm, were closed by the tourniquet, and so the blood could only come from the more deeply lying arteries through the capillaries joining the veins to the arteries which completed the circulation under normal conditions.

Criticizing the traditionally received view of Galen, Harvey pointed out that blood could not pass through the septum of the heart, not only because it was too thick, but also because both ventricles of the heart contracted and expanded together, so that there was at no time a pressure tending to drive blood through the septum. Moreover, the septum had its own system of arteries and veins which would not be required if the blood passed through it. Finally, Harvey cut away the left ventricle from the heart of a dog and showed that no blood came through the septum from the right ventricle.

Harvey established the generality of the circulation of the blood in sanguineous creatures by examining the blood vessels of some forty different species of animals, including worms, insects, and fish. The study of the cold-blooded animals was particularly important, for their hearts beat slowly, permitting the detail of the heart movement to be observed, and continued to beat for some time when removed from the body. Harvey observed that if the heart of an animal was held in the hand it was felt to harden when it contracted just as a muscle did. He therefore regarded the heart as a hollow muscle, the contraction of the muscle being responsible for the movement of the blood. Previously in the theory of Galen, and even in that of so recent a figure as Bruno, it was the spirits in the blood which were regarded as the cause of the blood's motion, the heart serving mainly as the organ preparing the vital spirits and responding passively to the movement they caused. Harvey was of the view that the heart manufactured vital spirit, such a spirit being akin to the soul of man. With Servetus he held that 'The soul itself is the blood', but the idea was not essential to his system, for the soul or spirit in the blood was no longer responsible for the circulation. Harvey in fact was the first to ascribe the motion of the blood to a mechanical cause, the muscular contraction of the heart. He first made his theory known in a lecture delivered to the Royal College of Physicians in 1616, the notes of the lecture being preserved. Here Harvey compared the heart to a water bellows or a pump. In a mixture of Latin and English he wrote: 'W.H. (William Harvey) constat per fabricam cordis sanguinem per pulmones in aorta perpetua transferri, as by 2 clacks of a water bellow to rayse water.' The use of a mixture of the scholarly Latin and the craftsman's vernacular was not uncommon at the time: Galileo, for example, wrote up his experiments in Italian and his theory in Latin, the practice reflecting the union which had taken place between the scholarly and the craft traditions.

In his book *On the Motion of the Heart and the Blood*, published in 1628, Harvey carried further his proof that the muscular contraction of the heart was the mechanical cause of the circulation of the blood. He pointed out that when the heart contracts the arteries show an immediate expansion which must be due to the ejection of blood from the heart, not to the self-propulsion of the blood under the impulse of the spirits within it. Moreover, the arteries near to the heart were very thick, so that they could withstand the direct shock of the heart beat, and the arteries in general were thicker than the veins since they were subject to the pressure of the blood ejected from the heart, while the veins were not. Again blood thrown out of the heart invariably passed into arteries, not into veins, even in the case where venous blood was so transported, namely from the right ventricle to the pulmonary artery. In the case of cold-blooded animals with slowly beating hearts, such as the frog, Harvey observed that the upper chambers of the heart, the auricles, contracted before the lower chambers, the ventricles, and from the arrangement of the valves in the heart he traced the detail of the circulation. He showed that the right auricle pumped blood from the veins to the right ventricle, which then ejected the venous blood through the pulmonary artery into the lungs. There the venous blood was transformed into the bright red arterial blood which was pumped by the left auricle from the lungs through the pulmonary vein to the left ventricle whence it was ejected into the arterial system. Thus it appeared that the right ventricle, which was less muscular than the left, was concerned entirely with the lesser circulation of the blood between the heart and the lungs. Such a view was corroborated by the fact that animals without lungs, such as the fishes, had only one ventricle, that corresponding to the left ventricle in animals with lungs, which was responsible for the general circulation of the blood round the body.

Harvey's theory was an important addition to the new mechanical philosophy for he had indicated that the heart, veins, and arteries, constituted a mechanical system for the transport of the blood. The treatment of living organisms and their parts as mechanical systems had begun notably with Leonardo da Vinci, who had shown that the bones of animals worked like levers, and it had been continued by Galileo, who had used his theory of the strength of materials to explain such scale effects as to why it was that elephants must have thick and stumpy legs in contrast to the fine and delicate limbs of insects. The idea that living organisms were machines was generalized by Descartes and it was applied in detail to a variety of examples by Alphonse Borelli in a posthumous work *On the Motions of Animals* (1680). Borelli dealt with the mechanical actions involved in walking, running, jumping, skating, and lifting weights. He treated similarly the flight of birds, the swimming of fishes, and the crawling action of worms. Turning to the mechanical action of the internal organs of man, Borelli calculated that if the heart functioned like a piston in a cylinder it must exert a pressure equal to the weight of about 135,000 pounds during the course of a heart-beat. Similarly he considered the lungs to be a pair of bellows and the stomach to be a kind of grinding machine.

On the practical side Harvey left a number of questions which were to be answered later in the seventeenth century. He had shown, notably, that the capillaries joining the arteries and veins must exist though he had not been able to see them. Robert Boyle in 1663 traced the course of the capillaries with injections of coloured fluids and coloured molten wax. With the microscope, Marcello Malpighi, 1628-94, succeeded in seeing the capillaries in the lungs of a frog in 1660, and finally in 1688 Antony van Leeuwenhoek, 1632-1723, saw the actual circulation of the blood through the capillaries in the tail of a tadpole and in the foot of a frog. With such discoveries the mechanics of the movement of the blood were finally settled and it now became possible for scientists to investigate what function the circulation of the blood served. In particular the chemists could now study how it was that there was a change from the dark red venous blood to the bright red arterial blood in the lungs and to examine the physiological significance of the change.

Chapter 20

From Alchemy to Medical Chemistry

THE ANCIENT Greek biologists and medical writers had never considered the physiology of the human body in specifically chemical terms. It was admitted that fundamentally living creatures were composed of the four elements, but for medical purposes the workings of the body were analysed in terms of the properties of distinctively biological substances, such as the four constitutional humours, the sanguine, choleric, melancholic, and the phlegmatic, and the three functional fluids, the arterial and venous bloods, and the nervous fluid, with their controlling spirits, the vital, natural, and animal. Diseases were not ascribed to external entities invading the human body but to internal disarrangements in the proportions of the four humours. Thus there were no such things as diseases in themselves; there were only disease states of the body. Early on, in the time of Hippocrates, it was held that the human body itself had the power of restoring the balance of the disarranged humours, and patients were not overtroubled with medicaments. But by the time of Galen it had become the practice to attempt to restore the balance of the humours with medicines, and since the humours were thought of as essentially organic substances, such medicines were mainly of plant or animal origin. The Muslims added to the drugs used by Galen and made medicines more complex, so that during the middle ages it was the standard practice to administer comprehensive cure-all remedies containing as many as sixty or seventy constituents. Such constituents were still of biological origin in the main, and not infrequently they were of a somewhat noxious character. The first London Pharmacopoeia of 1618 listed, for example, bile, blood, cock's combs, and wood lice as medicaments to be taken orally.

Few of these medicines were derived from mineral sources for the ancient Greek philosophers and the medieval scholars were not greatly interested in chemical substances and their properties. However, chemical problems had been studied by the alchemists during later antiquity and the middle ages, and some of them became interested in the application of alchemy to medicine, as we have seen. Such a movement culminated in the work of Paracelsus, c. 1493-1541, a Swiss physician who endeavoured to bring into being a new science of medical-chemistry, or iatrochemistry as it was called, by uniting medicine with alchemy. Throughout the ages alchemy had been associated with the chemical crafts on the one hand and mystical religion on the other, and so in a period of religious ferment when, at the same time, the craft tradition was coming to maturity it is not surprising perhaps that such a development should occur in the field of alchemy, paralleling the contemporaneous Copernican revolution, the Protestant Reformation, and the first moves towards the theory of the circulation of the blood.

Paracelsus was the son of a Zurich physician named von Hohenheim, but he called himself Paracelsus because he considered himself to be greater than the Roman physician, Celsus. About 1514 he worked in the mines and metallurgical workshops of Sigismund Fugger, the south German financier and alchemist, and then he studied medicine at Basle, where he subsequently taught for two years. He broke with tradition by lecturing in the vernacular German instead of the Latin of the scholars; indeed he was the first to do so at a university, and he brought together the craftsmen and the scholars of the medical profession by inviting the apothecaries and barber-surgeons of Basle to his lectures. Furthermore, he inaugurated his course of lectures by burning the books of Galen and Avicenna, the generally received medical authorities, just as Luther had burnt the Papal Bull. Paracelsus admired Luther as a man who was 'abundantly learned', and Luther was much attracted by alchemy as it embodied the quasi-religious conception of ennobling metals by a process of death and resurrection.

Paracelsus defined alchemy as the science of transforming the raw materials of nature into the finished products useful to humanity. It was a definition which included all the chemical and biochemical crafts. The smelter transforming minerals into metals was an alchemist, and so too were the cook and the baker preparing food from flesh and grain. Paracelsus himself was particularly interested in making medicines from natural substances, either from minerals or plants, for according to his definition the apothecary and physician were equally alchemists. He adopted the basic view of the alchemists that minerals grew and developed into more perfect forms underground, and that man in the laboratory could imitate artificially what occurred naturally below the surface of the earth. His interpretation of such an idea was wider, for he held that all substances were alive and grew naturally and that man could speed up or adapt such natural processes for his own ends.

In medicine Paracelsus rejected the view that the health of the body was determined by the four constitutional humours, and he put forward the theory that the human body was essentially a chemical system composed of the two principles of the alchemists, mercury and sulphur, and a third principle, salt, which he himself added. Salt had been regarded as a fundamental substance earlier, though it was not generally regarded as a basic chemical principle before the time of Paracelsus. Illnesses, according to Paracelsus, could arise from a lack of balance between the principles, just as the Galenical doctors thought that they arose from a lack of harmony between the humours, but the theory of Paracelsus indicated that the balance was to be restored by mineral medicines, not organic remedies. The followers of Paracelsus, the iatrochemists as they were called, occasionally found good inorganic medicines, though usually for quite fanciful reasons. Thus they administered iron salts to anæmic patients with a poor blood supply as iron was associated with the red planet, Mars, and with Mars the war-god of blood and iron. However, the general view that the human body is a chemical system was ultimately more useful than the older humoral theory.

Another useful element in the theories of Paracelsus was the doctrine that diseases were highly specific in their action and that for each disease there was a specific chemical cure. Thus Paracelsus opposed the old cure-all remedies containing numerous constituents and advocated the administration of single substances as medicaments. Such a reorientation gave an impulse to the study of particular diseases and helped to differentiate between useful and harmful medicaments. So long as the complex cure-all remedies were used such a differentiation was difficult, since any one of the many constituents might be useful or harmful. Paracelsus appears to have derived his doctrine of the specific character of diseases from the idea that all things in nature were autonomous living beings. He was of the view that in the beginning God had created a primordial matter and then numerous seeds to grow in it. Each seed developed into a particular entity within a predestined period of time, whereupon the entity itself passed away but its seed lived on to begin a fresh cycle of growth. 'God has created all things,' wrote Paracelsus. 'He has created something out of nothing. This something is a seed, in which the purpose of its use and function is inherent from the beginning.' Thus the development of each entity was shaped by an immanent pattern and a power contained within its seed. The growth of all beings was autonomous, free from the influence of external forces, for 'Never can anything develop from what is not in itself.' The growth-promoting power within each seed was a vital or spiritual force, the Archeus as Paracelsus termed it. The Archeus in the human body separated the useful from the useless in the ingested food and transformed nutriment into body tissue, like a miniature alchemist inside the laboratory of the human frame. Different Archei transformed and arranged matter in different ways, giving rise to an immense diversity of living organisms, each with its own specific individuality: 'For God has carefully differentiated all His Creation from the beginning, and has never given to different things the same shape and form.'

Paracelsus supposed that a disease was a force of a specific and vital character, like an Archeus or a seed. It battled with the Archeus of the body which it invaded, or the subsidiary Archei of particular organs, and it could be quelled by the Archeus of the mineral or plant that provided the specific remedy. In this way Paracelsus approached towards the idea that diseases were entities in themselves, in contrast to the traditional view which recognized only diseased states of the human body. Paracelsus made known his views on disease in 1531, and a few years later, in 1546, Girolamo Fracastoro, 1484-1553, put forward a similar theory that diseases were seed-like entities in themselves. Fracastoro was an adherent of the atomic theory and so, to explain the long-known fact that certain diseases were contagious or infectious, he suggested that there were atoms or seeds of disease which reproduced themselves and were transferred from person to person by contact or through the air. In this way Paracelsus and Fracastoro anticipated in a general way the later germ theory of disease, though their views were vague and were quite unsupported by experimental evidence.

Paracelsus did not owe much to the atomic theory for his views appear to have been connected more closely with the ideas of the Reformation than with the ancient philosophies revived during the Renaissance. He regarded his system as a religious revelation which was destined to restore the purity of Hippocratic medicine, just as the Reformers regarded their theologies as revelations restoring the purity of primitive Christianity. The Galenical doctors, Paracelsus wrote, 'are completely ignorant of the great secrets of nature which in these last days of grace have been revealed to me from on high'. As we have seen, the general ideas of the Calvinists and the mechanical philosophers had certain elements in common, and it seems that the vitalistic philosophy of Paracelsus had elements in common with the ideas of Luther's German Reformation. Paracelsus associated himself specifically with the Germanic lands, earning for himself the title of the 'Luther of Chemistry', and his system of iatrochemistry was more popular in Germany than elsewhere.

'For even as Avicenna was the best physician of the Arabs, and Galen of the men of Pergamum,' Paracelsus wrote, 'so also most fortunate Germany has chosen me as her indispensable physician.'

Alchemy, and now iatrochemistry, were connected with mystical religion, and the Germanic lands were particularly notable for their mystics from the time of Eckhart, c. 1260-1327, to that of Boehme, 1575-1624. Indeed, in the works of Boehme Protestant mysticism and iatrochemical theory were fused, as we shall see later.

The men of the scientific revolution and the Protestant Reformation reacted against the corporate-hierarchical world view of the middle ages in two ways. The Calvinists and mechanical philosophers, as we have seen, moved away from the hierarchical conception of the government of the universe towards an absolutist theory of cosmic rule in which all beings were to be mechanically uniform, enjoying parity of powers under the Supreme Governor. The Lutherans and the vitalist philosophers, notably the iatrochemists, rejected the traditional connections of domination and servitude which had been presumed to exist between the various kinds of creatures inhabiting the world, and they supposed that all of the beings in the universe were free and independent, deriving their autonomy from an inward principle. Paracelsus held such a view, supposing that each entity in nature developed under the impulse of its own internal vital force, independently of other entities. The mystics always had been individualists, giving primacy of place to the private visions experienced within their own minds. The mystical religious tradition of the Germans coloured their Reformation, and in common with the mystics Luther stressed the spiritual autonomy of man. For Luther it was through the justification of his inner spiritual faith that man was to be saved, not by the externality of Divine predestination as Calvin taught. Man was a little world unto himself, free and autonomous in the spirit, comprehending all things from within.

'In all other creatures God is known as by his footsteps only,' wrote Luther, 'but in man, especially in Adam, He is known truly and fully; for in Adam is seen that wisdom and righteousness, and knowledge of all things, that he might be rightly called a *microcosm*, or little world in himself; for he understands the heavens, the earth, and the whole creation.'

The iatrochemists likewise understood man to be an autonomous microcosm, thereby differing from the earlier alchemists who had been of the view that man, the microcosm, was ruled over by the heavenly bodies of the macrocosm. According to Paracelsus, it was in the realm of the spirit that the freedom and liberty of man resided.

'Just as the firmament with all its constellations forms a whole in itself,' wrote Paracelsus, 'likewise man in himself is a free and mighty firmament; and just as the firmament rests in itself and is not ruled by any creature, the firmament of man is not ruled by other creatures, but stands for itself and is free of all bonds. . . . Thoughts are free and are subject to no rule: on them rests the freedom of man. . . . Everything we invent has its origin in the spirit. Therefore let us not concern ourselves with how things have come to us, but trust that everything is done at God's command.'

From such a standpoint the iatrochemists, like Luther, were led to distrust methods of enquiry involving the use of logic. 'Such is the physical science of the Greeks,' wrote Paracelsus, 'deduced only from what is seen, recognizing nothing occult by mental experiment.' Knowledge for the iatrochemists was to be obtained by the empirical exploration of mystic insights and analogies, particularly the analogies obtaining between man, the microcosm, and the world as a whole, the macrocosm. By the analogies between them, Paracelsus wrote, 'he that knoweth the origin of thunder, wind, and storms, knoweth where colic and torsions come from'. However, the failure of such insights and analogies to work in practice could only be due to the incompetence of the experimenter, not to a lack of reality in the ideas. 'All the fault and cause of difficulty in Alchemy,' wrote Paracelsus, 'is wholly lack of skill in the operator.' Another source of knowledge for the iatrochemists were the Scriptures and the religious concepts of the day. Paracelsus felt that the primary matter of the world must have a threefold complexity as the Creator was three persons in one. Accordingly he added a third primary principle, that of salt, to the two principles of sulphur and mercury postulated earlier by the alchemists. 'These three principles,' wrote Paracelsus, 'are the prime matter and have only one name: the first matter is God, and as in the Deity there are three persons, so here each species is separate as to its office, but the three offices are comprehended under the one name of the first matter.'

The theories of Paracelsus were of great influence during the sixteenth and seventeenth centuries, and they came to rival those of Galen. In general the works of Paracelsus were forbidden in the universities but his theories seem to have been quite popular there, for in the late sixteenth century there were student riots at Paris and Heidelberg in protest against the prohibition of Paracelsian doctrines. Iatrochemistry appealed much more to the apothecaries than to the academic physicians, however, as it gave a theory to their art and provided a basis on which they could go into medical practice on their own account. In England the status of the apothecaries rose considerably during the seventeenth century. In 1608 they separated themselves off from the original drug merchants, the Company of Grocers, and set up their own Company of Apothecaries. More and more they went into medical practice, and during the Great Plague of 1665-66 they made good by staying at their posts in London whilst many of the Galenical doctors fled the city. Finally in 1703 a test case gave the apothecaries the full right to practise medicine, a right which was not revoked until the nineteenth century.

Iatrochemistry itself was developed further by John Baptist van Helmont, 1580-1644, a nobleman of Brussels. His main work, *On the Development of Medicine*, was published posthumously in 1648, though he brought out several minor works expounding the same views in his lifetime. Helmont followed Paracelsus in rejecting the Aristotelian logic and the deductive reasoning of the schools. 'Logic is unprofitable', he wrote, as 'the rules of Mathematicks or Learning by demonstration do ill square with nature'. He regarded knowledge obtained by the vision and insight of the mystic as divine, and knowledge obtained by logic and demonstration as merely human, and much inferior. Like Paracelsus, Helmont argued much by analogy, basing his views in particular upon the supposed similarity between terrestrial and celestial objects. 'Sublunary things do express themselves an analogy or proportion of things above,' he wrote. Helmont was more of an experimentalist than Paracelsus and more consistently followed up his ideas in a practical way. He thought that chemistry ought to be introduced into the universities, and taught 'not by naked discourse, but by handicraft demonstration of the fire . . . by distilling, moistening, drying, calcining, resolving, as nature works'.

Helmont did not accept the view of Paracelsus that the primary matter consisted of the three principles, salt, mercury, and sulphur, but like Paracelsus he argued in a theological way and held that water must be the primary matter, since the waters were mentioned in the Bible as the primeval chaos antecedent to the rest of the Creation. Such a view he thought confirmed by an ingenious experiment. He planted a willow cutting, weighing five pounds, in a tub containing two hundred pounds of earth and watered the cutting for five years after which it had become a tree weighing one hundred and sixty-nine pounds, the weight of earth in the tub remaining unchanged. Helmont argued that as water only had been added to the tree, its increase in weight must be due to assimilated water which had become transformed into wood. Further experiments led him to believe that water could be converted into earth, so that earth was a derivative and not a primary element. Water boiled in the crude glass vessels of the time gave a sediment, owing to some of the glass dissolving and some precipitating. Helmont took the sediment to be earth, and he understood the phenomenon to be a case of the transformation of water into earth. Fire, another of the traditional four elements, was just burning smoke in Helmont's view and not an element at all. However, he thought that air might be an independent element as he found that no matter how much he compressed a volume of air it would not condense to a 'water', or the liquid state, unlike steam and other vapours.

Like Paracelsus, Helmont thought that the various entities of the natural world derived from seeds which grew in the primary matter. Each seed or ferment contained a vital force or spirit which laid down a unique pattern of development and behaviour for the entity growing from that seed. Thus nature consisted of a multiplicity of autonomous beings, each self-determined by its own inner vital force.

'There are two chief causes or first beginnings of Bodies,' wrote van Helmont, 'to wit, the element of water or the (material) beginning, and the ferment or leven or seminal beginning. . . . The ferment is a formal created being, framed from the beginning of the world in the place of its own monarchie. (It) containeth the types or patterns of things to be done by itself, the figure, motions, houre, respects, inclinations, fitness, equalisings, proportions, alienations, defect, and whatsoever falls under the succession of days, as well in the business of generation as of government.'

Even the parts of a composite entity, such as the separate organs of the human body, had a life and being of their own, deriving from the spirit or vital impulse which governed that part. Diseases were living beings which entered man from outside and settled upon some particular organ, monopolizing its vital processes. 'A disease is an unknown guest,' wrote van Helmont, for each disease contains 'its own efficient cause and its own matter within itself'. Thus a disease was specific as to its cause, effect, and location, and it was to be treated by its specific remedy—a simple inorganic or plant substance— not the old complex cure-all medicaments. Helmont was of the view that the ancient doctors had confused diseases with their symptoms and had tried to remove the symptoms but not the causes. 'Curing is seated ofttimes in the removal of the occasional cause,' he wrote, 'but never in the removing of symptoms.'

In the field of chemistry Helmont was the first to distinguish between gases and the element of air. Before his time, and indeed for some time after, gases were thought to be different forms of the element of air, or air mingled with various impurities. Helmont, however, held that gases were substances differing fundamentally one from the other, and from air and condensable vapours. The alchemists had thought that bodies were made up of 'matter' and 'spirit', and they supposed that in some cases they could isolate the spirit by heating the body and condensing the vapour given off. Thus they obtained alcohol, or 'spirit of wine', and hydrochloric acid, or 'spirit of salt'. Similarly, Helmont thought that each entity was made up of primary matter, which was common to all bodies, and a vital force or spirit, which was specific to that entity. He endeavoured to isolate such spirits, not only by the traditional pyrotechnical methods, but also by the action of acids on substances, and in some cases he found vapours or spirits given off which could not be condensed. These vapours he termed 'gases', and as he thought that the spirit of a substance was specific to it, he held that gases differed one from the other according to their origin. Such a point of view led Helmont to isolate and characterize several gases, notably the oxides of carbon, nitrogen, and sulphur. Since gases were the pure spiritual energies or vital forces of substances, according to Helmont's theory, he thought that gases in themselves were potent agents and he instanced the explosion of gunpowder and the shattering of vessels by compressed gases to illustrate the contention.

Medical and chemical problems were studied not only by the iatrochemists, the vitalists, but also by the philosophers of the mechanical school of thought during the seventeenth century. In general the iatrochemists were of the view that inorganic substances were alive, changing by virtue of inner vital forces, whilst the mechanical philosophers considered matter to be dead and inert, undergoing change only when subject to external mechanical forces. Both points of view were of limited application to chemistry during the seventeenth century, though the mechanical philosophy was ultimately the more useful. The laws of mechanics were highly generalized, prescribing the same kinds of change to all the varieties of matter, so that they could not readily explain the specificity of chemical reactions, whilst iatrochemistry offered only a pseudo-explanation of chemical specificity, ascribing it to particularized vital forces.

One of the earliest chemists to put forward a mechanical theory of chemical change was Jean Rey, a French metallurgist of the first half of the seventeenth century. It had been known for some time that metals increased in weight when they were heated in air and formed a calx. To explain the phenomenon, Rey suggested in 1630 that air had weight and that it was taken up by metals on heating. He did not think of the process as a chemical combination of the air with the metal but as a mechanical mixing, like dry sand taking up water and becoming heavier.

'The increase in weight comes from the air,' wrote Rey, 'which air mingles with the calx (frequent agitation aiding) and becomes attached to its most minute particles; not otherwise than water makes heavier sand which you throw into it and agitate, by moistening it and adhering to the smallest of its grains.'

A more famous chemist who described himself as a mechanical philosopher was Robert Boyle, 1627-91, a younger son of the second Earl of Cork who was a *nouveau riche* of the early Stuart period. Boyle was interested in the work of the iatro-chemists, particularly in their empirical observations, but he was of the view that those observations should be explained in terms of the mechanical philosophy. For Boyle the mechanical philosophy was the theory that matter consists of particles of corpuscles in motion. He averred that he did not expect to 'see any principles proposed more comprehensive and intelligible than the corpuscularian'. Descartes, who had generalized the mechanistic point of view, rejected the atomic theory as he believed that a vacuum could not exist, but Boyle thought that a vacuum could be produced by means of the air-pump. Moreover, the atomic theory had been reviving thoughout the sixteenth century and it was expounded by several authors in the seventeenth, the most notable being Pierre Gassendi, 1592-1655, a professor of mathematics at the Collège de France, so that the atomic theory gradually became part of the mechanical philosophy.

Such a view Boyle found readily applicable to his work in physics. He discovered that the pressure of a gas varied inversely with its volume, and he saw that this law could be explained by assuming the particles of a gas to be either minute stationary springs or small spheres in random motion. Later Newton showed that Boyle's law of gases followed from the first supposition while Bernoulli showed that it followed from the second. But when he came to chemical problems Boyle found that the mechanical philosophy could be far less readily applied. Writing of the individual properties of chemical substances and the specificity of their reactions, Boyle spoke of 'the grand difficulty objected against the corpuscularian philosophy proposed by me about the origin of qualities: viz., that it is incredible that so great a variety of qualities as we actually find to be in natural bodies should spring from principles so few in number as two, and so simple as matter and local motion'. Boyle pointed out that salt will dissolve in water but not in oil or mercury, whilst gold will dissolve in mercury but not in water or oil, and again sulphur dissolved in oil but not in water or mercury. Such properties might be explained, he thought, in terms of certain 'principles of variation in bodies', whereby it was supposed that the fundamental atoms of substances had various shapes and sizes, and moved in different ways, or were fixed in various orders and arrangements relative to one another and retained certain 'subtle emanations' of 'effluviums' within their pores. According to Boyle, these 'principles of variation in bodies', like the letters of an alphabet, could be linked together in various ways to give numerous different combinations, each combination representing a possible set of properties possessed by a chemical substance. However, Boyle failed to give a fruitful interpretation of the observed chemical properties of bodies in terms of his 'principles of variation', for his theory did not systematize the then scattered facts of chemistry, nor was it open to experimental verification.

In other ways Boyle contributed to the foundation of modern chemistry by improving upon the existing procedures and general axioms of the subject. As a follower of Bacon he demanded that chemistry should be founded upon a substantial body of experimental observations, and he called in particular for the quantitative study of chemical changes. Boyle pointed out the importance of working with pure homogeneous substances and in this connection he formulated the definition, though perhaps not the conception, of a chemical element.

'I mean by elements', Boyle wrote, 'certain primitive and simple, or perfectly unmingled bodies which not being made of any other bodies . . . are the ingredients of which all those called perfectly mixed bodies are immediately compounded, and into which they are ultimately resolved.'

However, Boyle inclined to the view that water, air, and fire were elementary substances. He thought that there existed a 'material of fire' which was responsible for the increase in the weight of metals when they were heated in the air and formed a calx. Thus Boyle failed to perpetuate Rey's theory that the calcination of metals involved the taking up of air and Helmont's conception that gases were individual chemical substances, not forms of elementary air or air contaminated with impurities.

In physiology Boyle, and more especially his followers, Robert Hooke, 1635-1703, Richard Lower, 1631-91, and John Mayow, 1645-79, went far to show that the change from dark red venous blood to bright red arterial blood in the lungs was due to the uptake of part of the air and that the air so absorbed served a process in the body akin to that of chemical combustion. Paracelsus, arguing in his characteristic way by analogy, had suggested that air was a form of nourishment absorbed by the lungs, just as food was absorbed by the stomach. Boyle concurred with such a view for he found that animals soon expired without a supply of fresh air and died almost immediately when placed in the vacuum produced by his air pump. A candle flame similarly required air, and so Hooke suggested that respiration served a physiological process similar to combustion. Boyle discovered that animals and candle-flames in closed vessels absorbed part of the air available to them, and he suggested that this part might be a 'vital quintessence' mixed with the elementary air. Hooke found that gunpowder would burn in the vacuum or under water, and so he presumed that the 'vital quintessence' in the air, required for respiration and combustion, was a 'nitrous spirit' which was contained also in the saltpetre of the gunpowder mixture. With Lower, Hooke discovered that dark red venous blood when shaken up with air became bright red arterial blood, and Lower himself found that all of the blood in a suffocated animal was dark red and that it was all bright red in an animal whose lungs had been aerated with a pair of bellows. Lower suggested therefore that the blood picked up the vital part of the air, the 'nitrous spirit', in the lungs and by so doing was transformed from the dark venous blood to the bright arterial blood which carried the 'nitrous spirit' round the body, supplying the various vital processes. In this way, for example, an embryo, which could not respire, received the vital 'nitrous spirit' from the arterial blood of its mother.

The work of Boyle, Hooke, and Lower, was summarized and extended by John Mayow in his *Five Medico-Physical Treatises*, published in 1674. Mayow suggested that the vital part of the air indispensable for respiration and combustion was composed of certain 'nitro-aerial particles'. These particles were present in saltpetre, as gunpowder would burn in the absence of air. They were contained also in nitric acid since antimony gave the same product when treated with nitric acid or when heated in air. The product in both cases was heavier than the original antimony, the increase in weight being due to the 'nitro-aerial particles' absorbed. Such particles were also taken up by the blood during respiration and they generated animal heat by uniting with the 'sulphureous' or combustible particles in the blood. Mayow did not think of the air as a mixture of two gases, a mixture of his 'nitro-aerial particles' with other inert particles. He conceived of the air as an elementary substance, the 'nitro-aerial particles' being attached to the particles of air. The latter were like small springs and when the 'nitro-aerial particles' were removed from them by combustion or respiration they lost some of their springiness, the air accordingly contracting in volume. Mayow's view of the part played by the 'nitro-aerial particles' in chemical and vital processes differs somewhat from the modern conception of the chemical and physiological role of oxygen, which he is sometimes said to have anticipated. He was an adherent of the mechanical philosophy and for him the actions of the 'nitro-aerial particles' were largely mechanical; they did not undergo what we would term specifically chemical combination. The rusting of iron was due to the mechanical attrition of the 'nitro-aerial particles', whilst muscular contraction derived from their rapid motions in the muscles, and hunger was the pain caused by those particles pinching the stomach walls. Heat and light were the 'nitro-aerial particles' in rapid motion while cold and the blue of the sky were those same particles at rest. In this way Mayow elevated his 'nitro-aerial' spirit or particles to the position of a general principle, akin to the principles of the alchemists and iatro- chemists. Indeed many of his views were mechanical versions of the vitalist iatrochemical theories. Like the iatrochemists he thought that there were three fundamental principles in nature, nitro-aerial spirit, sulphur, and salt, his nitro-aerial spirit taking the place of the mercury of the alchemists. Just as the alchemists had thought that the metals were generated by the interaction of mercury and sulphur, so Mayow supposed

242 / A History of the Sciences

that chemical change in general derived from the interaction of nitro-aerial spirit and sulphur.

'Nitro-aerial spirit and Sulphur are engaged in perpetual hostilities with each other,' he wrote, 'and indeed from their mutual struggle when they meet, and from their diverse states when they succumb by turns, all the changes of things seem to arise.'

With Mayow the work of the seventeenth-century English school of medical chemists came virtually to an end. The tradition of the school was continued to some degree by Stephen Hales, 1671-1761, but it did not survive and modern chemistry was founded elsewhere, in France at the end of the eighteenth century. Boyle had arrived at a reasonable definition of a chemical element and at a promising conception of method in chemistry. He and his school had come to the conclusion that the air played a vital role in combustion and respiration, part of the air being absorbed in these processes. Such views, together with Helmont's conception of gases and Rey's theory of the calcination of metals, were somewhat eclipsed, if not entirely lost, in the subsequent period when a fresh wave of iatrochemistry spread over Europe from Germany in the form of the phlogiston theory.

Chapter 21

Some Early Modern Applications of Science

THE FIRST notable applications of science had been the utilization of astronomy and mathematics for the keeping of accounts, surveying, and the making of maps and calendars during the bronze age. Such applications were improved upon by the Greeks and the Arabs, but science was not applied to radically new fields until early modern times when the great geographical discoveries posed novel practical problems in navigation for which tradition had no ready-made solutions. Since the end of the great Mongol empire, direct landward trading with the Orient had become more and more difficult while the trading potential of Europe had increased. Accordingly alternative routes to the east were sought for by the Europeans, who now had a sailing ship capable of ocean-going voyages. The first explorers were the Genoese, who had been ousted from the Levant trade by the Venetians, but they accomplished little and they were soon replaced by the Portuguese. However, the Genoese retained for some time a considerable reputation for their instrument makers and their pilots, amongst whom were Columbus and the Cabots.

From the beginning of the fifteenth century the Portuguese pushed down the coast of West Africa, still following the coast-wise navigation of the middle ages. The Infante of Portugal, Henry the Navigator, 1394-1460, set up a navigation research institute and an astronomical observatory at Sagres on the Cape St. Vincent about 1420. The works of Ptolemy and other ancient writers on astronomy and geography were brought to the institute and were used on the explorations. But, as one of Henry's captains observed, 'With all due respect to the renowned Ptolemy, we found everything the opposite of what he said.' The observatory and institute were staffed with German mathematicians and Italian map makers who prepared charts of the lands and seas explored from the 1450's. The mathematicians recalculated the circumference of the earth and, using the small estimate of Posidonius and Ptolemy for the length of a degree, they obtained an optimistically short estimate. With this estimate a course for the westward circumnavigation of the earth was charted in 1474.

The Portuguese explored eastward, Bartholomew Diaz rounding the Cape of Good Hope in 1486 and Vasco da Gama reaching India by that route in 1497. The Spaniards explored westwards, Columbus reaching the West Indies in 1492. Both Columbus and da Gama had to cross extensive stretches of ocean out of sight of land and, in order to make their journeys repeatable, ocean charts and methods of determining a ship's position at sea were required. The making of ocean charts and maps of large areas in general involved the problems of ascertaining the relative positions of places on the earth's surface and of representing the spherical earth on a plane map. The determination of a ship's position at sea and the positions of the newly discovered lands required procedures for the measurement of the latitudes and longitudes of places. The latitude of a position north or south of the equator was easily ascertained by measuring the altitude of the sun in the daytime or the altitude of the pole star at night in the northern hemisphere, as was well known to the Greeks. The longitudinal distance between two places was more difficult to determine, however. Such distances could be estimated by measuring the differences between the local times of the places in question: for example, at midday in England it is sunset in Siberia and dawn in America, and from the time differences the relative separations of these places may be determined. The local time at a given position could be measured quite easily from the sun, as with the sundial, or at night from the apparent rotation of the circum-polar stars, such as the stars of the Great Bear constellation which were known as the 'clock stars'. But in order to obtain the time differences between places it was necessary that the local times should be measured relative to some standard event perceptible to observers at different places at the same instant, such as an eclipse of the sun or the moon, or to some arbitrary scale of standard time, like the present Greenwich mean time. The simultaneous observation of an eclipse and the local times at different places was the only method of determining the longitude known in antiquity and the middle ages, but solar and lunar eclipses were too infrequent to make the method of much use to the navigators of the sixteenth century.

Such problems, and the older ones of surveying and calendar making, were studied during the fifteenth and sixteenth centuries, notably by the men of south Germany, the Netherlands, and northern Italy. As we have seen, Johannes Muller, 1436-76, and his patron, Bernhard Walther, 1430-1504, a wealthy merchant of Nuremberg, observed the heavens at Walther's observatory in order to obtain data for the reform of the calendar and for the preparation of nautical almanacs which were used by the Spanish and Portuguese navigators. A little later, in 1524, Peter Bennewitz, 1495-1552, a professor of mathematics at Ingolstadt and a friend of the Spanish Emperor, Charles V, suggested that standard time for the determination of the longitude could be obtained by observing the moon's position amongst the fixed stars. Bennewitz also invented a projection suitable for maps of large land areas. He represented the earth's latitudes as parallel straight lines and the meridians of longitude as portions of circles increasing in curvature from the prime meridian, which was linear, so that the map of the world was divided into thirty-six segments having equal widths at the equator. Another south German, Martin Waldseemuller, a mathematician of Strassburg, developed an early prototype of the theodolite about 1522 for surveying and observing the heavens.

In the Netherlands the Flemings had long rivalled the Genoese as the best compass makers of Europe, correcting their compasses for the deviation of the magnetic needle from the true north while the Genoese did not. Here Gemma Frisius, 1508-55, a professor of mathematics at Louvain and Cosmographer Royal to Emperor Charles V, suggested in 1530 that standard time for the measurement of longitude might be obtained by means of accurate mechanical clocks set to the local time of a prime meridian, like the present one of Greenwich. In 1533 Frisius made known his method of surveying by triangulation which replaced the current laborious method of pacing out distances. He suggested that the length of an arbitrary base line be measured, and the angles subtended by various distant objects ascertained from the ends of that line, such measurements giving trigonometrically the distances of the objects sighted. His pupil, Gerard Mercator, 1512-94, another Fleming, was one of the most able of the instrument makers, surveyors, and map makers of the sixteenth century. In 1569 he published a map of the world, based on a projection he had invented, which was ideal for navigators' charts. Mercator represented the meridians of longitude by equally-spaced parallel lines and the latitudes by parallel lines perpendicular to the meridians. The latitude lines were gradually spaced out towards the polar regions so that the degrees of latitudes were exaggerated in exactly the same proportion as the degrees of longitude. Such a projection simplified considerably the charting of a course, for on a Mercator map the course of a ship along a constant compass bearing always appeared as a straight line, not a complicated curved line as with other projections.

The problem of representing the spherical earth upon a plane map for the purposes of navigation was thus solved by Mercator. However, the question of determining the longitude at sea remained, as the methods proposed by Bennewitz and Frisius were not yet practicable. In 1598 Philip III of Spain offered a prize of 1,000 crowns for a method of ascertaining the longitude and about the same time the States General of Holland offered similarly a prize of 10,000 florins for a solution to the longitude problem. Shortly afterwards Galileo observed with his telescope the four moons of Jupiter and he noted that their frequent eclipses provided a good method of determining standard time at different places. Accordingly Galileo drew up tables of the motions of Jupiter's moons, and he offered the tables to both Holland and Spain. However, Galileo's tables were not very accurate and the method proved to be impracticable at sea as the motion of the ship prevented the making of steady observations with a telescope. Galileo made another discovery which later promised to solve the longitude problem by Frisius' method. He found that the beats of a swinging pendulum took the same period of time if the amplitudes of swing were small, and he left behind him designs for a mechanical clock regulated by such a pendulum. Subsequently attempts were made by Christian Huygens of Holland to construct pendulum chronometers which would keep standard time at sea, but they were not a success, as the motion of the ship interfered with the swing of the pendulum.

A notable feature of the developments described hitherto was the lack of significant contributions from the men of the Iberian peninsula to the solution of the problems of navigation. The Spaniards and the Portuguese had been foremost in making and exploiting the great geographical discoveries and they experienced the problems of navigation most acutely, yet they did little to solve these problems either by applying science or by advancing science. During the sixteenth and early seventeenth centuries it was the men of northern Italy, Germany, and the Netherlands who contributed the most to the advance of science. The Germans and the Italians, with their long-standing cultural and commercial traditions, developed both fundamental and applied science, though their science stagnated during the mid-seventeenth century after their prosperity had gone. The scientific activity of the Netherlanders, who possessed a more recent cultural and technical tradition, was at first mainly of an applied character. Science in the Netherlands was continued by the Dutch when the contribution of the Flemings declined, Simon Stevin, 1548-1620, of Bruges transferring his allegiance to Holland during the Dutch struggle for independence from Spain. His study of mechanical problems, though mainly of an applied nature, marked the beginnings of Dutch theoretical science which came to fruition with Christian Huygens, 1629-95, and the late seventeenth-century school of Dutch microscopists. Other countries which had profited technically and commercially from the great geographical discoveries, notably England and France, began the development of pure and applied science about the same time, their scientific tradition coming to maturity also in the mid-seventeenth century.

Spain and Portugal, the countries which had made the great geographical discoveries, did not develop much of a scientific tradition, while their commerce and industries declined. The Spaniards were deceived into thinking that they had discovered wealth in the gold and silver which they found in the New World, but the precious metals they brought from America served only to stimulate the crafts and commerce of other lands and to ruin the native Spanish industries. Such industries were further retarded by the expulsion of native commercial elements, such as the Jews and Protestants, and by the disabilities placed upon foreign craftsmen and foreign investors. The Portuguese scientific navigation movement inaugurated by Henry the Navigator began to produce positive results in the sixteenth century, but the achievement proved to be ephemeral. The Portuguese Jew, Pedro Nunez, 1502-78, designed a conical projection for ocean charts and invented a device for reading fine subdivisions on the scales of astronomical and other instruments. But the position of Jews became untenable and he left the country. After Nunez little of importance for either fundamental or applied science appeared from the Iberian peninsula. Some good technical handbooks were produced, such as the *Art of Navigation* (1550) by Cortez and the *Art of Metals* (1640) by Barba, but they did not contain much that was novel. Thus the commerce, arts, and sciences of the Iberians were still-born, and as the might of Spain declined, so too did the scientific achievement of the men who had been stimulated by their connections with Spain to study applied scientific problems in the early sixteenth century, notably the Flemings and the south Germans.

Elsewhere, particularly in England, converse policies were adopted; industries were planted and foreign craftsmen welcomed. Commerce prospered, and England soon developed a tradition of first applied, and then fundamental science. Like Spain, England relied at first upon Genoese pilots, notably the Cabots, and foreign mathematicians, such as the Frenchman, Jean Rotz, who became Royal Hydrographer to Henry VIII in 1524. But as navigation and trade developed native pilots and mathematicians soon appeared. The earliest English mathematicians of note were Robert Recorde, *c.* 1510-58, Leonard Digges, died 1571, and John Dee, 1527-1606. Recorde published several books on astronomy and mathematics, notably the *Castle of Knowledge* (1556), which was written for the English Cathay voyagers, and the *Whetstone of Witte* (1557), which was dedicated to the Governors of the Muscovy Company. When Recorde died in 1558 his work as technical adviser to the Muscovy Company and the Cathay voyagers was taken over by John Dee. Mathematics and commerce in England were closely connected during the sixteenth century and applied science was promoted by the merchants and trading companies. Writing of the art of navigation, Dee spoke of two compasses 'of me Invented, for our two Muscovy Master Pilots, at the request of the Company'. The Muscovy Company also authorized one of Dee's pupils, Richard Eden, to translate the Spanish work of Cortez on the art of navigation.

Dee studied the mathematical sciences at Louvain under Gemma Frisius and Gerard Mercator, and in turn he taught most of the important English pilots of the sixteenth century. Richard Chancellor was instructed by Dee prior to his search, on behalf of the Muscovy Company, for a north-east passage to Russia and China in 1553. Dee gave instruction to Martin Frobisher prior to his search for a north-west passage to China in 1576-78, and Dee also taught Chancellor's successors, Stephen and William Borough. Dee was not only a teacher of mathematics, but also a practising astronomer and a notable author. He and Chancellor constructed a large cross staff with which Dee made astronomical observations for the reform of the calendar, and his ward and pupil, Thomas Digges, died 1595, observed the new star of 1572 with the instrument, Digges' observations of the star being second in accuracy only to those of Tycho Brahe. In 1570 Dee wrote an important preface setting forth the philosophical rôle and practical applications of mathematics for the first English translation of Euclid's geometry which was made by his friend, Henry Billingsley, a wealthy haberdasher who was Lord Mayor of London in 1595.

'All thinges which are, and have beyng', wrote Dee in this preface, 'are found under a triple diversitie generall. For either they are demed Supernaturall, Naturall, or of a third being . . . which by a peculiar name also, are called Thynges Mathematicall. For, these, beyng (in a maner) middle, betwene thinges supernaturall and naturall are not so absolute and excellent as thinges supernaturall, nor yet so base and grosse as thinges naturall: but are thinges immateriall and neverthelesse by materiall things hable somewhat to be signified.'

Accordingly, Dee went on to say, by means of things mathematical we can comprehend something of things supernatural and order things natural. We can apply mathematics to the computations of commerce, to the problems of architecture, astrology, music, geography, astronomy, and to the arts of navigation, medicine, and war. Such applications, it seems, were not practised very competently in England during the sixteenth century.

'Of these feates (of mathematics)', continued Dee, 'is sprong the feate of Geodesie, or Land Measuring: more cunning to measure and survey Land, Woods, and Waters, afarre off. More cunning, I say: But God knoweth, hitherto, in these Realmes of England and Ireland, whether through ignorance or fraude, I can not tell, in every particular, how great wrong and injurie hath in my time bene committed by untrue measuring and surveying.'

The money saved by improved surveying, Dee averred, would suffice to found a mathematical readership in each of the two universities.

The English merchants, like Dee's friend, Billingsley, were particularly concerned to promote the study of mathematics, notably in connection with the problems of navigation and the surveying involved in the great land transfers of the period. The merchants of London, led by Thomas Smith who became the first governor of the East India Company in 1600, set up a mathematical lectureship in London to train navigators and militia captains at the time of the Armada threat in 1588. The merchants appointed Thomas Hood, the son of one of their members, to the post which was continued for two years. Later the East India Company sponsored the mathematical lectures of Edward Wright, 1560-1615, who had worked out in 1599 the mathematics of Mercator's map projection. Such lectureships did not last for long but a more important and permanent educational institution was established by Sir Thomas Gresham, 1519-79, a wealthy member of the Company of Mercers who left his estate and mansion in London for the foundation of a College in which the teaching of the sciences was to be prominent. His will vested the management of Gresham College, as the institution was called, not in hands of churchmen like the universities, but in the charge of the Company of Mercers

and the Mayor and Aldermen of London who put Sir
Thomas' will into effect upon the death of his widow in 1597.
Gresham's bequest provided for seven professors who were to
give two lectures each per week, one in Latin and one in English, upon the subjects of divinity, music, astronomy, geometry, medicine, law, and rhetoric, the lectures being free and
open to all citizens of London. In his will Gresham laid down
the duties of the astronomy professor as follows:

'The astronomy reader is to read in his solemn lectures, first
the principles of the sphere, and the theoriques of the planets,
and the use of the astrolabe and the staff and other common
instruments for the capacities of mariners; which being read
and opened, he shall apply to use, by reading geography and
the art of navigation in some one term of every year.'

Once founded, Gresham College became the main focus
of scientific activity in England during the first half of the seventeenth century. There were groups outside of the College,
such as that centring round Sir Walter Raleigh and the Earl of
Northumberland which consisted notably of Thomas Harriot,
1560-1621, Robert Hues and Robert Warner, who were
termed 'the earle of Northumberland's three Magi'. Of these
men Harriot was the most important. He went out to America
with Raleigh and there mapped out Virginia. Harriot corresponded with Kepler upon astronomical and optical questions,
and in the subject of mathematics he anticipated some elements
of Descartes' analytical geometry. Another group centred
round John Wells and his son who held the post of Keeper of
H.M. Naval Stores at Deptford during the first half of the seventeenth century. At the house of the Wells gathered the
master shipwrights, Phineas Pett, Edward Stevens, Hugh Lyi-
ard, and Henry Goddard, the father of Jonathan Goddard
who became a professor of medicine at Gresham College.
However, the Deptford group was associated fairly closely
with the Gresham professors of astronomy and geometry and
it did not constitute a group markedly distinct from the Gresham circle.

The first professor of geometry at Gresham College was
Henry Briggs, 1556-1630, whose circle of friends included Ed-

ward Wright, William Gilbert of Colchester, Marke Ridley, one of Gilbert's medical colleagues who had been physician to the English merchants in Russia, and William Barlow, a Prebendary and subsequently Bishop of Winchester. All of these men made experiments on magnetism, the most important being those made by Gilbert. The English navigators were much concerned with possible north-east or north-west passages to Asia, such explorations involving the navigation of polar regions where the deviation of the magnetic compass from the true north was large and uncertain. The science of magnetism was therefore of particular importance to the English and they were foremost in this field of study. Columbus had found that the deviation of the magnetic compass from the true north varied from place to place on the earth's surface. Sebastian Cabot and Jean Rotz, in the service of the English, suggested that the phenomenon might be used to determine the longitudes of places but the compass deviation was found to vary too irregularly from place to place to serve this purpose. Moreover, Gilbert was led to believe, from his experiments with spherical lodestones, that the deviation of the compass from true north at a given position was caused by the presence of land masses and was quite independent of the longitude of that position. Gilbert suggested that a northerly sea route to India was more likely to be found by the north-east passage than by the north-west as there was a smaller deviation of the compass along the former route and thus fewer land obstacles according to his theory. Whilst he rejected the possibility of a connection between the compass deviation and the longitude at any position, Gilbert thought that there might be a correlation between the dip of the magnetic needle and the latitude at various places on the earth's surface. Hence he suggested that the magnetic dipping needle could be used to discover the latitude under cloudy skies when the heavenly bodies could not be seen. Gilbert predicted that the dipping needle would assume a vertical position in the polar regions, and Hudson found that such was the case on his voyage to North America in 1609.

Another of Briggs' friends, John Napier, 1550-1617, of Merchiston in Scotland, invented logarithms sometime towards the end of the sixteenth century in order to facilitate the technique of arithmetic computation. Napier's logarithms were of angular measures for the use of astronomers and Briggs extended the invention by introducing the base of ten and by calculating the logarithms of 30,000 common numbers to fourteen decimal places. Briggs' colleague in the chair of astronomy at Gresham College, Edmund Gunter, 1581-1626, constructed rulers with a logarithmic scale for further ease of calculation, and in 1632 an amateur mathematician, William Oughtred, 1575-1660, obtained the slide-rule by arranging two of Gunter's rulers so that they would move side by side.

Briggs, Gunter, and Gunter's successor to the chair of astronomy at Gresham College, Henry Gellibrand, 1597-1637, were all associated with the shipwrights, navigators, and merchants of the time. In fact Briggs was himself a member of the Virginia Company. They were connected in particular with the group of shipwrights and others interested in navigation which centred round John Wells, the Keeper of Naval Stores at Deptford. Briggs in 1609 judged between two factions of shipwrights in a controversy concerning some innovations which Phineas Pett had introduced. Gunter between 1626 and 1628 worked with Wells to find better methods of calculating the tonnage of ships. Later, in 1635, Gellibrand discovered that the deviation of the magnetic compass from true north varied with time as well as with place by comparing a measurement of the magnetic declination he made in the garden of Wells' house at Deptford with similar measurements made by Gunter twelve years before and by William Borough some half a century before. Gellibrand's discovery finally put an end to all hope of ascertaining the longitude from the phenomena of terrestrial magnetism as it was now apparent that positions on the earth's surface could not be characterized by time-invariant values of the declination or other magnetic measures.

During the second quarter of the seventeenth century the scientific movement in England became more coherent and organized, culminating, as we shall see, with the foundation of the Royal Society in 1660. The interests of the English scientists in that period also became wider in scope, spreading beyond the utilization of the mathematical sciences for surveying, navigation, and so on, to embrace the intellectual aspects of the scientific revolution developed on the continent, and the possibility of applying science to craft processes and industrial techniques. With certain notable exceptions, like William Gilbert of Colchester, the English scientists of the sixteenth century had been concerned more with the technical rather than the ideological aspects of the new astronomy, and with the application of science more to navigation and commercial problems than to industrial problems of a mechanical or chemical character. Robert Recorde, for example, had been not only technical adviser to the Muscovy Company and the Cathay voyagers but also Comptroller of the Royal Mines and Monies. In the former capacity his work was unexceptionable, but in the latter he was accused of incompetence in connection with an attempt to plant the alum industry in England. Again Recorde's discussion of the Copernican theory was technical rather than philosophical. Half a century later, however, Gilbert devoted considerable attention to craft problems and to the question of providing a rational physical explanation for the movements of the heavenly bodies, rather than merely 'saving the appearances' mathematically. Gilbert devoted one tenth of his *De Magnete* (1600) to cosmological theory and another tenth to the mining, smelting, and fashioning of iron, while a quarter was concerned with navigation and nautical instruments and the rest with his work on magnetism.

After Gilbert, Francis Bacon was particularly concerned with the application of science to the arts and crafts, rather than to commerce and navigation, and with the construction of a new natural philosophy, rather than theories 'feigned and assumed for ease and advantage of calculation'. For Bacon the arts and crafts were the basis of human civilization.

'Let anyone consider', Bacon wrote, 'what an immense difference there is between man's life in any highly cultivated part of Europe, and in some very wild and barbarous region of the New Indies: and this, not the soil, not the climate, not bodily powers, but the arts provide.'

England had already assimilated many of the arts and crafts practised on the continent. Under Elizabeth I the immigration of skilled craftsmen from Spain, Italy, and the Netherlands was encouraged; the industries of wire-drawing and brass-founding were planted, and the pumping methods of the Germans were introduced into the mines. Bacon wished to give a further impulse to such a trend by applying scientific knowledge, derived with his new method, to the development of novel industrial and craft techniques. For the purpose of advancing and applying science Bacon proposed the foundation of a scientific academy, a 'House of Solomon' as he called it, in his posthumous *New Atlantis* (1627). The academy was to be not just a learned society but a research and teaching institute equipped with laboratories, gardens, a library, workshops, and power houses. The members of the academy were to collect information from foreign lands, from books, from craftsmen, and from their own experiments and observations. The information so collected was to be arranged into the form of an encyclopedia from which a new system of natural philosophy could be derived, a system that would be of great use when applied to the common needs of mankind.

'The end of our Foundation', wrote Bacon, 'is the knowledge of causes, and the secret motion of things, and the enlarging of the bounds of the Human Empire, to the effecting of all things possible.'

The suggested foundation of a scientific society, like Bacon's other projects, did not attract a great deal of attention in his own day but Bacon's writings became popular in the mid-seventeenth century, influencing in particular the men who founded the Royal Society.

Chapter 22

The Scientific Societies of the Seventeenth Century

DURING THE second quarter of the seventeenth century the scientific movement in England gradually became more coherent and broader in scope, as we have noted. Even the popular almanac-makers in England responded to the new intellectual currents, and from about 1640 the more important of them dropped the Ptolemaic system in favour of either the Tychonian or the Copernican theory, the latter being generally adopted by about 1680. A work which helped considerably to change the view-point of the almanac-makers was the *Discourse concerning a New Planet* (1640) by the Puritan clergyman, John Wilkins, 1614-72, who did much to popularize the new cosmology and to harmonize it with Calvinist theology. Wilkins was also concerned to promote the crafts and industries through the applications of science in contrast to the orientation of the earlier English scientists towards commerce and navigation. His *Mathematical Magic* (1648) was concerned mainly with the mechanical principles of machines, from which, Wilkins wrote, 'there is much real benefit to be learned; particularly of such gentlemen as employ their estates in those chargeable adventures of Drayning Mines, Cole-pits, etc, who may from hence learn the chief grounds and nature of engines'. Wilkins again was foremost in the organization of science in mid-seventeenth-century England. He was the leading spirit of a group of young scientists, calling themselves the 'Philosophical College', who began meeting regularly in London about the end of 1644. The group included the Puritan divines, John Wilkins and John Wallis, 1616-1703, the Gresham professors of astronomy and medicine, Samuel Foster, died 1652, and Jonathan Goddard, 1617-75, four medical men, and from 1646, Robert Boyle, 1627-91, and William Petty, 1623-87. These men met weekly to make experiments and to discuss scientific theories, first at the Bull Head Tavern in Cheapside and then at Gresham College in Bishopsgate. Of the ten known members of the 'Philosophical College', six were definitely Puritan and Parliamentarian during the Civil Wars while only one was a certain Anglican and Royalist. Goddard was Cromwell's physician, and Wilkins became Cromwell's brother-in-law in 1656. Wallis decoded Royalist ciphers for the Parliamentary forces, while Foster had been removed from his chair at Gresham College in 1636 for refusing to kneel at the communion table. It seems that Gresham College had been a meeting place for Puritans as well as scientists, for Antony Wood stated that Foster's predecessor, Henry Gellibrand, 'suffered conventicles (being himself a Puritan) to be kept in his lodgings' at Gresham College. Gellibrand in fact had been arrested at the instigation of Laud, then Bishop of London, for bringing out an almanac in 1631 which left out the list of Catholic saints and martyrs usually included in such works.

Since 1642, Charles I had made Oxford his capital and he had removed several university Puritans and Parliamentarians from their posts. At the instigation of the professor of astronomy, John Greaves, an ardent Royalist, he removed Sir Nathaniel Brent from the Wardenship of Merton College and replaced him with William Harvey, the royal physician. Oxford fell to Cromwell in 1646 and two years later a Parliamentary Commission was set up to reform the university, ejecting Royalists and replacing them with Parliamentarians. The London group of scientists composing the 'Philosophical College' was heavily called upon to fill vacant appointments. John Wilkins was made Warden of Wadham College, John Wallis, professor of geometry, William Petty, professor of anatomy, and Jonathan Goddard became Warden of Merton College in 1651 when Brent, who had been reinstated, resigned. Most of these changes were made for political reasons but one seems to have been made on ideological grounds. The ejected professor of astronomy, John Greaves, was a Royalist, and so too was the new one, Seth Ward, who had been removed from Cambridge for his Royalist sympathies. But Greaves was the last professor of astronomy at Oxford to support the Ptolemaic theory while Ward was the first to support the Copernican.

John Wilkins attracted several notable students to Oxford and he founded there a scientific club, the 'Philosophical Society', which lasted until 1690. At Wadham resided Christopher

Wren, Thomas Sydenham, John Mayow, Laurence Rooke, and Thomas Sprat, all enthusiasts of the new 'experimental philosophy'. Christopher Wren, besides his work as an architect, distinguished himself as an anatomist and an astronomer, being elected to the chair of astronomy at Gresham College in 1657. Thomas Sydenham did much to reform medical practice and earned for himself the title of the 'English Hippocrates'. John Mayow, the chemist, worked with Robert Boyle, who had come to Oxford with the other members of the 'Philosophical College'. Laurence Rooke, an astronomer, came over from Cambridge to join Wilkins' circle, leaving in 1652 to become professor of astronomy, and then of geometry, at Gresham College. Thomas Sprat wrote the first history of the Royal Society, published in 1667, which detailed the events and the causes leading to the foundation of the institution.

With the restoration of Charles II in 1660, London again became the main centre of scientific activity in England as many of the scientists appointed by the Commonwealth left Oxford or were removed from their posts. The number of persons interested in science had increased greatly during the Commonwealth period and it was now felt that an official scientific organization should be founded in England. Accordingly the scientists in London met at Gresham College in November 1660 after one of Christopher Wren's lectures, and proposed formally the foundation of a 'College for the promoting of Physico-Mathematical Experimental Learning'. John Wilkins was elected chairman and a list of forty-one persons 'judged willing and fit to join with them in their design' was drawn up. Shortly afterwards the courtier, Sir Robert Moray, brought the King's verbal approval of the 'College' and Moray was made president of the assembly. Two years later Charles II sealed the charter which formally incorporated the institution as 'The Royal Society for the Improvement of Natural Knowledge'. Another courtier, Lord Brouncker, was made the first President of the Royal Society and the first joint Secretaries were John Wilkins and Henry Oldenburg, a business man with extensive connections on the continent. The members of the Royal Society increased in number from about a hundred at its foundation in 1660 to over two hundred in the 1670's, but interest in science began to decline towards the end of the century and the Society had only a hundred and twenty-five Fellows in 1700. Membership thereafter increased again, reaching five hundred by 1800, but less than half of that number were scientists properly speaking, the rest being honorary Fellows. There were of course members of the Royal Society in the seventeenth century who added little that was new to the theories or the experiments of science. We do not associate the names of Wilkins and his pupil Sprat with any notable discoveries but they were most active in promoting the organization, application, and dissemination of science whereas the honorary Fellows of the eighteenth century, generally speaking, did little in these directions.

The Fellows of the Royal Society in its early days were greatly influenced by the writings of Francis Bacon and they followed the policies he had suggested fairly closely. William Petty compiled 'histories' or descriptions of shipping, clothing manufacture, and the dyeing trade, while Robert Boyle made a general survey of the methods used in the crafts involving chemical processes. The Baconian influence pervaded the proposed statutes of the Royal Society drawn up by its Curator, Robert Hooke, in 1663.

'The business and design of the Royal Society', wrote Hooke, is 'to improve the knowledge of natural things, and all useful Arts, Manufactures, Mechanick practices, Engynes, and Inventions by Experiments, —(not meddling with Divinity, Metaphysics, Moralls, Politicks, Grammar, Rhetorick, or Logick). To attempt the recovering of such allowable arts and inventions as are now lost. To examine all systems, theories, principles, hypotheses, elements, histories, and experiments of things natural, mathematical, and mechanical, invented, recorded, or practiced, by any considerable author, ancient or modern. In order to the compiling of a complete system of solid philosophy for explicating all phenomena produced by nature or art, and recording a rationall account of the causes of things.'

The influence of Bacon was reflected again in the number of Fellows joining the eight committees which the Royal Society set up in 1664 to examine specialist problems. The committee for the study of mechanical questions was the most popular with sixty-nine members. Next came the committee concerned with the 'histories' of trades with thirty-five members and the agricultural committee with thirty-two. In contrast the astronomy committee had only fifteen members. However, the influence of Bacon upon the Fellows of the Royal Society declined during the 1670's and it was supplanted by a 'Galileian' trend as manifest above all in the work of Newton who became a Fellow of the Society in 1671. For example, the percentage of the total number of papers in the *Philosophical Transactions* of the Royal Society devoted to applied science declined from 10·3 per cent in the years 1665-78 to 6·6 per cent in the years 1681-99. Later, during the eighteenth century, the mathematical trend of the 1680's in turn declined and the work of the members of the Royal Society and British scientists in general became more empirical and largely experimental.

The committees set up by the Royal Society in 1664 indicate that the English interest in applied science by now had spread beyond the mercantile problems of navigation to include the problems of crafts, industry, and even agriculture. However, it seems that the merchants still gave a considerable impetus to the scientific movement as Robert Hooke in the preface to his *Micrographia* (1664) stated that the merchants had played a considerable part in the foundation of the Royal Society. The Fellows of the Society, wrote Hooke,

'have one advantage peculiar to themselves, that very many of their number are men of converse and traffick, which is a good omen that their attempts will bring philosophy from words to action, seeing the men of business have had so great a share in their first foundation.'

Thomas Sprat in his *History of the Royal Society* (1667) wrote in a similar vein that 'the noble and inquisitive genius of our merchants had done much to advance science and establish the Royal Society, the members of which had 'principally consulted the advancement of navigation'. The nobility, Sprat suggested, were interested in science to the degree to which they had become commercialized, either through intermarriage with city families or by commoners making fortunes and becoming ennobled.

'And still we have reason to expect that this change will proceed farther for the better,' Sprat wrote, 'if our Gentlemen shall more condescend to engage in commerce and to regard the Philosophy of Nature.'

The earlier Puritan affiliation of many of the older members of the Royal Society was a somewhat delicate matter in the uneasy religious and political atmosphere of the Restoration period. As we have seen, divinity and politics were specifically mentioned in the statutes proposed by Hooke as subjects with which the Royal Society would not meddle. Of the sixty-eight Fellows of the Royal Society in 1663 for which information is available, some forty-two had been Puritan and Parliamentarian during the Civil War period and twenty-six had been Royalist. Such a conspicuity of former Puritans was widely advertised by the opponents of the Royal Society, the most notable of whom was Henry Stubbe, 1632-76, a Warwick physician who attacked the Society and all that it stood for in a series of pamphlets published 1670-71. Stubbe had been a Puritan experimental scientist appointed to Oxford university under the Commonwealth and removed at the Restoration. Thereupon he performed a *volte-face* and attacked his former associates, defending the systems of Aristotle and Galen against the new science. Stubbe described the Fellows of the Royal Society as his 'Oliverian adversaries', a reference to Cromwell, and he suggested that they were undermining the universities and destroying established religion. 'To Preserve our old Religion 'tis absolutely necessary that we retain our old Learning,' wrote Stubbe. However, the religion of England had changed and so too had the natural philosophy generally received by educated opinion. The Church of England was no longer an autonomous Catholic Church, as it had been in the days of Henry VIII, for it had become permeated with Calvinism, particularly under Elizabeth I and the Commonwealth. Now the new science and the new religion were felt to be fully congruent and, for the most part, they replaced the old religion and the old learning in England.

Most of the foundation members of the Royal Society accepted the Anglican Church of the Restoration. John Wilkins, the one-time Puritan, ended his days as Bishop of Chester and his pupil, Sprat, became Bishop of Rochester. Some held unorthodox opinions, like Isaac Newton and John Locke who were Unitarians, but they did not allow their views to become widely known. Few English scientists of the seventeenth century were to be found amongst the Nonconformists, who had been excluded from the universities and the professions by the Act of Uniformity (1662). Notable exceptions were the medical-chemist, Richard Lower, who was a Quaker, and William Whiston, who joined the Baptists when removed from the chair of mathematics at Cambridge for his heterodox beliefs. Since they were excluded from the universities, the Nonconformists set up their own educational institutions in which the teaching of science had an important place, and during the eighteenth century the Nonconformists became prominent amongst the scientists of Britain, as we shall see.

As in England, scientific societies were set up on the continent during the sixteenth and seventeenth centuries. Those formed in the old sixteenth-century regions of scientific activity, Germany and Italy, were ephemeral and did not last, while those set up in the new seventeenth-century centres of science, the Royal Society of London and the Paris Academy of Sciences, were more stable and enduring. The earliest scientific societies were those of Italy. The first notable society was the Academia Secretorum Naturae which met in Naples at the house of its president, Baptisa della Porta, 1538-1615, during the 1560's, but it was soon closed down on the charge of meddling with witchcraft. Next there was the Accademia dei Lincei at Rome which was active between 1601 and 1630 under the patronage of Duke Federigo Cesi. The society had a membership of thirty-two, including Galileo and della Porta, but it split up with the condemnation of the Copernican theory in 1615 and ended with the death of its patron in 1630. The last of the important Italian scientific societies of the time was the Academia del Cimento, the academy of experiments, which met in Florence between 1657 and 1667 under the patronage of the Medici brothers, Grand Duke Ferdinand II and Leopold. The academy had about ten members of whom the most notable were Viviani, 1622-1713, Borelli, 1608-70, and Redi, 1618-76. Viviani had worked with Galileo's pupil, Torricelli, 1608-47, with whom he had constructed the first barometer.

Redi was a biologist and showed, contrary to common belief, that insects were not generated spontaneously from decaying matter. Owing to the condemnation of the Copernican theory, the academy worked mainly upon experimental topics though Borelli ventured into the field of theoretical mechanics. Other members determined the velocity of sound and developed various scientific instruments, such as the thermometer for the measurement of temperature and the barometer for the estimation of atmosphere pressure. In 1667 Leopold Medici was made a cardinal and thereupon the academy was dissolved.

The German scientific societies of the time were even less important than those of Italy. The Societas Ereunetica was established at Rostock in 1622 by the botanist, Joachim Jung, 1587-1657, and the Collegium Curiosum sive Experimentale was founded in 1672 by Christopher Sturm, a professor of mathematics at Altdorf. However, neither of these societies outlived their founders and it was not until the eighteenth century that a stable scientific academy was set up in Germany. Mainly through the exertions of the scientist and philosopher, Leibniz, 1646-1716, the Berlin Academy was established by Elector Frederick I of Prussia in 1700. Through his influence also the St. Petersburg Academy was founded by Peter the Great in 1724. Neither of these societies was an immediate success as science did not have deep roots in Germany and Russia at the time. The first notable members of these societies were foreign scientists brought to Berlin and St. Petersburg respectively in order to staff the academies. During the eighteenth century the Berlin Academy was staffed notably with French *savants,* such as Maupertuis, 1698-1759, who was made secretary of the Academy by Frederick II of Prussia, Lamettrie, 1709-51, Voltaire, 1694-1778, and later Lagrange, 1736-1813. Indeed French was made the official language of the Berlin Academy in 1745. Similarly the St. Petersburg Academy was staffed at first with Swiss scientists, notably Nicolas Bernoulli, 1687-1759, Daniel Bernoulli, 1700-82, and Leonard Euler, 1707-83.

In France the development of scientific institutions followed a course similar to that pursued in England, though there were some important differences. French science was much more dependent upon patronage than English science during the seventeenth century and it was less concentrated geographically upon the metropolis. In Paris there was an institution somewhat similar to Gresham College in London, the Collège de France which had been founded in 1518 by Francis I to establish a home for humanism, then much opposed by the university of Paris. A professor of mathematics at the Collège de France, Orontius Finaeus, 1494-1555, prepared maps of the newly discovered lands and advised the explorers, just as his contemporary, Recorde, did in England. Later professors at the Collège de France, such as Gassendi, 1592-1652, and Roberval, 1602-75, were prominent in the scientific movement, as were the Gresham professors. The earliest notable scientific group in France met at Aix about 1620 in the house of Claude de Peiresc, 1580-1637, a wealthy ecclesiastic and a member of the *parlement* of Provence. Before he moved to Paris, Gassendi was a professor at Aix and a member of the Peiresc group. At Paris the cell of Marin Mersenne, 1588-1648, a Minorite friar like Gassendi, provided a meeting place for scientists and a centre for the interchange of scientific correspondence. Mersenne corresponded with Galileo, Descartes, and Hobbes, while at his cell met Fermat, Roberval, Gassendi, and Pascal. Later meetings were held in Paris at the house of Habert de Montmor, 1600-79, a Counsellor of state, the meetings becoming formalized about 1654.

The Montmor Academy gradually sank into financial difficulty and in 1663 Louis XIV's minister, Colbert, 1619-83, was asked for aid on the ground that the advance of science would benefit France economically. Colbert appreciated that the application of science would further his policy of expanding the industry and commerce of the country but he decided to set up a new scientific society under the patronage of the French crown. In 1666 the Paris Academy of Sciences was founded, the society consisting of about twenty members each of whom was paid a salary by the King. The Academicians were professional scientists working as a body upon problems set to them by the royal ministers. In contrast the Royal Society was a self-supporting association of amateur scientists who worked upon their own research problems. Colbert adopted an enlightened policy towards the Paris Academy of Sciences, issuing only general directives, but his successor, Louvois, ordered the work of the academicians towards narrowly practical, and even trivial ends, such as problems connected with the royal fountains and court games of chance. Louvois was also responsible for the Revocation of the Edict of Nantes (1685) which drove the Huguenot scientists out of the country. Many of the Huguenots fled to Switzerland but some came to England, notably the mathematician, de Moivre, and the physicist, Papin. In 1692 Louvois was succeeded by Pontchartrain who placed the Academy of Sciences under the direction of his nephew, Bignon. The Academy had grown considerably since its foundation, another thirty academicians being appointed between 1666 and 1699. Now it was further enlarged and reorganized. Bignon increased the number of members to seventy and arranged their posts hierarchically so that they enjoyed very different privileges and rights. As before, the academicians were salaried by the Crown, and the Academy as a whole was controlled by the royal ministers. The Paris Academy of Sciences retained such a form of organization until the French Revolution when it was again reorganized and given an egalitarian constitution.

During its early days the Paris Academy of Sciences, like the Royal Society, was influenced by the writings of Bacon, notably through the Dutch member of the Academy, Christian Huygens. The academicians adopted Bacon's suggestion of compiling 'histories' of natural phenomena and craft processes, producing collectively a natural history of animals and plants and later a large catalogue of machines and inventions. They also began the construction of a map of France and devoted much attention to the problem of determining the longitude at sea. As with the Royal Society, the influence of Bacon upon the Academy of Sciences declined fairly rapidly, notably in the time of Louvois who suspended work on the map of France and the longitude problem. The views of Descartes then became more popular, and the French interest shifted from the practical to the literary and philosophical aspects of science. Such a trend was first marked in the work of Fontenelle, 1657-1757, who from 1699 was secretary to the Paris Academy of Sciences for forty years.

Several literary and scientific societies were set up in the provinces of France during the seventeenth and early eight-

eenth centuries, some thirty-seven important provincial academies being founded by 1760. The societies mainly concerned
with science were those of the south of France; the Montpellier
society, founded in 1706, that of Bordeaux, set up in 1716,
and that of Toulouse, established in 1746, all of which were
affiliated to the Paris Academy of Sciences. Valuable scientific
work was carried out by the provincial French societies during
the eighteenth century but the events of the French Revolution
ended their importance and made Paris the main centre of
French science during the nineteenth century. In Britain the
converse was true. London became less and less the main focus
of British science, provincial centres appearing in the industrial
midlands and the north, the shift finding an organizational
expression in the rise of the provincial literary and philosophical societies during the late eighteenth and nineteenth centuries.

PART FOUR

EIGHTEENTH-CENTURY SCIENCE: THE DEVELOPMENT OF NATIONAL SCIENTIFIC TRADITIONS

Chapter 23

The Application of Science During the Eighteenth Century

AT THE end of the seventeenth century the problem of ascertaining the longitude of a ship at sea was still unsolved. In England the Fellows of the Royal Society devoted considerable attention to the problem but it was felt, as the Council of the Society put it in 1663, that 'the matter of navigation being a state concern was not proper to be managed by the Society', a principle which was restated in 1685 when Samuel Pepys requested that the records of the Society be examined for all references to navigation. The longitude problem and other navigational matters were of important national concern and accordingly a state institution was set up to deal with them, namely the Greenwich observatory built 1675-76. The foundation of the Greenwich observatory arose from the work of a Commission which had been set up to examine the possibility of determining the longitude by the method of Bennewitz from the position of the moon amongst the fixed stars. The Commission, composed of Sir Jonas Moore, the Surveyor of the Ordnance, Christopher Wren, Robert Hooke, and others, consulted in 1675 a young Cambridge astronomer, John Flamsteed, 1646-1719, on the feasibility of the method. Flamsteed expressed the opinion that the contemporary star catalogues and lunar tables were too inaccurate for the purpose required. Upon hearing of this opinion Charles II declared that he would have the positions of the stars and moon 'anew observed, examined, and corrected for the use of his seamen', and he appointed Flamsteed to the post of Astronomer Royal. Flamsteed had to equip Greenwich observatory with instruments before he could begin his work, some £2,000 of his own money being spent on such equipment according to his own estimate. During his forty-four years at Greenwich, Flamsteed prepared tables of the moon's motions and a catalogue of star places which were published posthumously in 1725. They were much more accurate than previous tables and catalogues but not accurate enough for the determination of longitude at sea.

The Paris Academy of Sciences, as a state institution, had its own observatory which was built 1667-72. Here the longitude problem was studied by Jean-Dominique Cassini, 1625-1712, an Italian civil engineer and astronomer who was placed in charge of the Paris observatory in 1669. He attempted to solve the longitude question by Galileo's method, preparing more accurate tables of the motion of Jupiter's moons, though his results were not of practical use. Another member of the Paris Academy of Sciences, the Dutch amateur, Christian Huygens, attempted to solve the problem by Frisius' method, namely the use of a mechanical clock for keeping standard time. Huygens was the first to construct a clock regulated by a pendulum, and in the course of so doing he studied the properties of various kinds of pendulum and the general question of motion along curved paths. In 1659 Huygens constructed a chronometer for use at sea but the motion of the ship interfered with the swing of the pendulum so that it did not keep accurate time. Later Huygens, and Robert Hooke independently, found that the oscillations of a spiral hair spring were isochronous and they suggested that such a spring could be used to regulate a time-keeper.

Neither Hooke nor Huygens seem to have made the suggested spring-regulated chronometer, and in general the seventeenth-century attempts to solve the longitude problem came to no practical conclusion. Accordingly the British government in 1714 set up a Board of Longitude which offered a series of prizes ranging from £10,000 to £20,000 for methods of finding the longitude to varying degrees of accuracy, and in 1716 the French government followed suit, offering a prize of 100,000 livres for a solution to the longitude problem. The British Board of Longitude made its first award to a German, Tobias Mayer, 1723-63, director of the observatory at Göttingen who in 1753 brought out tables of the moon's motions relative to the fixed stars which were accurate enough for approximate determinations of the longitude at sea by the method of Bennewitz. Such tables were printed in the Nautical Almanac which was published by the Greenwich observatory from 1767 on. The Bennewitz method as improved by Mayer was only accurate to within twenty miles at the best and each measurement of the longitude involved many hours of calculation, so that it was soon superseded when accurate mechanical clocks became available.

The construction of accurate chronometers for finding the longitude at sea was attempted by many persons during the eighteenth century, the most notable being John Harrison, 1693-1776, a Yorkshire watchmaker, and Pierre Le Roy, 1717-85, clock-maker to the King of France. Between 1728 and 1770 Harrison constructed five time-keepers, making them successively more accurate and gradually minimizing their inherent defects by sheer mechanical skill. The first four chronometers were massive machines. The fourth weighed sixty-six pounds, but it was accurate to within one-tenth of a second a day and it secured for Harrison a prize from the Board of Longitude in 1765. However, his fifth chronometer was no bigger than a large watch but it was just as accurate as his fourth machine, that is, to within one and a third miles of longitude in terms of distance. About the same time Le Roy in France finished his marine chronometer which was tested in 1763. Le Roy worked in a very different way from Harrison. Instead of minimizing defects by mechanical skill as Harrison had done, Le Roy attempted to eliminate them altogether by examining the fundamental principles of watch construction and devising new mechanisms which did not involve the inherent shortcomings of the current chronometers' movements. Harrison's achievement was that of a skilled craftsman, while Le Roy's work was more scientific, embodying both experiments and theoretical calculations.

With the work of Harrison, and more particularly Le Roy, the problem of finding the longitude at sea was finally solved, though it was some time before marine chronometers were produced in quantity. Now most of the fundamental technical problems relating to commerce and navigation had been cleared up, but meanwhile problems in the field of industry had become more acute. One of the most important of the industrial problems was the drainage of the mines. Coal had become more and more important as a fuel during the sixteenth and seventeenth centuries, owing to the scarcity of wood, which hitherto had been the main source of fuel used by the pyro-technical trades, such as soap-, and glass-making, and the metallurgical industries. Thus firewood prices in England increased nearly eight times between 1500 and 1640 while prices in general rose only threefold. The ship-builders had greater priority to the wood available during the sixteenth and seventeenth centuries, and so the pyro-technical trades switched over to coal fuels as far as possible in that period and completed the transition during the eighteenth century. Such a change from wood to coal fuels had the effect of bringing the pyro-technical trades from the south to the midland and northern coalfields of England and of greatly stimulating the mining of coal. In consequence coal mines were multiplied and they were pushed to greater and greater depths which made the removal of water more and more difficult. Three-quarters of the patents issued in England between 1561 and 1668 were connected directly or indirectly with the coal industry, 14 per cent of them being devoted solely to methods of draining the mines.

The solution to the problem of removing water from the mines was fundamentally a matter of finding a cheap source of power. Water-raising devices, such as the chain of pots, the suction and the force pumps, had been known from ancient times. Water could be raised from great depths by placing several pumps in series, but considerable power was required to drive such a series of pumps. During the sixteenth and seventeenth centuries the mines were drained mainly by horsepower as wind and water mills were not always reliable and they depended upon a suitable location—a river or windy place —which only occasionally coincided with the location of a mine. Georg Bauer in 1556 described a pumping device used in the German metal mines which required ninety-three horses to work it, and by the end of the seventeenth century as many as five hundred horses were being used to operate the pumps in some English mines.

The force of fire and heated substances had long been known and had been utilized as a minor source of energy. In antiquity Hero of Alexandria had used the power of hot air and steam to drive his mechanical toys, and Leonardo da Vinci during the Renaissance had left behind him a design for a steam-driven cannon. Towards the end of the sixteenth century

Baptisa della Porta described a device for raising water by means of steam pressure. The apparatus consisted of a closed vessel fitted with a pipe for the admission of steam which ejected water in the vessel through another pipe, so placed that one end was below the surface of the water while the other was open to the atmosphere. Solomon de Caus of France described a similar device in 1615, and David Ramseye of England obtained a patent from Charles I in 1630 for an invention to 'raise water from low pitts by fire', using the same method. About the same time Edward Somerset, the Marquis of Worcester, began his work on a 'water-commanding engine' which was said to have thrown water forty feet high when tested in the 1650's. Sir Samuel Moreland, Master Mechanic to Charles II, designed a similar machine in the 1670's, and finally Captain Thomas Savery, an engineer of Dartmouth, constructed in 1698 a steam pump based on della Porta's principle which was tested out in the mines. Savery's engine consisted of a closed vessel fitted with one pipe leading down to the water supply, a second connected to a steam boiler, and a third leading upwards to the delivery outlet. Steam was admitted into the vessel and allowed to condense. The partial vacuum produced then drew up water through the lower pipe, which was fitted with a one-way valve so that the water could not return. More steam was now admitted to the vessel, and the steam pressure drove the water up to the outlet through the delivery pipe, which was also fitted with a one-way valve. Such a cycle of operations was then repeated by the engineer in charge of the boiler. The Savery steam pump was not a success as high steam pressures had to be used in order to force the water up the delivery pipe with the consequence that there was considerable danger of the boiler bursting. Furthermore the pump had to be situated near to the bottom of a mine, at least within thirty feet of the low-water level, so that if the pump failed it quickly became flooded and inaccessible to repair.

Another line of attack upon the problem of producing mechanical power from heat commenced with the observation made by the sixteenth-century mining engineers that water could not be raised to a height greater than about thirty feet by means of suction pumps. Galileo thought it curious that nature should abhor a vacuum in this way only up to a certain limit, and he suggested that if this limit were constant other liquids should be sucked up to a limiting height which depended on their density. His pupils, Torricelli and Viviani, discovered in 1643 that such was the case, mercury with a density fourteen times as great as that of water rising to a height of about twenty-nine inches. Torricelli and Viviani filled a tube which had been closed at one end with mercury and inverted the tube with its open end immersed in a dish of mercury. They found that the level of the mercury in the tube sank until it was about twenty-nine inches above the level of the dish and that this height remained constant whether the tube were vertical or inclined to one side. Viviani suggested that the rise of mercury in the tube to the height of twenty-nine inches was due to the pressure of the atmosphere upon the free surface of the mercury in the dish, a view for which evidence was provided by Blaise Pascal, 1623-62, in 1648. Pascal ascended the Puy de Dome in southern France with the apparatus of Torricelli and Viviani and he found that the height of the mercury column decreased as he went up the mountain. From his experiment, Pascal presumed that the pressure of the mercury column was balanced by the pressure of the air on the surface of the mercury in the dish, the atmospheric pressure falling off as the mountain was ascended.

Independently of the Italian and French work, Otto von Guericke, 1602-86, a brewer and engineer of Magdeburg in Germany, followed a similar line of investigation between 1635 and 1645. Guericke attempted to produce a vacuum by pumping water out of a barrel with a suction pump but he found that air leaked into his barrel. Next he endeavoured to pump air out of a copper globe directly and this time he succeeded in creating a vacuum. Guericke found that the vacuum, or rather the pressure of the atmosphere, could exert a tremendous mechanical force. He showed that when two metallic hemispheres were fitted together and evacuated they could not be pulled apart by a team of sixteen horses, and that the pressure of the atmosphere on a piston closing an evacuated cylinder could not be overcome by the force exerted by twenty men. Thus it appeared from Guericke's experiments that if a vacuum could be produced by some non-mechanical means it should be possible to obtain considerable mechanical power from the pressure of the atmosphere.

Guericke's work was repeated and extended in England by Robert Boyle and Robert Hooke, who made numerous experiments with the air pump, and showed that the pressure of a given amount of air varied inversely with its volume. Later the French Huguenot physicist, Denis Papin, 1647-1712, came to England and acted as Boyle's assistant for a time. During that period Papin invented the pressure cooker, or 'digester' as he called it, and the safety valve, which was an essential part of his 'digester'. Papin suggested that mechanical power might be transmitted over considerable distances by means of a vacuum acting through a pipe. Hooke thought that the vacuum required for the purpose could be produced by the condensation of steam, and towards the end of his life (died 1703) he got into touch with Thomas Newcomen, 1663-1729, a blacksmith of Dartmouth, who was working on the matter. The details of Newcomen's work are not known, but some time in the first decade of the eighteenth century he developed his atmospheric steam engine, which was the first effective machine to convert heat into mechanical energy on a large scale.

The Newcomen engine, as generally adopted for the purpose of draining a mine, consisted of a cylinder closed by a piston which was fastened by means of a rod and chain to one end of a rocking beam, the other end of the beam being connected to the drainage pumps. Steam at low pressure was admitted into the cylinder and it was then condensed by means of a small spray of water. In this way a partial vacuum was produced in the cylinder so that the pressure of the atmosphere acted on the piston which pulled down the rocking beam and thus worked the pumps. The cycle of operations was then repeated, the opening of the requisite valves, and so on, being performed automatically by mechanisms connected to the rocking beam. The Newcomen engine was much superior to Savery's pump. It was placed at the pit-head, not at the bottom of the mine, and thus it could not be flooded. It required low-pressure steam, the mechanical work being performed by the pressure of the atmosphere, so that there was no danger of boiler explosions. Moreover, the Newcomen engine was a generalized prime mover, converting heat into mechanical energy, not a specialized steam pump like Savery's machine. The use of New-comen engines spread rapidly. From 1712 they were set up in collieries and metal mines all over Britain and after 1720 they were exported abroad.

The work of Newcomen on the steam engine, like that of Harrison on the marine chronometer, seems to have been largely empirical. These men had the skill required to put into practical effect the suggestions made by the gentleman-amateur scientists. Later in the eighteenth century such engineers and instrument makers absorbed more of the scientific method and then applied science for themselves. The earliest important figure in this movement was John Smeaton, 1724-92, who has been described as the first of the modern engineers. Smeaton designed and built bridges, harbours, canals, mills, and steam engines all over Britain, and he performed laboratory experiments with models of machines in order to improve the performance of the large-scale versions he was concerned with as a consulting engineer. Between 1752 and 1754 Smeaton made small-scale models of the old prime movers, the wind-mill and the water-wheel, and he varied the component parts from which they were constructed to discover upon what factors their efficiency depended. In this way he found experimentally that the overshot water-wheel was twice as efficient as the undershot wheel while various theoretical scientists had stated that it should be one-sixth to ten times more efficient.

Later Smeaton studied the Newcomen steam engine in a similar fashion. In 1769 he drew up a list of the sizes and performance of about a hundred of the atmospheric steam engines then used in Britain. He then constructed a small model of the engine and between 1769 and 1772 performed some hundred and thirty different experiments with it, varying each factor which affected the performance of the engine while the others were kept constant. By these means Smeaton obtained the best values for the length and diameter of the cylinder, the number of piston strokes per minute, the size of the boiler required, and the probable coal consumption, for an engine of any horse-power from one to seventy-six. By thus applying the quantitative experimental method, Smeaton greatly improved the machines he studied. However, he did not introduce anything that was new in principle. He did not apply any new scientific knowledge, while the next great engineer, James Watt, did.

James Watt, 1736-1819, was the son of a Greenock merchant. He studied the trade of instrument-making in London and took up the post of instrument-maker to the university of Glasgow in 1757. During the course of his duties, about 1763, he discovered that small models of the Newcomen engine were less efficient than the large-scale machines, a phenomenon which he discussed with Joseph Black, 1728-99, the professor of medicine at Glasgow. Shortly before, Black had found that different substances have different capacities for taking up heat, their so-termed specific heats, and that substances undergoing a change of state from the solid to the liquid or the liquid to the vapour absorb a great deal of heat, their so-called latent heats, though the temperature remained unchanged. Black and Watt saw that a considerable amount of steam was wasted in the Newcomen engine as each fresh charge of steam had to warm up the cylinder which had been cooled down during the previous condensation. From the latent heat of the conversion of steam into water and the specific heat of the cylinder materials, Watt calculated the amount of steam wasted by engines of various sizes and he showed that the loss was greatest in small models where the ratio of the surface area of the cylinder to its volume was larger than in the full-scale engines.

To prevent the wastage of steam, Watt kept the main cylinder of the steam engine at the same elevated temperature all the time, and condensed the steam in a separate chamber which was maintained at a low temperature, so that no steam was lost in rewarming the cylinder at each stroke. Black put Watt into contact with one of his friends, John Roebuck, 1718-94, who had set up the first large iron works in Scotland at Carron in 1760 and who was interested in manufacturing the new steam engine. Roebuck had connections with Birmingham where he had set up a sulphuric acid works in 1740, using the lead-chamber process which he had invented, and when Roebuck went bankrupt in 1775 his share in the Watt steam engine was taken over by Mathew Boulton, 1728-1809, a Birmingham manufacturer of light metal goods. From 1776 Boulton and Watt manufactured the new engine in partnership at Birmingham and Watt improved the machine further, notably by increasing its efficiency and adapting it to give rotary mechanical motion.

The Newcomen engine was still used in the collieries long after Watt's improvements as the saving of coal fuel was not an important consideration at the pit-head. But the Cornish tin miners, who had to bring coal from south Wales, used the Watt engine from the time of its inception and so too did the iron masters who required a powerful prime mover to work the blowing machines providing the blast for their furnaces. A reliable and efficient source of mechanical power was also required to drive the complex textile machinery which had been developed during the eighteenth century, and this too was provided by the Watt engine when it had been adapted to provide rotary motion. The new textile machinery and the iron-smelting innovations of the eighteenth century were primarily technical inventions, but the development of the steam engine was greatly dependent upon both the content and the method of science, of which it was perhaps the most important single application prior to the nineteenth century.

Chapter 24

The Background to Eighteenth-Century Science

A NOTABLE feature of the work on the longitude and mine-drainage problems was that the gentlemen-amateur scientists of the seventeenth and early eighteenth centuries failed to bring their suggestions to a successful practical conclusion, men who belonged to the craft tradition producing both the marine chronometer and the steam engine. The scientists of the seventeenth century had been all-round men, interested in pure and applied science. In both fields they had had considerable success, developing on the applied side such instruments as the thermometer, barometer, telescope, microscope, air-pump, electrical machine, and pendulum clock. But on the two important questions of the steam engine and the marine chronometer there seems to have been a slackening of effort towards the end of the seventeenth century, just when the solutions to those problems were in sight. Hooke and Huygens did not construct in a practical form the balance spring chronometer which they had designed, while the development of the atmospheric engine was left to the blacksmith, Newcomen. The scientists of the time were aware that they had failed to realize their earlier Baconian programme. In the opening year of the eighteenth century the Council of the Royal Society regretfully placed it on record that,

'the discouraging neglect of the great, the impetuous contradiction of the ignorant, and the reproaches of the unreasonable, had unhappily thwarted them in their design to perpetuate a succession of useful inventions'.

Such a slackening of effort was not confined only to the sphere of applied science. It extended also to the development of natural philosophy itself, for we find that the first half of the eighteenth century was a singularly bleak period in the history of scientific thought, compared with the period which had gone before and the period that was to follow. The malaise was noted at its inception. In the *Philosophical Transactions* of the Royal Society for 1698 the German philosopher, Leibniz, and John Wallis, now the sole surviving member of the original 'Philosophical College', discussed, as they put it, 'the cause of the present languid state of Philosophy'. They noted a marked fall in the standard of scientific discussions, and it seemed to them that amongst their younger contemporaries 'nature had few diligent observers'.

Scientific activity seems to have slackened in the period between the commercial expansion of the sixteenth and seventeenth centuries and the agrarian and industrial revolutions of the eighteenth. As we have noted, the merchants and overseas trading companies in England did much to promote science in the earlier period, while the industrial revolution was not without its influence upon science in subsequent times. In between there was a period of readjustment and reorientation in which science subsisted largely upon its own established tradition without much stimulus from external agencies. By the eighteenth century all was not well with the ventures which had profited from the great geographical discoveries. In England there was a series of crises leading to the spectacular collapse of the South Sea Company in 1720. In France John Law's Company of the Indies failed about the same time, while in Holland there was a series of similar collapses somewhat earlier. Agriculture and industry provided alternative and more stable forms of enterprise, and in Britain about the middle of the eighteenth century activity in these fields quickened. The gentlemen-landlords, who were usually of an Anglican persuasion, introduced the crop rotations and the improved livestock breeds of the agrarian revolution, while in the towns comparatively humble men, with the Nonconformists strong amongst them, engineered the industrial revolution. In France such ventures were much discussed, but much less implemented, owing to the rigid controls exercised by the French government. Thus it was that such discussions turned to criticisms of the *ancien régime,* which prepared the way for the French Revolution.

With the new movements of the second half of the eighteenth century science revived again, though now it had new characteristics. One of the most pronounced was the methodological division of science along national lines. Most of the natural philosophers of the seventeenth century had been concerned with the experimental, theoretical, and the applied aspects of science. Newton, for example, had made notable experiments in optics, developed the theory of universal gravitation, and invented a reflecting telescope and a nautical sextant. But during the eighteenth century English scientists were mainly experimentalists, and the French were primarily theoreticians, while applied science passed over from the gentlemen-amateur scientists to the instrument-makers and engineers of England and, to a lesser degree, of France. The English Astronomers Royal, Bradley, 1692-1762, and Maskelyne, 1732-1811, made notable empirical observations and discoveries, while the French scientists, Lagrange, 1736-1813, and Laplace, 1749-1827, developed the theory of mechanics and astronomy. In the same way Lavoisier, 1743-94, worked out the theory of the chemical revolution, using the experimental discoveries of the English scientists, notably Joseph Priestley, 1733-1804. The division was reflected to some degree even in the work of Harrison and Le Roy on the marine chronometer, as we have noted.

Such a division along national lines had been foreshadowed in the philosophies of Bacon and Descartes who had emphasized respectively the empirical and the deductive aspects of the scientific method. The difference between the work of the English and the French scientists had already become perceptible in the late seventeenth century though then it was not yet marked. Oldenburg, on a continental tour, noted in a letter to Boyle, 'Ye French naturalists are more discursive than active or experimental.' During the eighteenth century the division became profound and it may perhaps be said to have reflected the national preoccupations of the century; in England the industrial revolution, in France the events leading up to the political revolution. By developing experimental and applied science the English scientists and engineers gave an immediate impetus to the improvement of industrial technique. As we have seen, Black's experimental investigation of latent and specific heats found an immediate practical application in the new steam engine developed by Watt. France on the other hand produced but little of the technical equipment of the industrial revolution except perhaps in certain luxury trades, such as the Jacquard draw loom for weaving patterned cloths. The French were preoccupied with criticizing theoretically the doctrines of the established church and state, and with disseminating the philosophy of 'enlightenment'. The theoretical and rather unpractical spirit of the movement infected French science, for the criticism of the church and state was conducted in the name of science, and the philosophy of the enlightenment was nothing other than the Newtonian system and its extensions.

The French movement for the dissemination of enlightened opinion began notably with Fontenelle, 1657-1757, who popularized the theories of Descartes, and it was developed by Voltaire, 1694-1778, who brought the Newtonian system to France. The most important product of the movement was the great French *Encyclopédie* which was published in twenty-two volumes between 1751 and 1777. Many encyclopedias appeared during the eighteenth century, such works in England and France expressing the division of interests between the two countries. The English encyclopedias were primarily technical at first, like Harris' *Lexicon Technicon* (1704) and Chambers' *Dictionary of the Arts and Sciences* (1714), but later they became wider in scope, leading finally to the *Encyclopædia Britannica,* the first edition of which was published at Edinburgh in 1771. The French encyclopedias on the other hand were at first primarily critical and theoretical, like the *Historical and Critical Dictionary,* published in 1695 at Amsterdam by Pierre Bayle, a French Huguenot refugee.

Besides the division of theory and experiment between the French and the English in the science of the eighteenth century, there were marked shifts in the centres of scientific activity and changes in the social background from which scientists came. Apart from Napier and the Gregorys in Scotland, most of the important British scientists of the seventeenth century had come from places south of the line joining the Severn to the Wash. Now in the second half of the eighteenth century scientists were appearing in Scotland, the industrial midlands, and the north country. Such men were no longer connected with the trading companies or the landed gentry, like Robert Boyle, who had been a director of the East India Company, or the first President of the Royal Society, Lord Brouncker, a *nouveau riche* who had been ennobled by Charles II, but they were now sons of weavers, like Joseph Priestley, and John Dalton, 1766-1844, or craft apprentices, like Humphry Davy, 1778-1829, and Michael Faraday, 1791-1867. Moreover, many of these new men were Nonconformists, not Anglicans. Joseph Priestley was a Unitarian minister, John Dalton a Quaker, and Michael Faraday a Sandemanian. Again these men were mainly interested in chemistry and electricity, and were concerned only to a lesser degree with the sciences of astronomy and mechanics which had been the main foci of attention during the sixteenth and seventeenth centuries.

As in Britain, scientific activity appeared in new regions on the continent during the eighteenth century, though there was not the same marked change in the social background from which the continental scientists came. French scientists still came notably from families connected with the state bureaucracy, though members of the French *parlements*, or law courts, were not so prominent in science as they had been, perhaps because the *parlements* by now had lost much of their importance. However, men from lawyer families or with a training in law, like Fontenelle and Voltaire, still had an important place in the scientific movement. Switzerland became an important centre of scientific activity during the eighteenth century, probably because it was the refuge of so many Protestant scientists who had left Catholic lands; men like the Bernoulli family of mathematicians who had left Flanders and settled at Basle, and the French Huguenot families, such as the Trembleys, the Saussures, and the Candolles, who settled at Geneva. Sweden, with her rich iron ore and extensive timber preserves, became an important centre of iron production in the eighteenth century, and at the same time she produced scientists who, like those of Britain, came from a comparatively humble background: men such as Linnaeus, 1707-78, and Berzelius, 1779-1848, who were pastors' sons, and Scheele, 1724-86, who was an apothecary.

The older centres of science, notably Paris and London, remained active during the eighteenth century, though the science of Holland stagnated with her prosperity. The Royal Society had lost the youthful vigour of its seventeenth-century days but it still contained eminent scientists of the gentleman-amateur tradition, such as Henry Cavendish, 1731-1810. Generally speaking, the new type of Nonconformist scientist from the industrial north was elected to the Society, indeed the increase in the number of Quakers who became Fellows of the Royal Society affords a good index to the rise of science amongst the Nonconformists. During the seventeenth century there were four Quaker Fellows of the Royal Society, in the eighteenth there were fourteen, and in the nineteenth thirty-six. The proportion of the Quaker population of Great Britain elected to the Royal Society was about the same as that of the Anglican during the seventeenth century, but it was four times as great in the eighteenth and thirty times as great during the nineteenth. The new type of British scientist coming from a Nonconformist or craft background found the atmosphere of the Royal Society somewhat uncongenial at first. Some, like Humphry Davy, adapted themselves to the traditions of the Society, striving at the same time to transform it from a London club into a national organization for the promotion of science in Britain. Others, like Joseph Priestley, turned away from the Society and set up their own scientific institutions in the midlands and the north to cater for their own local needs.

The various Protestant sects which had refused to conform to the requirements of the 1662 Act of Uniformity had formed, or had derived from, the radical wing of the earlier Puritan movement. At that time their secular activities had been mainly of a political nature, and their composition had been heterogeneous, though even then it had inclined towards the less wealthy classes. In the late seventeenth and early eighteenth centuries the Nonconformist sects, particularly the Quakers, assumed a composition which was more sharply lower middle class in character, and they reorientated their secular activities to the development of the crafts, taking a leading part in the industrial revolution. As the universities were closed to them, the Nonconformists set up their own educational establishments, some of which reached a university standard of teaching during the eighteenth century. Many Anglicans chose to go to the Nonconformist Academies rather than the universities, as more modern subjects were taught in the former, including, in particular, a great deal of science. Some of the tutors at the Nonconformist Academies became notable sicentists, such as Joseph Priestley, who taught at the Warrington Academy, and

John Dalton, who was a tutor at the Manchester New College.

The Nonconformists were particularly prominent in the development of the heavy industries, such as the iron industry. Here the Quakers all but dominated the field during the eighteenth century, as the Anglican ironmasters tended to buy estates and become landed gentlemen, leaving the iron industry to the Quakers, who, rejecting the principle of hereditary aristocracy, did not aspire to join the landed nobility. One of the more important of the Quaker iron-master families was the Darbys of Coalbrookdale, who invented the method of smelting iron with coke in place of the traditional wood charcoal, which by now had become scare and costly. The Darby family also developed a method of casting iron in moulds of sand, and they possessed the monopoly of casting steam engine cylinders from 1724 to 1760. Other Nonconformist inventors were the Presbyterian, James Watt, his first partner, John Roebuck, who was an Independent, and John Wilkinson, whose precision boring machine made the Watt engine a commercial success.

The men of the industrial regions with their scientific education at the Nonconformist Academies and their technical interests founded institutions to promote the arts and the sciences in their own localities. One of the earliest was the Lunar Society, founded at Birmingham by Mathew Boulton about 1766. Its membership included Boulton and Watt; the chemist, Joseph Priestley; Erasmus Darwin, 1731-1802, a physician, poet, and naturalist, who anticipated some of the evolutionary theories of his grandson, Charles Darwin; James Keir, 1735-1814, the owner of a chemical works at Bromwich; Samuel Galton, a manufacturer of iron goods who had been ejected from the Quakers for making guns; John Baskerville, 1706-75, who had improved printing type; William Murdoch, 1754-1839, an engineer of the Boulton-Watt works who invented gas lighting and a steam locomotive; two physicians of Birmingham, Dr. William Withering, 1741-99, and Dr. William Small, 1734-75; and two literary men, Thomas Day, 1748-89, and Richard Lovell Edgworth, 1744-1817. The Lunar Society met at the house of one or other of its members once a month on the night of the full moon, whence it derived its name, so that the members could find their way home with facility. The Society flourished until 1791 when riots were fomented in

Birmingham against Nonconformists and supporters of the French Revolution, a category which included several members of the Society. Priestley's house was wrecked and his scientific apparatus and library were destroyed. Shortly afterwards he emigrated to America, and others having moved away from Birmingham the Lunar Society came to its end.

Another important provincial scientific institution which was established in the same period was the Manchester Literary and Philosophical Society. The Manchester Society arose from meetings of scientists and industrialists which were held at the Nonconformist Academy in Warrington. The meetings of the Society were regularly recorded from 1781 and papers read at the meetings or submitted to the Society were published from 1785. It was a larger organization than the Lunar Society, possessing some forty foundation members, and it was more stable, thriving to the present day. The main founder and first President of the Manchester Society was Thomas Percival, 1740-1804, who had studied at Warrington Academy under Joseph Priestley and practised as a physician at Manchester. Another founder of the Society and its first Secretary was Thomas Henry, 1734-1816, a Manchester apothecary, whose son, William Henry, 1774-1836, discovered in 1803 the gas law named after him.

In its early days the Manchester Society was greatly concerned with the science of chemistry, a subject which was of considerable importance locally in connection with the bleaching and dyeing of textiles. The earliest eminent scientist associated with the Society was a chemist, namely, John Dalton, who presided over the Society from 1817 to 1844. The importance of chemistry was stressed even in the constitution of the Society where it was listed as a topic for discussion separate from natural philosophy in general. Law VIII of the Society runs:

'That the subjects of conversation comprehend natural philosophy, theoretical and experimental chemistry, polite literature, civil law, general politics, commerce, and the arts.'

The constitution also indicates that the Society was not averse to the discussion of law, politics, and commerce, subjects with which the Royal Society had declared it would not meddle.

Like the Lunar Society the Manchester Society contained several members who supported the French Revolution, indeed two of them, Thomas Cooper and James Watt junior, were elected deputies of the French National Assembly. In consequence the Manchester Society also experienced some difficulties in the 'King and God' riots of 1791, but unlike the Lunar Society it weathered the storm. During the nineteenth century the Manchester Society became an important institution of British science, and it served as the prototype of the other provincial literary and philosophical societies which were formed at the rate of one, five, and even ten per decade in the Victorian period.

In Scotland the Philosophical Society of Edinburgh was founded in 1732, the Society becoming important from about 1783 when it received a royal charter. Amongst the members of the Society were the philosopher, David Hume, 1711-76, Adam Smith, 1723-90, the economist, Joseph Black, 1728-97, professor of medicine first at Glasgow and then at Edinburgh, James Hutton, 1726-97, who first put forward the theory of geological evolution in a paper read to the Society in 1785, and John Playfair, 1748-1819, a professor of natural philosophy at Edinburgh who, together with the scientific amateur, Sir James Hall, 1762-1831, extended and illustrated the Huttonian theory.

The universities of Presbyterian Scotland, like the Nonconformist Academies of England, were noted for the teaching of the sciences during the eighteenth century: in fact Edinburgh became famous for her medical school from that time. The Scots scientists, again like the Nonconformist scientists of England, were in contact with the industrial development of their time. Hutton became a chemical manufacturer, while Black advised Watt and placed him in contact with the industrialist, Roebuck. However, in one respect the Scots scientist of the late eighteenth century differed from their English colleagues; they concerned themselves more with theoretical science. Such is apparent in the work of Hutton on the theory of geological development, while Playfair criticized the lack of British response to the French theoretical work on mechanics and astronomy. But at the turn of the century British science as a whole became more theoretical again. From 1801 Thomas

Young, 1773-1829, revived the wave theory of light and John Dalton in 1803 introduced the atomic theory into chemistry, both men incidentally coming from Quaker families. In a complementary way the French scientists became more empirical about the same time, spurred on to experimental and applied science by the needs of Napoleon's armies.

Chapter 25

Astronomy, and the Newtonian Philosophy in the Eighteenth Century

AFTER NEWTON had published his *Principia* in 1687, there was little development of the theory of universal gravitation for another half century, and it was another hundred years before men of his calibre appeared again. Some of the men who had been working towards the theory at the same time as Newton, notably his friend, Edmund Halley, 1656-1742, applied the idea to particular cases—Halley to the comet which bears his name—but in the succeeding generation there were few important theorists in astronomy. The Newtonian theory was generally accepted by the English scientists of the period, and it was taught at Cambridge by Newton's successor, William Whiston, and at Edinburgh by David Gregory. However, the theories of Descartes were still popular at Cambridge, for Whiston wrote that he went up to Cambridge in 1693 in order to study

'Particularly the Mathematicks, and the Cartesian philosophy: which was alone in Vogue with us at that time. But it was not long before I, with immense Pains, but no Assistance, set myself with the utmost Zeal to the study of Sir Isaac Newton's wonderful Discoveries.'

Newton's theories for the most part were accepted by the Anglican ecclesiastics, though some of his followers were regarded with suspicion. Edmund Halley in 1694 submitted a paper to the Royal Society in which he suggested that the body which we now call Halley's comet had at one time approached near the earth, and had raised a huge tidal wave that constituted Noah's Flood. He did not publish his paper until 1724 as he was 'Apprehensive lest he might incur the censure of the sacred order'. Whiston in 1696 put forward a similar theory, and for this and other heterodoxies he was removed from his chair in 1701. The philosopher, Bishop

Berkeley, 1685-1753, considered these suggestions offensive to religion, and he advanced the opinion that the Newtonian theory was a piece of grammatical pedantry.

'As in reading other books', Berkeley wrote in 1709, 'a wise man will choose to fix his thoughts on the sense and apply it to use, rather than lay them out in grammatical remarks on the language: so in perusing the volume of nature, it seems beneath the dignity of the mind to effect an exactness in reducing each particular phenomenon to general rules, or showing how it follows from them.'

Newton's *Principia*, Berkeley thought, was one of the worst examples of this 'grammatical' study of nature.

In 1734 Berkeley attacked the differential calculus discovered by Newton in a work entitled *The Analyst: or, a Discourse addressed to an Infidel Mathematician*, that is, to Halley. The calculus dealt with such problems as finding the slope of a curve, by taking a small section and reducing it to a point mathematically, from which the slope of the curve at that point could be obtained. Berkeley considered that such a device was unsound for he maintained that a point could not have a slope. Comparing mathematics with religion he wrote:

'Do not mathematicians submit to authority, take things upon trust, and believe points inconceivable? Have they not their mysteries, and what is more, their repugnancies and contradictions?'

Such an attack was not without its positive side, for it stimulated British mathematicians to enquire into the logical foundations of the calculus, and to place what was obviously a useful mathematical technique upon a more secure foundation. Within a year of the publication of Berkeley's *Analyst*, several of Newton's followers had published answers to the bishop, and indeed had criticized each other's answers. Finally Colin Maclaurin, 1698-1746, a professor of mathematics at Edinburgh, found a reasonably sound basis for the calculus in 1742.

On the continent the acceptance of Newton's discoveries was much slower. In the Paris Academy of Sciences even the theories of Descartes were suspect until the late seventeenth century. However, when the Academy was reorganized in 1699 Fontenelle, the reigning popularizer of the Cartesian philosophy, was made secretary, a post he held for forty years. During that period Descartes' theories were dominant in the Academy of Sciences: the first time a prize was awarded for a Newtonian memoir by the Academy was in 1734, and the last time a prize was given for a Cartesian work was in 1740. With Fontenelle we find the beginnings of a tendency to extend the mechanical philosophy from the physical world to the world of man, a tendency which was to become marked in France later during the century.

'The geometrical spirit', Fontenelle wrote in 1699, 'is not so tied to geometry that it cannot be detached from it and transported to other branches of knowledge. A work of morals, or politics, or criticism, perhaps even of eloquence, would be better, other things being equal, if it were done in the style of a geometer.'

The 'geometrical spirit', as Fontenelle understood it, was the viewpoint of the mechanical Cartesian philosophy.

The movement begun by Fontenelle was extended by Voltaire, 1694-1778, who was more critical of the established institutions of Church and state than Fontenelle, and who brought to France and popularized the more 'enlightened' philosophy of Newton. After his second sojourn in the Bastille for his critical activities, Voltaire stayed in England for the period 1726-29, during the course of which he became acquainted with English thought and institutions. Upon his return to France, Voltaire published in 1734 his famous *Letters on the English*, in which he dealt with such diverse topics as the philosophy of Bacon and Locke, Quakerism, the character of the British Government, and the practice of inoculation, then recently introduced from the Middle East by Lady Mary Whortley Montague. But what was more important for French science, Voltaire discussed extensively in four chapters the natural philosophy of Newton which he compared favourably with the Cartesian system. Voltaire's praise of all things English was a criticism of French thought and institutions, and ac-

292 / A History of the Sciences

cordingly his *Letters on the English* were condemned and burnt. Voltaire himself fled to the independent Duchy of Lorraine with Mme de Chatelet, with whom he wrote a popular account of the Newtonian theory, which was published in 1738.

The Newtonian theory raised a storm of controversy in the Paris Academy of Sciences during the 1730's, for it was vigorously opposed by the directors of the Paris observatory, the Cassini family. The first of them, Jean Cassini, 1635-1712, had been brought from Italy in 1668 at the invitation of Louis XIV. He was a supporter of the old geocentric theory of the solar system, and both his son and grandson opposed aspects of the Newtonian theory. A French expedition to Cayenne in 1673 had found that the length of a pendulum beating out seconds near the equator was shorter than the seconds pendulum in Paris: they were 990, and 994 mm. respectively. Newton had interpreted this to mean that the fôrce of gravity was smaller at the equator than near the poles, which implies that the earth was an oblate spheroid, flattened at the poles, and bulging at the equator. He suggested that such a shape had resulted from the rotation of the earth on its axis when it was in a plastic condition. In the 1730's when Newton's natural philosophy came to France the second Cassini, Jacques, 1677-1756, opposed his theory of the shape of the earth. From the lengths of a degree of latitude which had been measured at Dunkirk in the north and Perpignan in the south of France, Cassini maintained on the contrary that the earth bulged at the poles and was flattened at the equator.

It was pointed out, when the question arose in the 1730's, that measurements of a degree of latitude were required at points much further north and south to find out accurately what the shape of the earth actually was. Accordingly Louis XV authorized the Academy of Sciences to send out two expeditions, one to Lapland, and the other to Peru, to ascertain the lengths of a degree of latitude in these regions. The Peru expedition set out in 1735, and the Lapland party a year later. When the leaders of the Peru group returned in 1744 it was found that Newton was essentially correct, the length of a degree in Peru being 110,600 metres, and in Lapland 111,900 metres. Maupertuis, the leader of the Lapland party, had been sceptical of Newton's views, but now he was quite convinced. Finally Clairault, also of the Lapland party, accounted quantitatively for the flattening of the earth at the poles using the Newtonian theory.

The son of a Paris mathematics professor, Alexis Claude Clairault, 1713-65, and the natural son of a French general, Jean le Rond D'Alembert, 1717-83, found abandoned as a baby on the steps of the church, Jean le Rond, were amongst the first to accept and develop Newton's theories in France. Others on the continent were the Basle group, Daniel Bernoulli, 1700-82, John Bernoulli, 1710-90, and Leonard Euler, 1707-83, who more particularly developed the calculus, discovered by Newton and Leibniz, and applied it in astronomical analysis using the Newtonian theory. Newton himself had used Euclidian geometry which was more readily comprehended by the men of his time. These men were particularly concerned with the analysis of the moon's motions in connection with the longitude problem: in fact it was with Euler's lunar theory that Tobias Mayer made the tables which obtained for him in 1755 the prize offered by the British government for finding the longitude.

With the calculus these men examined in detail problems that had only been touched upon by Newton. One such was the three body problem: what are the expected motions of, say, the earth, sun, and moon, under the mutual influence of each other's gravitational fields? For two bodies the problem was simple, but for three and more bodies it was complex, and methods of successive approximation had to be used. Clairault at one time(1747) had his doubts about the theory of gravitation, for he found that the nearest distance the moon approached to the earth was theoretically only a half of what was observed, but on improving his calculations he found in 1749 that the theory agreed with observation.

The successors of Clairault and D'Alembert, Lagrange, 1736-1813, and Laplace, 1749-1827, carried the analysis of the Newtonian theory of the solar system to much greater detail. Laplace was the son of a Normandy farmer. In 1767 he went to Paris and presented D'Alembert with a treatise on the principles of mechanics. D'Alembert was impressed by the work and obtained Laplace a post at the Paris military school, where he taught mathematics to Napoleon Bonaparte, amongst others. Lagrange was the son of a French officer in Turin. He obtained a post first as a member of the Berlin Academy, and then of the Paris Academy of Sciences. The work of Lagrange and Laplace covers the period from 1770 to their deaths, Lagrange contributing notably to the theory of mechanics and Laplace to the theory of astronomy.

The main result of their work was to show that the solar system was a perfectly self-regulating mechanism, in which irregularities were self-correcting. Newton had observed anomalies in the movements of Saturn and Jupiter, which he could not completely explain, and for this and other reasons, he had thought that God was always present in the universe to correct such irregularities. Laplace, however, declared that he had no need for such a hypothesis. He showed first that the apparent acceleration of the moon's motion was a self-rectifying phenomenon, the acceleration reversing in about 24,000 years. Then he demonstrated that the large variations in the velocities of the planets, Jupiter and Saturn, were the result of their gravitational interactions. Finally he showed that all such irregularities were self-correcting, for, as he put it,

'From the sole consideration that the motions of the planets and satellites are performed in orbits nearly circular, in the same direction, and in planes which are inconsiderably inclined to each other, the system will oscillate about a mean state, from which it will deviate but by very small quantities.'

It appeared from the work of Laplace that the universe had no history: it was a perfectly self-regulating machine which had been working over indefinitely long periods in the past, and which would persist indefinitely in the future. However, Laplace devised another and more qualitative picture of the universe which incorporated the idea of evolution that was lacking in his quantitative model. He noted that all of the planets rotated in the same direction round the sun in orbits which were more or less in the same plane. Moreover, the spins of the sun, the seven planets, and their fourteen satellites, were all in this same direction, and all of these bodies were close together compared with their distances from the nearest stars.

Laplace was of the view that such an arrangement of the bodies in the solar system could not have arisen by chance, and he endeavoured to account for it by assuming that the sun, planets, and their satellites had all originated from a primordial mass of gas. He supposed that the hot mass of original nebulous gas had rotated round its centre from the beginning and that it had gradually cooled and contracted. As the mass contracted its rate of rotation increased, because the total angular momentum of the nebula remained constant. The increasing rate of rotation caused a nebulous ring to separate off from the equator of the mass, a process which happened repeatedly giving several concentric rings lying in the same equatorial plane. Each ring broke up and was reformed into a spherical body, namely a planet, whilst the residual central core of the nebula formed the sun. Each planet in turn threw off one or more rings, which condensed to form spherical satellites except in the case of the planet Saturn where the rings still remain.

The nebula hypothesis, as the theory was termed, encountered the objection that nebulous rings split off in the way Laplace had suggested would condense into several bodies, not just one, and furthermore the distribution of momentum in the solar system was not accounted for satisfactorily by the theory. Even so it was very popular throughout the nineteenth century, perhaps because it was agreeable to the general evolutionary conceptions of the period. For Laplace, and for scientists in general, his work on the stability of the solar system was the more important. Newton's inverse square law of the variation of gravitational force with distance came to be considered as the fundamental law of the universe and the archetype of other laws. The intensity of light and radiant heat seemed to vary inversely with the square of the distance from the source, and towards the end of the eighteenth century it was discovered that magnetic and electrical attractions and repulsions obeyed the inverse square law. 'The inverse square law of force', wrote Laplace, 'is that of all the emanations which come from a centre like that of light.'

Such a pattern of inverse square law forces acting across empty space between the particles of matter, seemed to be ubiquitous in nature at the turn of the eighteenth century.

The world appeared to be entirely comprehensible in terms of the Newtonian type of laws, and completely determinate. Thus Laplace in 1812 put forward his famous conception of a Divine Calculator who, knowing the velocities and positions of all the particles in the world at a particular instant, could calculate all that had happened in the past, and all that would happen in the future. Such a conception illustrates the theoretical tendency of French science at the time. The Divine Calculator was the mathematical physicist writ large, a being who appeared to be outside of the system he was investigating, and who contemplated the world as though it were a play, deducing from the events of the moment the preceding and the subsequent action. Indeed Laplace himself declared that the problems of astronomy must be considered 'as a great problem of mechanics, from which it was important to banish as much as possible all empiricism'.

The biologist, Bonnet, 1720-90, somewhat earlier had put forward a similar conception in the sphere of psychology. He suggested that if an Intelligence could have analysed the workings of all the fibres in Homer's brain, he would have been able to picture the *Iliad* as it was conceived of by the poet. However, the mathematician, Condorcet, 1743-94, perceived that there were drawbacks to the idea when it was extended to the human sphere. The physical, organic, and human worlds would be identical in principle, he wrote in 1782,

'for a being who, as a stranger to our race, studied human society as we study the beaver and the bee . . . But here the observer is part of the society he observes, and truth can only be judged, imprisoned, or bribed.'

Curiously enough it was becoming apparent that observations depended on the observer as well as the object observed about the same time as Laplace propounded the conception of the Divine Calculator. From the seventeenth century on it had been known in astronomy that the observation of the same object by different observers often gave slightly differing results. The British Astronomer Royal, Maskelyne, 1732-1811, dismissed one of his assistants because their observations always differed by the same small amount, and he thought the assistant was wrong. By the end of the eighteenth century it was appreciated that observers as well as instruments had their own errors, and techniques were devised for obtaining the most accurate measurement by averaging several observations. The most notable method was that of least squares, which was developed by Legendre, 1752-1833, in 1806, and Gauss, 1777-1855, in 1809. Gauss and Laplace tried to prove that this method gave truly 'objective' measurements which would be independent of any observer, but they established nothing that was conclusive on the matter.

Whilst the French were developing the theory of astronomy the English were engaged upon important observational and experimental work. The French of course had their observers but their work was not so important as that of the English. The third British Astronomer Royal, James Bradley, 1692-1762, made numerous observations of the positions of the fixed stars in an attempt to discover the shift in the apparent relative positions of the stars with the movement of the earth, which was to be expected according to the Copernican theory. He did not find such an effect, but he did discover in 1728 what is termed the aberration of light, that is, the apparent annual movement of each star in a small ellipse, owing to the fact that the earth moves and that the light from the stars takes some time to reach the earth. Later, in 1748, he discovered what is called the nutation of the earth's axis, a rotational wobble of the earth's axis in space, superimposed upon the larger precessional wobble, which again gave rise to an apparent change in the positions of the fixed stars.

The fifth Astronomer Royal, Nevil Maskelyne, measured the mass of the earth in 1774. On either side of a fairly steep granite mountain in Perthshire, he measured the deviation of a plumb line from the vertical caused by the gravitational attraction of the mountain. Maskelyne estimated the mass of the mountain from its volume and the density of granite, and thus he was able to find the mass of the earth by comparing the pull of the earth and that of the mountain on his plumb lines. He found that the mean density of the earth was roughly 4·5 times that of water. The same problem was studied experimentally by the wealthy amateur scientist, Henry Cavendish, 1731-1810, in 1798, using a method suggested by a

Cambridge clergyman, John Michell, about 1750. Cavendish fastened two small lead balls to a right rod, one at either end, and suspended the rod by means of a fine wire attached to its middle. He then brought up two large lead spheres towards the small lead balls and measured the force of gravitational attraction between them by means of the twist imparted to the fine suspension wire. Comparing the force so measured with the force of gravity at the earth's surface, Cavendish obtained values for the mass and the density of the earth. He found that the mean density of the earth was 5·5 times that of water, a figure which is near to the modern estimate.

Hitherto astronomers had concerned themselves primarily with the motions and arrangements of the bodies composing the solar system, the fixed stars being regarded as a kind of spherical map or background relative to which the motions of the planets could be conveniently measured. However, more and more attention was devoted to the study of the nature, arrangement, and possible motions of the fixed stars during the early modern period which culminated in the opening up of the new field of sidereal astronomy in the eighteenth century. Copernicus had been forced to place the traditional eighth sphere of the fixed stars at a very great distance from the bodies of the solar system in order to account for the lack of stellar parallax. Thomas Digges, expounding the Copernican theory in 1576, did away with the notion of an eighth sphere and supposed the sidereal region of the heavens to be infinite, being 'garnished with lights innumerable and reaching up in Sphaericall altitude without end.' Giordano Bruno then suggested that each of the innumerable stars in the infinite void was the sun of a planetary system, a view which was quite widely accepted in the seventeenth century. The generally received theory of the universe, wrote William Derham, a canon of Windsor, in his *Astro-Theology* (1715),

'is the same as the Copernican as to the system of the sun and planets. . . . But then whereas the Copernican hypothesis supposeth the Firmament of the Fixt stars to be the bounds of the Universe, and to be placed at an equal distance from its centre, the Sun; the new system supposeth that there are many other systems of suns and planets, besides that in which we have our residence; namely, that every Fixt star is a sun and encompassed with a system of planets, both primary and secondary.'

If the planets moving round the sun formed a system, then it seemed possible that the stars spread throughout space might form a system or a series of systems. Such a view was first put forward by the amateur astronomer, Thomas Wright of Durham, 1711-86, in 1750. He suggested that the sun and the stars of the Milky Way formed a giant sidereal system, all moving round a common centre. Wright thought that there might be a large heavenly body at the centre of the sidereal system just as, on a much smaller scale, the sun was the centre of the solar system. The theory of Wright was extended by the German philosopher, Immanuel Kant, 1724-1804, who in 1755 suggested that the nebulae were vast systems of stars like the Milky Way and all the sidereal systems in the universe might move round a common centre and compose an even vaster system. Kant supposed that in the beginning there was a chaos of material particles which were endowed with the properties of attraction and repulsion, and that through the property of attraction these particles conglomerated into the heavenly bodies, while through the property of repulsion the heavenly bodies were set in vortex motion, forming solar systems, sidereal systems, and ultimately the vast cosmic system.

Such speculations attracted the attention of some observational astronomers, the most notable of whom was William Herschel, 1738-1822. Herschel was a native of Hanover and he came over to England, where he settled in 1757, as a musician in the Hanoverian Guards. He quickly gained a reputation as an astronomer and in 1782 George III gave him a salaried post as the King's Astronomer. Herschel was a skilled telescope maker, constructing what were for that time giant reflecting telescopes, some of which he sold and some of which he used himself. With such instruments Herschel systematically studied the motions and the distribution of the stars in space, a field that was then quite new. He divided up the celestial hemisphere visible to him into a number of areas which he explored in detail, noting the number and the character of the stars in each. Herschel made four of such studies which he termed 'reviews of the heavens'. During the course of the second review he discovered in 1781 another planet beyond Saturn, namely Uranus. He also noted in this review some 269 pairs of stars revolving round one another which seemed to indicate that the force of gravity operated in the sidereal system as well as in the solar system. In the most complete of his reviews, the fourth, which was started in 1783, Herschel assumed that the intrinsic brightness of all the stars was the same, so that the apparent brightness of a star could be taken as a measure of its distance from the solar system. By observing the number and the brightness of the stars in various areas of the celestial hemisphere, Herschel then found that the stars were not symmetrically distributed throughout space either in distance or direction. There were local clusterings of the stars in space, the most notable being that of the Milky Way which had the form of a bun or disc. Herschel presumed that the stars of the Milky Way system were revolving round a common centre, as he discovered that the sun was moving relative to the other stars by observing that the distribution of stars in front of the solar system was opening up while the stars behind were closing in upon one another.

Herschel thought that most of the stars, with the exception of the star-pairs, possessed planetary systems and that most of the nebulae were sidereal systems like the Milky Way or such systems in the process of formation. Following Kant, Herschel arranged the two thousand nebulae he had observed into four classes showing an evolutionary sequence. According to Herschel, the first formed were the irregular, bright, and diffuse nebulae, composed of 'some shining fluid unknown to us', from which the star clusters had condensed out. The star clusters then evolved into the bun or disc shaped nebulae, like our Milky Way, and these in turn would become globular ultimately, owing to gravitational attraction.

William Herschel worked very much alone in the field of sidereal astronomy. For the most part, the men who bought his telescopes did not study the stars, perhaps because they were mainly gentlemen-amateurs who belonged to a tradition which had been laid down in the seventeenth century, and who worked in the field of telescopic planetary astronomy which had been opened up in that period. From the middle of the nineteenth century sidereal astronomy attracted an ever increasing number of observers, Herschel's work in the meantime being continued notably by his son, John, 1792-1871, who between 1834 and 1838 examined at Cape Town the stars of the southern celestial hemisphere not visible to his father.

Chapter 26

The Phlogiston Theory and the Chemical Revolution

WHEN NATURAL philosophy began to show signs of stagnation towards the end of the seventeenth century the nascent science of chemistry was particularly affected. Boyle and the English school of medical chemists which flourished in the 1660's and '70's failed to establish much of a tradition, and their achievement did not attract the attention it merited in the period immediately subsequent. It was noted by William Wotton in his *Reflections upon Ancient and Modern Learning* (1694) that medical chemistry had been the most affected by the decline of science at the time.

'The Humour of the Age as to these things (the sciences)' wrote Wotton, 'is visibly altered from what it was Twenty or Thirty years ago. . . . The sly insinuations of the Men of Wit. . . . have so far taken off the edge of those who have opulent fortunes and a love to learning that Physiological studies begin to be contracted amongst Physicians and Mechanics.'

The 'Men of Wit', the literary figures of the time, were scarcely responsible for the decline of science, for their ridicule of the Royal Society was a far less serious matter than Stubbe's earlier political and ideological attacks, which had had little effect. Moreover, the decline of science was just as marked in France where literary men, such as Fontenelle, supported the scientific movement.

As the English school of chemistry declined towards the end of the seventeenth century the German iatrochemical school revived, producing the phlogiston theory. The iatrochemists had supposed that chemical substances contained three essences or principles; Sulphur, the principle of inflammability; Mercury, the principle of fluidity and volatility; and Salt, the principle of fixity and inertness. Joachim Becher, 1635-82, a professor of medicine at Mainz, modified the iatrochemical doctrine somewhat, suggesting in 1669 that solid earthy substances contained in general three constituents; firstly, a *terra lapida*, a fixed earth present in all solids, corresponding to the Salt principle of the earlier iatrochemists; secondly, a *terra pinguis*, an oily earth present in all combustible bodies, corresponding to the Sulphur principle; and thirdly, a *terra mercurialis*, a fluid earth, corresponding to the Mercury principle. Becher held that all bodies which would undergo combustion contained the sulphureous, oily, *terra pinguis* which escaped from its combination with the other earths during the process of burning.

The processes of burning and calcination involved therefore the decomposition of a compound body into its constituent parts, namely, the sulphureous *terra lapida* and the fixed *terra lapida* in the simplest cases. In theory simple bodies could not undergo combustion for substances containing *terra pinguis* and another earth were necessarily compound. In 1703 Georg Ernst Stahl, 1660-1734, a professor of medicine and chemistry at Halle, renamed Becher's *terra pinguis* 'Phlogiston', this being 'the motion of heat', or 'the motion of fire', as well as 'the Sulphureous Principle' and the 'Oily Principle'. A metal was a compound of a calx and phlogiston, heat liberating the phlogiston and leaving the calx. In general phlogiston was the essential element of all combustible bodies, oils, fats, wood, charcoal, and other fuels containing particularly large amounts. The phlogiston escaped when those bodies were burnt and it entered either into the atmosphere or into a substance which would combine with it, like a calx which formed a metal.

Such a theory of calcination and combustion contained a large element of the view of the early iatrochemists and the alchemists before them that substances in general were composed of a matter and a spirit which could be separated by pyrotechnical methods, the spirit escaping from the matter when the substance was subject to heat. During the sixteenth century it had become apparent in some cases, notably in the calcination of the metals, that the residues of matter, or 'dead bodies' as they were termed, were heavier than the original substances. The phenomenon was explained by assuming that the spirits of substances had no weight, or even possessed a positive lightness, so that a substance became heavier when it lost its volatile or spiritous part. Biringuccio in his *Pyrotechnia* (1540) held such a view:

'Lead, when the watery and aerial parts have been removed by the fire, . . . falls like a thing left to itself and altogether dead, and so comes to be greater in weight, as also the like is shown to occur in the dead animal body, which indeed weighs much more than when alive.'

In a more sophisticated form the phlogiston theory offered a similar explanation of the calcination of metals. The phlogiston theorists presumed that metals were compound substances and that upon calcination they broke up into their constituent parts, the heavy calx and the aerial phlogiston. Stahl, with whom the phlogiston theory came into vogue, did not make explicit the notion, which was implied by the theory, that phlogiston had a negative weight or a positive lightness, like the elements of air and fire in the Aristotelian cosmology. However, some of his followers did. The phlogiston theory was much discussed by the scientific societies of the south of France, notably those of Dijon and Bordeaux, and it was particularly influential at the medical school of Montpelier. In the 1760's Gabriel Venel, 1723-75, professor of medicine at Montpelier, laid it down explicitly that phlogiston had a positive lightness.

'Phlogiston', he is reported to have said, 'is not attracted towards the centre of the earth, but tends to rise: thence comes the increase of weight in the formation of metallic calces, and the diminution of weight in their reduction.'

Such an idea illustrates how far chemistry was separated from physics and the mechanical philosophy in the mid-eighteenth century, for in physics it was generally accepted that all bodies on the earth were attracted by the force of gravity towards the earth's centre.

During the second half of the eighteenth century the phlogiston theory was generally received by the chemists of Britain, notably Joseph Black, 1728-99, Henry Cavendish, 1731-1810, and Joseph Priestley, 1733-1804, though they carried out the experimental work which was destined to overthrow that theory and the ancient Greek doctrine that natural substances were composed of the four elements, earth, water, air, and fire. Earth was not widely thought of as an element any more as many kinds of 'earths' had been recognized. But water, air, and fire, were still generally considered to be elements, indeed phlogiston was sometimes regarded as the element of fire, or more generally as the activating agent of fire. Early in the seventeenth century Helmont had conceived of gases as elementary substances distinct from air but his successors considered them to be merely forms of elementary air, 'factitious airs' as Boyle had called them. Now, in the mid-eighteenth century, Black demonstrated the existence of a gaseous substance, carbon dioxide, or 'fixed air' as he termed it, which was different from air in its chemical properties. In 1754 he showed that magnesium carbonate lost weight and a considerable amount of gas on heating, and that the same weight and the same gas was lost if an identical amount of magnesium carbonate were dissolved in an acid. Black demonstrated further that the residue left after heating, namely, magnesium oxide, gave the same salts as magnesium carbonate with acids but, unlike the carbonate, evolved no gas. Moreover, he indicated that the solution of magnesium oxide in acids gave a precipitate with soluble carbonates, such as soda, which was identical in weight and composition with the magnesium carbonate from which the magnesium oxide had been prepared in the first place. Thus it appeared that magnesium carbonate, and carbonates in general, were compounds of a base, such as magnesium oxide, with the ponderable gas, 'fixed air'. On heating, carbonates lost not the weightless and intangible phlogiston, but a well-defined chemical substance, 'fixed air', which had weight and which could be isolated and studied. Examining the properties of 'fixed air', Black found that it was absorbed by caustic alkalis while air was not absorbed, and that it did not support combustion and respiration while air did.

The work of Black served to direct the attention of the other British chemists towards the subject of the chemical nature of gases. In 1766 Henry Cavendish published an account of the preparation of hydrogen, or 'inflammable air' as he termed it, by means of the action of dilute acids on the metals, and of 'sulphureous vapours' and 'nitrous vapours' by the action of strong sulphuric and nitric acids respectively on the metals. In order to isolate such gases, Cavendish developed the pneumatic trough which had been introduced earlier by others, notably Hales, 1677-1761. Filling a bottle with water, he inverted it in a water trough and bubbled the gas into the bottle so that the gas displaced the water and filled the bottle which then could be sealed. If the gas were soluble in water Cavendish used mercury as the liquid in his trough and bottle, a technique which was developed notably by his contemporary, Joseph Priestley. In the 1770's Priestley discovered several gases and isolated them by means of the pneumatic trough, namely, ammonia, hydrochloric acid gas, nitrous oxide, nitric oxide, nitrogen dioxide, oxygen, nitrogen, carbon monoxide, and sulphur dioxide. Simultaneously and independently the Swedish apothecary, Carl Scheele, 1742-86, was working along the same lines, discovering the gas, oxygen, a little before Priestley. Scheele was amongst the first to recognize the importance of the discovery. In 1777 he pointed out that air could not be an elementary substance as it was composed of two gases, 'fire air' or oxygen, and 'foul air' or nitrogen, in the ratio of one to three parts by volume according to his estimate. However, Scheele adhered to the phlogiston theory. The function of 'fire air', or oxygen, in Scheele's view was to take up the phlogiston given out by burning substances. The amount that could be so absorbed was limited so that when the oxygen in a confined space was saturated with phlogiston it could no longer support combustion.

Meanwhile, in France, Antoine Lavoisier, 1743-94, was working along rather different lines, systematically criticizing the traditional chemical theories. Lavoisier was a typical scientist of eighteenth-century France. He took a degree in law and was well versed in several of the sciences. He became a member of the *Ferme Generale*, the French tax-collecting organization, and later he was appointed director of the state gunpowder industry. Lavoisier's earliest chemical work dates from 1769 when he showed that water was not converted into earth, contrary to the belief of van Helmont and other of his predecessors. The belief had been based on the fact that an earthy sediment collected in glass flasks used to boil water. Lavoisier demonstrated that the glass of the flask lost weight when water was boiled in it, the loss in weight being equal to the weight of sediment produced. Thus the earthy sediment came from the glass and not from the water. Late in 1772 Lavoisier repeated some of the work which had been carried out on combustion, showing that nonmetals, such as phosphorus, and metals, such as tin, increase in weight when burned in air. It seemed to Lavoisier that the increase in weight might be due to the absorption of air. He then read through and reviewed the work of his predecessor dealing with experiments which involved the absorption or the liberation of gases. He noted that different authors often gave different explanations for the same set of facts and he came to the view that the time had come for a critical repetition of many previous experiments in order to decide between the different explanations or replace them by an entirely new theory.

'These differences (of explanation) will be exhibited to their full extent', wrote Lavoisier in his *Journal Book*, 'when I shall give the history of all that has been done on the air that is liberated from substances and combines with them. The importance of the end in view prompted me to undertake all this work, which seemed destined to bring about a revolution in physics and chemistry. I have felt bound to look upon all that has been done before me merely as suggestive; I have proposed to repeat it all with new safeguards in order to link our knowledge of the air that goes into combination, or that is liberated from substances, with other acquired knowledge and so to form a theory.'

Lavoisier's note, which was written in 1772 or the following year, indicates that he had planned an entire revolution in chemical theory some time before he began his main experimental work on the subject. The note also shows that he intended to bring about the revolution by repeating past experimental work, not by venturing into new fields as Cavendish, Priestley, and Scheele had done. Lavoisier had shown already that Helmont was wrong in the belief that water could be converted into earth. Now in 1773-74 he set out to disprove Boyle's contention that the increase in the weight of metals on calcination was due to the absorption of fire particles. He repeated the experiment in which Boyle had heated tin in a flask, the flask and its contents being weighed before and after heating. Lavoisier, however, sealed the flask before the experiment and found that there was no change in weight after heating though the tin had calcinated. The process of calcination, therefore, could not be the absorption of ponderable fire particles as Boyle had supposed. On opening the flask, Lavoisier found that air rushed in, the flask and its contents then weighing more. The increase in weight due to air entering the flask was equal to the increase in weight of the tin on calcination, and so it appeared that the calx was a combination of the metal and the air.

Priestley had discovered in 1772 that metals on calcination absorbed at the most one-fifth of the volume of air in which they were enclosed. Lavoisier similarly found that only part of the air was taken up when metals were calcined, and he supposed that this part had properties which were different from those of the fraction which was not absorbed. The only gas known to Lavoisier at the time which differed from air in bulk and which was absorbed by chemical compounds was Black's 'fixed air', or carbon dioxide. Lavoisier found that lead when heated absorbed part of the air to give litharge, and that the litharge on heating with charcoal was converted back to lead and gave off a gas which he showed to be 'fixed air'. He presumed that the gas given off by the litharge was the same as that absorbed by the lead so that the active part of the atmosphere supporting combustion was 'fixed air'. However, Lavoisier found that phosphorus would not burn in 'fixed air' and that the gas would not support combustion in general. Accordingly he had to abandon the hypothesis that 'fixed air' was the portion of the atmosphere responsible for combustion and calcination. Further than this Lavoisier did not go by himself, though the problem of what role the atmosphere played in combustion remained.

The problem was not solved by Lavoisier's initial method of merely repeating the old experiments with refinements:

some of the new discoveries made by Cavendish, Priestley, and Scheele, were brought in. Priestley visited Paris in 1774 and acquainted Lavoisier with his discovery of a new gas, 'dephlogisticated air' as Priestley termed it, obtained by heating mercuric oxide. 'Dephlogisticated air', or oxygen as it is now called, was the active constituent of the atmosphere for which Lavoisier had been searching. In oxygen, candles burned more brightly and animals lived longer than in ordinary air. Metals on calcination absorbed the whole of a volume of oxygen, but only a fraction of a volume of air. Lavoisier at first, in 1775, thought that oxygen was the pure element of air itself, free from the impurities which normally contaminate the atmosphere. However, Scheele in 1777 showed that air consisted of two gases, oxygen which supported combustion, and nitrogen which was inert. Lavoisier accepted Scheele's view and he suggested in 1780 that the atmosphere was composed of one-quarter of oxygen and three-quarters of nitrogen by volume. Priestley, from his experiments on the fraction of a volume of air absorbed by metals during calcination, gave the more accurate ratio of one-fifth oxygen to four-fifths of nitrogen.

Finally in 1783 Lavoisier announced the renovation of chemical theory which he had planned a decade before. Mme Lavoisier ceremonially marked the beginning of the new chemistry by burning the books of Stahl and the phlogiston theorists, just as Paracelsus had opened the iatrochemical era some two and a half centuries before by burning the books of the medieval medical authorities. Combustion and calcination, Lavoisier held, involved in all cases the chemical combination of the combustible substance with oxygen, as the weight of the products formed invariably equalled the weight of the starting materials. The processes of burning and oxidation could not be ascribed to the escape of the so-called phlogiston, for the old theory required that phlogiston should possess weight in some cases, be imponderable in others, and possess a positive lightness in yet further cases. Heat and light, which had been regarded as manifestations of the escape of phlogiston, were often liberated during the processes of combustion and calcination, but their emission was external to the chemistry of those processes, as revealed by the weight changes during the reactions, since heat and light were imponderable.

The change in weight of a substance undergoing combustion or calcination was due entirely to its reaction with oxygen.

Lavoisier's chemical revolution was not quite complete for, in the manner of the iatrochemists, he elevated oxygen to the position of a general explanatory 'principle' by ascribing to it properties which were not warranted experimentally. According to Lavoisier, oxygen was the acidifying principle, all acids being composed of oxygen united to a non-metallic substance. Such a supposition was proved to be incorrect by Humphry Davy, 1778-1829, who showed in 1810 that hydrochloric acid did not contain oxygen. In 1784 Lavoisier extended his revolution by reforming chemical nomenclature, introducing the modern names of chemical substances. Similarly his reform of the nomenclature was carried only half-way, the old alchemical symbols for chemical substances being retained even though their names were modernized.

In Britain the new chemistry was taught from its inception by the Scot, Joseph Black, but the English chemists, Cavendish and Priestley, supported the phlogiston theory to the last. In 1781 Cavendish made an accurate measurement of the composition of the atmosphere, his result being very near to the modern estimate, but he continued to regard the air as an element, not a mixture of oxygen and nitrogen 'as Mr Scheele and La Voisier suppose'. However, Priestley, after his work on oxygen, accepted the modern view. Although the elementary nature of the air was a philosophical axiom, Priestley wrote in 1775, his experiments had convinced him 'that atmospherical air is not an unalterable thing'. In 1781 Priestley exploded together a mixture of hydrogen and oxygen, and he found that the gases were used up, leaving a dew. Cavendish repeated the experiment, discovering that water was produced by the union of one volume of oxygen with 2:02 volumes of hydrogen. His result implied that water was a compound of hydrogen and oxygen, not an element, but Cavendish refused to accept the implication. He presumed that water was an element, while oxygen was water deprived of phlogiston, and hydrogen was either phlogiston itself or water with an excess of phlogiston. On the basis of Priestley's experiments, James Watt put forward a similar theory of the nature of water independently and about the same time.

Meanwhile Lavoisier had been searching for the acid which should be formed, according to his theory, by the union of hydrogen, a nonmetal, with oxygen. He did not find such an acid, but in 1783 Cavendish's assistant, Blagden, visited Paris and informed Lavoisier that the English chemists had obtained water from hydrogen and oxygen. Lavoisier performed a rough experiment confirming the main result of the work carried out by Cavendish and Priestley on the composition of water. Lavoisier did not show that the weight of the combining gases was equal to the weight of water produced, remarking that since 'it is no less true in physics as in geometry that the whole is equal to its parts . . . we think we are right in concluding that the weight of this water was equal to that of the two airs which served to form it'.

From his experiment Lavoisier drew the modern conclusion that water was not an element but a compound of hydrogen and oxygen. Now Lavoisier was able to meet a serious objection which his new theory had encountered at its inception. A metal, such as tin or iron, dissolved in an acid liberating hydrogen and forming a salt. The calx of the metal dissolved in the acid forming the same salt without liberating any gas. It was generally thought therefore that hydrogen was phlogiston or perhaps phlogiston combined with water. The acid liberated phlogiston from the metal, but not from the calx, as the metal was presumed to be composed of the calx and phlogiston. Lavoisier could not account for these phenomena on the basis of his new theory at first, but as soon as it became apparent that water was composed of hydrogen and oxygen he provided an alternative to the phlogiston theory explanation. A metal, dissolving in dilute acid, took the oxygen from the water present forming its calx, or oxide, which united with the acid giving a salt, while the hydrogen of the water was set free.

Now Lavoisier's theory covered the known facts of chemistry much more satisfactorily than the phlogiston theory, and the latter rapidly lost ground. Earth, water, air, and fire, were no longer regarded as elements as there were many kinds of earths, and fire was resolved into heat, light, and smoke, whilst air was shown to be composed of oxygen and nitrogen, and water of hydrogen and oxygen. In place of the traditional view Lavoisier defined a chemical element, more precisely than Boyle, as 'the actual term whereat chemical analysis has arrived'. In his *Elements of Chemistry* (1789), which was the first of the modern chemical text-books, Lavoisier listed on this basis some twenty-three authentic elementary substances. However, he also included a substance which he termed 'caloric', the supposed imponderable matter of heat, amongst the elements of the inorganic world.

As regards the nature of heat the English chemist, Cavendish, was more modern than Lavoisier for he was old-fashioned and preferred the seventeenth-century doctrine, 'Sir Isaac Newton's opinion' as he put it, that heat was not a material substance but the mechanical motions of the particles composing material bodies. In general Cavendish was the most conservative theoretically of the late eighteenth century British chemists and, being unable to adapt himself to the new chemistry which his experiments had helped to establish, he abandoned his chemical work about 1785. Unlike Cavendish, Priestley did not belong to the by now largely conservative tradition of the gentleman-amateur scientists: he was of the new and more radical Noncomformist-industrialist tradition, but he never entirely abandoned the phlogiston theory. Priestley accepted the view that air was not an element but a mixture, and for a time he thought that water might be a compound and not an elementary substance, though James Watt, another phlogistonist, won him back to the traditional opinion. At one point, in 1785, he wavered in his support for the phlogiston theory as a whole when he found that water was produced by the action of hydrogen, or phlogiston as he thought it to be, on metal calces. However, he assimilated the discovery into the phlogiston theory by assuming that hydrogen was a compound of phlogiston and water, not phlogiston itself. Finally, during the last years of his life in America, he wrote a treatise defending the phlogiston theory in its entirety.

Much of the theoretical conservatism of the late eighteenth-century English chemists seems to have been due to marked empiricism of their work. They made the experiments which demonstrated to Lavoisier that air and water were not elements, and that combustion and calcination were essentially the reactions of bodies with oxygen. But they did not derive such conclusions from their work, and when the theoretical implications of their experiments were presented to the English chemists they found them unacceptable. Lavoisier on the other hand made no qualitatively novel experimental discoveries and he planned a whole revolution in the theory of chemistry before he had done much work on the subject. He repeated the experiments of Helmont and Boyle and deduced more from them than his predecessors had done, but he was held up until Priestley introduced him to the discovery of oxygen and again until he was acquainted with the work of Cavendish on the composition of water. In repeating the work of his English contemporaries, Lavoisier used practical methods which were markedly inferior to those employed by Cavendish and Priestley, but with the results he built up a new edifice of theory.

These men perceived the one-sided character of each other's work but it seems that they could not overcome their own limitations. Lavoisier described Priestley's *Observations upon Different Kinds of Air* (1772) as 'a web of experiments almost uninterrupted by any kind of reasoning'. Priestley, in a perhaps more gracious rejoinder, described Lavoisier as a man 'to whom in a variety of aspects the philosophical part of the world has very great obligations'. Priestley was perhaps the most curious figure in the episode of the chemical revolution. From an early age, Priestley wrote, 'I saw reason to embrace what is generally called the heterodox side of almost every question.' Yet in chemistry he adhered to the traditional phlogiston theory to the last and he did not espouse the new doctrines which his discovery of oxygen did so much to establish. Indeed it is somewhat ironical that his heterodoxies outside of chemistry, in politics and religion, were such that he had to leave England for America in 1794, the same year that Lavoisier, the chemical revolutionary, was led to the guillotine for having been associated with the conservative elements in French affairs, the *Ferme Generale* and other institutions of the *ancien régime*.

Chapter 27

The Idea of Progress in the Mechanical World of the Eighteenth Century

THE THEORY that biological organisms have evolved from simple to more complex forms, and the idea that man has progressed from rude beginnings to civilized society, appear to have been associated with one another historically. Both conceptions were particular aspects of more general philosophies of development, which conceived of man and other living creatures as having originated from the same simple organic forms, or that saw human progress as a continuation of biological evolution, as in the theories of Anaximander and Democritus amongst the pre-Socratic Greeks, or the Epicureans in later antiquity. Similarly the view that man has reached the limit of his development, has been associated historically with the idea that the organic species are more or less fixed, having been created in their present forms by an intelligent First Cause. The two general view-points were not necessarily opposed, for Democritus and the Epicureans thought of each world in the universe, and its inhabitants, as evolving, decaying, and then reforming, so that the cosmic process as a whole was an oscillation about a fixed level, with no overall evolution.

Elements from both viewpoints in fact crystallized out in later antiquity to give an opinion which was generally received by the men of the scholarly tradition down to modern times. Man and nature, it was thought, were much the same throughout the ages, both moving in a cycle about a standard mean. Man had reached the limit of his achievement, whilst animals and plants had been created in their present forms by an intelligent First Cause. The Roman emperor, Marcus Aurelius, A.D. 121-180, wrote that, the rational soul of man,

'goeth about the whole universe and the void surrounding it, and traces its plan, and stretches forth into the infinitude of time, and comprehends the cyclical regeneration of all things, and takes stock of it, and discerns that our children will see nothing afresh, just as our fathers too never saw anything more than we.'

More than a thousand years later the Florentine philosopher, Niccolo Machiavelli, 1469-1527, expressed exactly the same sentiment when he wrote:

'It is ordained by Providence that there should be a continual ebb and flow in the things of this world: as soon as they arrive at the utmost perfection and can ascend no higher, they must of necessity decline; and on the other hand, when they have fallen to the lowest degree, they begin to rise again.'

In popular thought the idea that mankind had degenerated was prominent, for the Greek word *presbiteros* and the Latin word *antiquior*, meant not only 'older' but also 'better', a conception that was given added force by the Christian doctrine of the Fall of Man. On the other hand, craftsmen, engineers, and scholars interested in the crafts, appear to have had a fairly clear idea of technological progress, and even of progress in scientific knowledge. Aristotle was of the view that the development of the mechanical arts had been completed already, but the Alexandrian engineer, Philo of Byzantium, noted that war machines had been improved considerably in his time, 'partly by learning from the earlier constructors, partly by observation of later trials'. The Roman philosopher, Seneca, A.D. 2-65, who objected to scholars depreciating the craftsmen—'Posidonius came very near declaring that even the cobbler's trade was the discovery of the philosopher,' he wrote—thought similarly that even the mystery of comets would be explained in time. Such views became prominent in early modern times when the craft tradition became more widely literate, and when a number of scholars interested themselves in the mechanical arts.

The most notable of such scholars was Francis Bacon, who pointed to the invention of gunpowder, printing, and the magnetic compass, like others before him, as examples of human technological progress, indicating an advance upon the achievement of antiquity. As we have seen, he wished to carry the process further by applying scientific knowledge derived by means of his new method. Bacon thought that the crafts could be progressively developed, but he did not think of scientific knowledge in the same way. The new natural philosophy was to be a closed and final system of knowledge, like the systems of antiquity which it was designed to replace. Given sufficient facts, Bacon thought that he himself could build up a definitive system of natural philosophy.

'So far as the work of the intellect is concerned', he wrote, 'I may perhaps successfully accomplish it by my own powers, but the materials for the intellect to work upon are so widely scattered that, to borrow a metaphor from the world of commerce, factors and merchants must seek them out from all sides and import them.'

Once this has been accomplished, 'the investigation of nature and of all the sciences will be the work of a few years'.

Descartes had the same view of the sciences: indeed he was of the opinion that he had formulated the new system of natural philosophy, for he concludes his *Principles of Philosophy* with the remark that 'No phenomenon of nature has been omitted in the explanations given in this treatise.' He also regarded his rules as the final statement of scientific method: Descartes wrote, that he did not think, 'there is any road by which the human intellect could ever discover better ones'. Bacon, who was a little nearer to the craft tradition with its sense of cumulative advance, did not regard his rules as the definitive method of science.

'I merely claim that my rules will make the process quicker and more reliable,' Bacon wrote, 'I do not mean to say that they cannot be improved upon. This would be utterly at variance with my way of thinking. My habit is to consider the mind, not only in its own faculties, but in close connection with things. It follows that I must admit that the art of discovery itself will advance as discoveries advance.'

Descartes too only had the idea of progress in connection with practical matters. He said that he wished, 'to induce intelligent men to try to advance farther by contributing each according to his inclination and ability, to the necessary experiments and also by publishing their findings. Thus the last would start where their predecessors had stopped, and by joining the lives and works of many people, we would proceed much farther together than each would have done by himself.'

In these ways, the finality of the old scholarly system-builders, and the cumulative empiricism of the craft tradition, were expressed together by the philosophers of the early modern scientific movement. Bacon had the idea of progress a little more than Descartes, being nearer to the craft tradition, whose empiricism he reflected in his method and also in the fact that he did not work out the new natural philosophy he announced. Descartes, with his orientation towards the scholarly tradition, did elaborate a new system of natural philosophy which in its content as well as its form expressed some of the values of the ancient philosophers. At first sight Descartes' universe appears to be a world in evolution: the giant cosmic vortices had fashioned the primordial matter of the universe, according to the laws of mechanics, until it assumed the arrangements we observe today. But Descartes was concerned to stress that any possible world of primordial matter would necessarily assume the present configuration of our world, and become stabilized in that form, as the laws of mechanics would always operate in the same way. Thus our present world was the predetermined end of any cosmic system: it was in fact the only possible world. Leibniz went further and insisted that this was indeed the best of all possible worlds, no improvement whatsoever could be envisaged. Bolingbroke and Pope, translating the sentiment into a moral idiom, averred that 'Whatever is, is right'.

When English philosophy spread to the continent during the eighteenth century, it was thought that Newton had finally constructed the new and definitive system of the world which had been advertised by Bacon and Descartes. Voltaire affirmed in a letter to Horace Walpole, that 'Newton pushed his work to the most daring truths which the human mind could ever reach'. Later in the century Lagrange wrote of Newton:

'There is but one universe, and it can happen to but one man in the world's history to be the interpreter of its laws.'

Newton had thought of the universe as a product of evolutionary development even less than Descartes. For Newton God had created the world fully fashioned in its present form, and only then had the laws of mechanics come into operation to sustain the cosmic machine. There were astronomical phénomena, Newton thought, which were not entirely explicable in terms of the laws of mechanics, but Laplace tied up these loose ends, as we have seen.

In this mechanical world of the eighteenth century nothing had developed historically, all the inhabitants and the creatures of the earth had existed in their present forms from the beginning. The London physician, Cheyne, remarked in 1715:

'If animals and vegetables cannot be produced from these (matter and motion) and I have clearly proved that they cannot, they must of necessity have existed from all eternity.'

Animals and plants were machines, but they could not generate themselves spontaneously from matter and motion. Like the world as a whole they were created in their present forms at the beginning of time, and so were all future generations of the creatures. It was widely thought that the first animals and plants had contained within themselves facsimile miniatures, like a series of boxes one inside the other, which constituted all the future generations of that species, each new-born creature being an enlargement of the preformed miniature.

Thus the formation of the world and its inhabitants, like the place of the scientific revolution in history, was seen during the seventeenth and eighteenth centuries as a single creative event, which once accomplished was eternally enduring and finished for all time. During the same period a similar view obtained concerning the formation of human society. For all their differences, Bodin, Hobbes, Locke, and Rousseau, thought that once upon a time isolated individual men had come together, and had contracted to live together with one another in human society for ever after. The mechanistic idea was extended to social theory in other ways by Bodin in 1577 and

Montesquieu in 1748. They held that the geographical location, or the climate of a region in which a nation lived determined its national character. The men of the north tend to be vigorous, but not very intelligent, whilst proceeding southwards, men became cleverer, but more feeble. Mankind could not avoid these things, for they were external determinations laid down by the environment.

In such a world neither the progress of mankind nor the evolution of the species could be envisaged. But the extension of the mechanistic viewpoint to another field, psychology, helped to generate the idea of progress and stimulated later writers to develop theories of evolution. The mechanical philosophy was applied to psychology in two ways. Firstly, it was thought that the mind of man was determined by the internal physiological mechanisms of the body, which implied that mankind might progress with the advance of medicine. Secondly, it was held that the mind of man was determined by external conditioning forces, such as education, so that man might progress if education were reformed.

The first school of thought was initiated by Descartes. In the abstract he held that there was one mechanical world of matter and motion, and another, spiritual world which included the soul of man, the two worlds being largely separate, connected only through the pineal-gland in the case of man. But when Descartes came down to detail, he thought that the human mind was largely determined by the internal mechanisms of the body. In his *Discourse on Method*, Descartes tells us,

'the mind is so intimately dependent upon the condition and relation of the organs of the body, that if any means can ever be found to render men wiser and more ingenious than hitherto, I believe that it is in medicine they must be sought for'.

Such a line of argument was taken up by one of the earliest French materialists, Lamettrie, 1709-51, a one-time army surgeon and member of the Berlin Academy. In his book, *Man a Machine*, published in 1748, he observed that the mental state of a patient was often dependent upon his bodily condition, and he reported that decapitated animals were capable of some sort of motion, like machines working imperfectly without one of their parts. He held therefore that the brain secreted thought, just as the liver secreted bile, and he believed that the progress of mankind depended upon the advance of medicine.

Such a theory was of limited application, and thus it was not so popular as the other mechanical theory of psychology during the eighteenth century. The second extension of the mechanical philosophy to psychological theory developed during the seventeenth century, finding its most important expression in John Locke's *Essay Concerning Human Understanding,* published in 1690. Locke pictured the mind of man at birth as a blank sheet of paper, upon which sensations and stimuli from the external world make their mark and impress, originating thoughts and ideas in this way.

'Let us then suppose the mind to be, as we say, white paper, void of all characters, without any ideas,' he wrote. 'How comes it to be furnished? Whence comes it by that vast store, which the busy and boundless fancy of man has painted on it with an almost endless variety? Whence has it all the materials of reason and knowledge? To this I answer in one word, From experience.'

Such a view was felt to be of a piece with the mechanical philosophy: Voltaire said of Locke that, 'no one has proved better than he that one can have the geometrical spirit without the aid of geometry'.

From Locke's theory of psychology were derived three important problems that were further investigated by the philosophers of the eighteenth century. Firstly, there was the question of how it was that the different impressions of vision, sound, taste, touch, and smell, from the five sense organs combined together to give a single sensation; secondly, there was the problem of how sensations were translated into ideas, and thirdly, how were ideas combined together. The first problem was discussed by one of Locke's friends, William Molyneux, a tutor of Trinity College, Dublin, and later by Molyneux's colleague at Trinity College, George Berkeley, in his *New Theory of Vision,* published in 1709. Molyneux and Berkeley studied case histories of men blind from birth who had recovered their sight, where it was found that the sense of sight was at first confusing, and was only slowly adapted to the determination of shapes, sizes and distances, which previously had been judged by the sense of touch. Thus they concluded that vision was not innately connected with touch, the two becoming linked and blended by experience.

In France similarly the case histories of men deprived of one or more senses were studied, notably by Denis Diderot, 1713-84, later the main editor of the *Encyclopédie*, who brought out his *Letter on the Blind* in 1749, and his *Letter on Deaf Mutes* in 1751.

'My idea', he wrote, 'would be to decompose a man, so to speak, and examine what he derives from each of the senses with which he is endowed.'

Like the English psychologists, Diderot thought that the sense of touch was fundamental in determining the spatial relations of objects, so that in viewing things at a distance, we see shapes that we have learned to recognize by touch in the past. Diderot's project of analysing each sense separately was carried out by his contemporary, Etienne Condillac, 1714-80, the son of a Grenoble government official, in his *Treatise on Sensations*, published in 1754. He thought of man as a kind of statue equipped with only one of the five senses. With the sense of smell, Condillac held that the statue would be able to receive sensations of smell, which would generate ideas of smell, and the persistence of such ideas would constitute a memory, whilst combinations of those ideas would form knowledge. Thus he concluded that all of the senses were of equal value in obtaining knowledge of the world, and that with one sense alone 'the understanding has as many faculties as with the five joined together'. The other senses did not add anything that was qualitatively new to the properties of the mind, they merely enlarged the sphere of its understanding. Impressions from the different sense organs were blended together by a kind of mechanical mixing which depended upon experience, but the resultant sensation and its corresponding idea differed in degree only, not in kind, from the sensation and idea produced by a single impression.

The question of how ideas combined together to give

thoughts was answered by supposing that ideas were associated together by reason of similarity in content, or because they appeared to be linked by the cause-effect relationship, or simply because they happened to stand next to each other in space and time. The philosopher, David Hume, 1711-76, wrote in his *Treatise of Human Nature* published in 1739, that,

'there appear to be only three principles of connection among ideas; namely Resemblance, Contiguity in time or place, and Cause or Effect'.

Such laws in psychology he thought to be the counterpart of the laws of mechanics in physics. The association of ideas, Hume wrote,

'is a kind of Attraction, which in the mental world will be found to have as extraordinary effects as in the natural, and to show itself in so many various forms.'

An attempt to solve the third problem of how impressions on the sense organs became transformed into ideas was made by the physician, David Hartley, 1705-57, in his *Observations on Man,* published in 1749. He held that stimuli applied to the sense organs excited vibrations in the nerves, which travelled up to the brain and there gave rise to ideas. The vibrations continued on in the brain long after the stimulus to the sense organ had died away, so that sensations and ideas were remembered. A number of residual vibrations in the brain were associated together if their corresponding sensations habitually came at the same time, or concurrently, so that after a while only one of those sensations was required to set going the residual vibrations of the group with which it was associated. In this way a single sensation could set off a whole train of associated ideas in the brain. Hartley thought that such a doctrine could be applied in everyday life to improve the human race.

'It is of the utmost importance to morality and religion', he wrote, 'that the affections and passions should be analysed into their simple compounding parts, by reversing the steps of the associations which concur to form them. For thus we learn how to cherish and improve good ones, check and root out such as are mischievous and immoral.'

This psychological theory enjoyed a considerable popularity in France, where it was applied to the project of improving the lot of mankind. The French philosophers were of the opinion that if the human mind was like a piece of blank paper at birth, as Locke had suggested, then it followed that all men were equal at birth, and became unequal through different environmental influences, such as the education they received, and the laws they had to obey. The philosopher Helvetius, 1715-71, a member of the French *Ferme Generale*, in his work *On the Mind*, published in 1758, said that he preferred the sensation psychology of Locke, to the medical psychology of Descartes, because the latter could not explain why there were different kinds of minds. The sensation psychology, on the other hand, he said, ascribes 'the inequality of minds to a known cause, and this cause is the difference of education'. Thus Helvetius thought that to secure the improvement of mankind it was only necessary to reform the legislative and educational systems: for man 'to be happy and powerful is only a matter of perfecting the science of education'.

In this way the French philosophers of the eighteenth century derived from the sensation psychology of the period the proposition that the opinions of men are governed by the legislative and educational institutions of their society. But they also entertained the opposite view that social institutions are fashioned and controlled by human opinion: in general 'opinion governs the world', as a phrase of the period put it. Such a view was a necessity, for if institutions completely governed opinion, all men would be perfectly adapted to their institutions, and the need for reform would not arise. But the French philosophers considered the reform of institutions desirable, and thus postulated the independent movement of opinion. In their theories the French philosophers never resolved the contradiction between the idea that opinion governs the world, and the view that the world governs opinion. Helvetius and Diderot at one point thought that there might be some third factor governing both institutions and opinion, though they did not find it. In general they got out of the difficulty by appealing to the idea of a *bon prince* who could simultaneously change his own opinions and the institutions of his land. Thus they addressed themselves to the so-termed enlightened despots of the age, Frederick the Great and Catherine II, but not the Bourbons of France, whom they considered despotic but not enlightened. Helvetius dedicated his second work, *On Man*, published in 1772, to Catherine II of Russia, whilst Diderot journeyed to St. Petersburg to advise her in person. In the same way, Voltaire, Lamettrie, and others, gathered round Frederick the Great at Berlin. In practice of course the French philosophers applied both of their views. They contributed to the *Encyclopédie* to enlighten the opinion that governed the world, and when they attained political power, as Turgot did when he became minister to Louis XVI in 1774, they attempted to reform the institutions that governed opinion.

The doctrine that opinion governs the world grew up in France during the eighteenth century together with the idea of progress. Both were an expression of the revolt against the notion that our world is the best of all possible worlds, and they were linked together in the project for the advance of mankind through the dissemination of enlightened opinion. As we have seen, the view that mankind could be advanced through the reform of educational institutions was deduced from the static, mechanical, psychology of the eighteenth century, and in the same way the opposite view that the progress of man resided in the advance and dissemination of enlightened opinion was deduced from the mechanistic sociologies of the period. Some, like Bodin and Montesquieu, believed that national characteristics were determined by geography and climate, others, notably Hume, thought that the character of a nation was determined by its institutions, but there was general agreement that mankind, like the world as a whole, was much the same at all times and places, apart from minor details. Hume averred that,

'mankind are so much the same in all times and places that history informs us of nothing new in this particular. Its chief use is only to discover the constant universal principles of human nature.'

From this point of view, it was argued by Fontenelle that men must have progressed by the sheer accumulation of knowledge throughout the ages. We have no reason to suppose that trees were larger in antiquity, he held, and thus both nature and mankind must have been constant and uniform throughout past time. These things being so, we must have men of the stature of Homer and Plato amongst us today, but they can proceed further, because they can start where their predecessors left off. Mankind, therefore, has become progressively enlightened for at least, he wrote,

'we are under obligation to the ancients for having exhausted all the false theories that could be formed'.

Voltaire, after Fontenelle, called for an active effort to advance mankind by criticizing traditional beliefs, and disseminating the newly acquired knowledge of the natural world. In his *History from Charlemagne to the death of Louis XIII*, published in 1756, Voltaire indicated that 'the progress of arts and sciences' was the principal part of his subject. We may believe, he wrote, that

'reason and industry will progress more and more, that useful arts will be improved, that the evils which have afflicted men and prejudices which are not their least scourge, will gradually disappear among all who govern nations.'

The most important publication in the French movement for the progress of man through enlightenment was the great *Encyclopédie*, which was published in twenty-two volumes between 1751 and 1777. In the prospectus issued in 1750, the editor Diderot stated that the aims of the work were—

'to bring together all the knowledge scattered over the surface of the earth, and thus to build up a general system of thought, so that the works of past ages shall not be useless, and our descendants becoming more instructed, shall become more virtuous and happier.'

The *Encyclopédie* was immensely effective, illustrating, the philosophers thought, the contention that opinion governs the world. The *Avocat Generale*, Seguier, confessed in 1770 that

'The philosophers have shaken the throne and upset the altars through changing public opinion.'

Finally when the political revolution came to France in 1789, it appeared to the philosophers that the days of the indefinite progress of mankind were at hand. Condorcet, who was destroyed by the revolution he supported, wrote his *History of the Progress of the Human Spirit* whilst in hiding during 1794. Here he stated:

'The result of my work will be to show, by reasoning and by facts, that human perfectibility is in reality indefinite: that the progress of this perfectibility henceforth independent of any power that might wish to stop it, has no other limit than the duration of the globe upon which nature has placed us. Doubtless this progress can proceed at a pace more or less rapid, but it will never go backwards; at least so long as the earth occupies the same place in the system of the universe, and as long as the general laws of this system do not produce upon the globe a general destruction, or changes which will no longer permit the human race to preserve itself, to employ the same powers, and to find the same resources.'

Laplace had already shown that the solar system was mechanically stable, and so it seemed that the indefinite progress of mankind was assured. The idea of progress, appearing to be firmly established upon the plane of social philosophy, found its way elsewhere in the guise of the theory of evolution. Laplace brought out this theory of the evolution of the solar system in 1796, and the physician, Pierre Cabanis, 1757-1808, developed a psychological theory from 1796 which envisaged the mental faculties of a man as the product of his developmental history. In 1809 Jean Baptiste Lamarck, 1744-1829, brought out a system of what he termed *Zoological Philosophy*, which was the first of the important modern theories of organic evolution. Lamarck was fully in the French tradition of the eighteenth century, indeed he has been called 'the last of the *philosophes*'. Animals he thought to be machines which had evolved to higher forms according to 'the law of progress'. However, his system was complex as we shall see, for it incorporated conceptions which had their roots in the natural philosophy of antiquity. In contrast the theory of organic evolution put forward by his older contemporary in England, Erasmus Darwin, 1731-1802, was much more a product of the immediate circumstances of the day.

The development of the idea of progress in England paralleled its growth in France, though here it assumed a theological form and was not at all anticlerical. The Christian doctrine of the Fall of Man was not conducive to the development of the idea that mankind progressed historically, but there was also the religious conception that there was a celestial world infinitely superior to the earth which was to be the residence of the blessed in their after life, and this doctrine when secularized provided the English idea of progress. Francis Bacon thought that some of the worse consequences of the Fall of Man could be repaired by the application of his method to the common uses of mankind, and had envisaged a considerable, if not indefinite progress in the arts and crafts. The Royal Society took up Bacon's projects in its early days, as we have seen, though by the end of the seventeenth century they had to confess themselves 'thwarted in their design to perpetuate a series of useful inventions'.

It was then that the conception of Leibniz that this is the best of all possible worlds became popular in England, as enshrined in the dictum of Pope that 'Whatever is, is right'. The idea of technological progress had largely evaporated, but the idea of progress in general did not disappear, it was spirtualized. The English essayist, Joseph Addison, 1672-1719, wrote in his journal, *The Spectator*, in 1711:

'A man considered in his present state seems only sent into the world to propagate his kind. He provides himself with a successor, and immediately quits his post to make room for him ... (but) would an infinitely wise Being make such glorious creatures for so mean a purpose? ... How can we find that wisdom which shines through all His works in the formation of man, without looking on this world as only a nursery for the next, and believing that the several generations of rational creatures, which rise up and disappear in such quick successions, are only to receive their first rudiments of existence here, and afterwards to be transplanted and flourish to all eternity? There is not in my opinion a more pleasing and triumphant consideration in religion than this of the perpetual progress which the soul makes towards the perfection of nature without arriving at a period in it . . . Nay it must be a prospect pleasing to God himself to see His creation for ever beautifying in His eyes, and drawing nearer to Him, by greater degree of resemblance. Methinks this single consideration of the progress of a finite spirit to perfection will be sufficient to extinguish all envy in inferior natures, and all contempt in superior.'

Such a conception was very popular in England at the time. The psychologist, David Hartley, whose theories we have mentioned, gave up the study of theology at Cambridge, and turned to medicine, because he could not accept the doctrine that sinners were eternally damned: they must at some point, he thought, partake of the universal progress of all souls after death. In the second, largely theological, volume of his *Observations on Man*, he worked out a mechanism for the temporary punishment of sinners after death. Bad habits, he thought, excite vibrations in the brain, which make an impress on the soul. These impressions persist for a while after death, so that souls 'receive according to the deeds done in the gross body, and reap as it is sowed'. However, the impressions eventually die away so that 'It is probable that all mankind will be made happy ultimately'.

Hartley had a considerable influence upon the members of the Birmingham Lunar Society, Priestley publishing an abridged version of his *Observations on Man*. The Lunar Society was, so to speak, the scientific general staff of the eighteenth-century industrial revolution in the Midlands, and its members were thoroughly imbued with the sense of technological progress, and the idea of the progress of mankind in general. Two of the members, Erasmus Darwin and Joseph Priestley, accordingly secularized the previous theological idea of progress, Darwin applying it to biology and Priestley to human affairs. Priestley in his book, *The First Principles of Government*, pubished in 1771, offered an idea of human progress similar to that put forward by Condorcet twenty years later.

'It requires but a few years', he wrote, 'to comprehend the whole preceding progress of any one art or science, and the rest of a man's life, in which his faculties are the most perfect, may be given to the extension of it. If by this means one art or science should grow too large for an easy comprehension . . . a commodious subdivision will be made. Thus all knowledge will be subdivided, and knowledge, as Lord Bacon observes, being power, the human powers will in fact be enlarged: nature, including both its materials and its laws, will be more at our command, men will make their situation more comfortable, they will probably prolong their existence in it, and will daily grow more happy and more able (and I believed more disposed) to communicate happiness to others. Thus whatever was the beginning of this world, the end will be paradisiacal beyond what our imaginations can now conceive.'

Erasmus Darwin in his *Zoonomia,* published in 1794, applied a similar conception to the development of the plant and animal species.

'The ingenious Dr. Hartley in his work on Man,' Darwin wrote, 'and some other philosophers, have been of the opinion that our immortal part acquires during this life certain habits of action or of sentiment, which become forever indissoluble, continuing after death in a future state of existence . . . I would apply this ingenious idea to the generation, or production of the embryon, or new animal which partakes so much of the form and propensities of the parent.'

Thus the habits acquired by an animal during its lifetime were inherited by its offspring, and led to an evolution of the species. Lamarck a few years later was to put forward a similar and much fuller theory of the inheritance of acquired characteristics, organisms being thought of as progressing through the accumulation of experience, like mankind. Like Lamarck also, Erasmus Darwin thought that there was an inner force within each organism driving it forwards towards higher forms.

'Would it be too bold to imagine', Darwin asked, 'that all warmblooded animals have arisen from one living filament which the great First Cause endued with animality, with the power of acquiring new parts, attended with new propensities, directed by irritations, sensations, volitions and associations; and thus possessing the faculty of continuing to improve by its own inherent activity, and of delivering down those improvements by generation to its posterity, world without end.'

Erasmus Darwin had one conception which Lamarck lacked, namely the idea that organisms evolved through the mechanism of competition and the survival of the more viable forms. Like his grandson, Charles Darwin, he believed that cocks had developed their spurs and stags their antlers through competing with one another for the females of their species. Plants, too, had changed by reason of 'their perpetual contest for light and air above ground, and for food and moisture beneath the soil'. Such a conception with its flavour of economic *laissez-faire* proved to be both popular and fruitful later in mid-Victorian England when it was elaborated by Charles Darwin.

Chapter 28

Evolution, and the Great Chain of Beings

THE BIOLOGISTS of modern times inherited from antiquity two rather contradictory views of the organic world, both of which had been elaborated by Aristotle. The one view conceived of the organic species as a hierarchy of creatures with comparatively large discontinuities between their ranks: there were, for example, only eleven classes in Aristotle's hierarchy of animals. The other view saw the various animals and plants as so many links in a great chain of creatures, the gradations between them being insensible and continuous. Philosophically, such conceptions could be reconciled by supposing that the highest creature of one class was directly contingent upon the lowest creature of the order above. In the practice of biological classification, however, the two conceptions could not be reconciled, and they gave rise to two different kinds of classificatory technique, the so-called 'artificial' and 'natural' systems of classification.

The artificial systems classified the organic species into discontinuous and well-marked groups, using a few, or even only one characteristic, such as the nature of the reproductive organs, for the purposes of the classification. The natural systems on the other hand aimed to bring the diverse organic species into natural families, in which there was a continuity of creatures, as many characteristics as could be found being studied in order to establish the affinity of the organisms within a family. Such methods of classification were very necessary to deal with the ever increasing number of animal and plant species which were being discovered. In antiquity Theophrastus knew of some 500 plant species, and Dioscorides later listed 600. The Swiss botanist, Casper Bauhin, 1560-1634, who introduced the modern binominal plant names, described some 6,000 plants known to him in 1623. In the next century the Swede, Linnaeus, 1707-78, classified about 18,000 species himself, whilst at the beginning of the nineteenth century Cuvier, 1769-1832, in France, declared that 50,000 different kinds of plants were then known.

During the sixteenth and seventeenth centuries, the artificial method of classification, which stressed the discontinuity and the hierarchical gradation of the organic species, was more popular in Catholic countries, as for example, Cesalpino and Malpighi in Italy, whilst the natural method emphasizing the continuity and the affinity of the species, was more popular in the Protestant lands, notably Lobelius, 1538-1616, in the Netherlands, Bauhin in Switzerland, and John Ray, 1627-1705, in England. In the eighteenth century these preferences were reversed, the Swedish Lutheran, Linnaeus, adopting the artificial method whilst the French naturalists, notably Buffon, 1707-88, adopted the natural one. Such a change may be connected with the fact that Lutheran thought gradually assimilated elements of the older theology with its concept of hierarchy, whilst the French philosophers of the eighteenth century adopted the mechanical philosophy, which levelled down all phenomena to the same mechanical uniformity.

The first important artificial system of classification was put forward by Andrea Cesalpino, 1524-1603, a professor of medicine at Pisa and later physician to Pope Clement VIII, in his treatise, *On Plants,* published in 1583. Cesalpino adopted the view of Aristotle that the most important feature of plants was the possession of a vegetative soul alone. The vegetative soul was responsible for the nutrition and reproduction of organisms, nutrition being mediated by the roots of plants, and reproduction by their fruit organs. Therefore, Cesalpino argued, the root and the fruit organs should serve as the main characters to be noted in the classification of plants, as they indicated the quality of the vegetative soul that the plant possessed, this soul placing the plant in its proper class of the hierarchy of organisms. Lichens and mushrooms, he thought, had no reproductive organs; they had roots only. Hence they possessed the most inferior of the vegetable souls, and they were to be placed at the bottom of the hierarchy of plants, providing a link between the plant and the mineral worlds.

In the seventeenth century, Marcello Malpighi, 1628-94, who was a professor of medicine at various Italian universities, and then physician to Pope Innocent XII, attempted to find a method of classifying all living creatures into a vertical scale. Malpighi was much interested in Harvey's theory of the circulation of the blood, and the problem of the role of respiration that arose from it. With the microscope, he was the first to see in 1660 the capillaries which connected up the veins with the arteries in the lungs of a frog. He then went on to examine the respiratory systems of other organisms, finding silk worms to be permeated by tubes through which they took in air. Plants he discovered to be filled with spiral air tubules, which he took to be part of the plant respiration apparatus. From these findings he suggested that the size of the respiratory organs in a living creature was inversely proportional to its grade of perfection in the scale of organic nature. Plants he placed at the bottom of the scale as they were filled with air tubules: next came insects with numerous air passages: then fishes with their smaller, though still complex, gill systems: and finally man and the higher animals came at the top of the scale as they had only a pair of lungs which were small compared with the size of their other organs. Malpighi's system of classification did not achieve a great deal of popularity, the first important modern classification of animals being carried out by John Ray in 1693 along the lines suggested by Aristotle.

On the other hand, Cesalpino's artificial system of classifying plants was most influential. It was attractive not only because it had a metaphysical justification in the conception that the species form a graded scale of beings, but also because it was simple and useful in practice. Cesalpino's system required the examination of only one or two organs in a plant, the roots and the fruit. The natural systems had the justification that they reflected more of the objective affinities of plants, but they were less simple, requiring the examination and comparison of all possible organs and characters—stems, leaves, and flowers, as well as roots and fruit organs. Artificial systems were more widely used, therefore, Cesalpino's classification being generally adopted after 1692 when a London physician, Nehemiah Grew, 1641-1712, found that plants reproduced sexually. Following Grew's discovery, Camerarius, 1665-1721, a professor of botany at Tubingen, indicated in 1694 that the stamens of plants were male organs and that the pistils were the female organs. Thus the nature and the number of the stamens and pistils possessed by plants allowed Cesalpino's method of classifying plants according to their reproductive organs to be carried out in greater detail.

Such a method was used the most extensively by the greatest plant classifier of the eighteenth century, Carl Linnaeus, 1707-78, the son of a Lutheran pastor, and professor of botany at Upsala in Sweden from 1741. Linnaeus' view of nature was that of a primitive Christian pietist, and he remained for the most part outside of the main mechanistic stream of scientific thought. His main work, *The System of Nature*, was published in 1735 and passed through twelve editions in his lifetime. The first edition was almost entirely devoid of mechanical conceptions, though by the last edition he had come to consider that animals in principle were machines. Linnaeus devoted his life to the classification of the great number of new plants which were being discovered and brought to Europe from all over the world. He grouped the plants known to him into classes, orders, genera and species, and he named them by means of the binominal nomenclature introduced by Bauhin, using one name for the genus and the other for the species. The number of its pistils determined the order to which a plant was assigned, whilst the number of its stamens determined its class, Linnaeus dividing up the known plants into twenty-four main classes by this method. While he used mainly the artificial method of classifying plants according to one characteristic, their reproductive organs, Linnaeus was attracted by the natural systems of classification based on many characteristics. In 1738 he published a fragment of such a natural system, though his influence was directed mainly towards the artificial method of classifying organisms.

Linnaeus had a passion for classification. His system of nature included not only the orders of the various animal and plant species, but also the different minerals and diseases. He even classified past and contemporary men of science according to military rank with himself as the general. Animals he divided into six classes, quadrupeds, birds, amphibia, fishes, insects and worms, following Ray in the seventeenth century and Aristotle before him. In the early editions of his work he classified whales amongst the fishes, though both Ray and Aristotle had realized that they were mammals. Minerals he thought to be living substances, which grew underground. In the last edition of his *System of Nature* he summed the essence of what he took to be the gradations in the hierarchy of organic nature with the words: 'Minerals grow, plants grow and live, animals grow, live and feel.'

Linnaeus was an influential figure. He attracted numerous pupils, and he and his pupils travelled widely exploring the flora and fauna of little known lands. It became usual for every exploration expedition to have its own naturalist. Captain Cook took Solander, who had studied under Linnaeus, with him on his first voyage of 1768-71. After the death of his son in 1783, the library and collections of Linnaeus were purchased by a London physician, J. E. Smith, who with two friends formed the Linnaean Society in 1788, which was devoted to the study of natural history, the society eventually acquiring Linnaeus' books and collections.

In France there was a strong reaction against the artificial methods of classification which was marked in the work of Georges Buffon 1707-88, Keeper of the King's Gardens at Paris, and Bernard de Jussieu, 1699-1777, Keeper of the Royal Gardens at Trianon. Bernard de Jussieu arranged the plants in the gardens under his supervision according to a natural system which he developed from the fragment published by Linnaeus. He himself published nothing, but his nephew, Antoinel Laurent de Jussieu, 1748-1836, carried on his uncle's work and published the details of the system in 1789, showing the arrangement of smaller groups of plants into natural families, such as the grasses, palms, lilies, and so on. The work of the Jussieus was continued in the nineteenth century by a Huguenot family of botanists who had been resident in Geneva, August de Candolle, 1778-1841, and his son, Alphonse.

Whilst the Jussieus were investigating the natural families of plants, Buffon and his assistant, Daubenton, 1716-1800, were studying the affinities of animals by comparing their anatomical structures. Buffon was of the opinion that all artificial classifications were an 'error in metaphysics'.

'The error consists', he wrote, 'in a failure to understand nature's processes, which always take place by gradations. . . . It is possible to descend by almost insensible degrees from the most perfect creature to the most formless matter . . . there will be found a great number of intermediate species, and of objects belonging half to one class and half to another. Objects of this sort, to which it is impossible to assign a place, necessarily render vain the attempt to a universal system.'

It was with these words that he introduced his main work, his *Natural History*, the first three volumes of which appeared in 1749, giving his general views. From 1753 to 1767 he brought out twelve volumes on the quadrupeds, and from 1781 to 1786, a further ten volumes on birds and minerals. Finally his pupil, Bernard Lacepede, 1756-1825, produced another eight volumes on serpents and fishes between 1788 and 1804 after Buffon's death.

Buffon was of the view that there were no discontinuous classes, orders, genera and species in nature—they were creations of the human mind and entirely artificial. In nature there were only individual organisms, and they showed very small and continuous gradations one from the other. However, he found that crossings between what were termed different species were generally infertile, and so he accepted the view that a species was a group of interfertile individuals. In studying the likenesses and the affinities existing between the animal species, Buffon was led to believe that what are now different species might have decended from a common ancestor. Discussing the two hundred different species of quadrupeds known to him, Buffon suggested that as many of them were so similar, it was possible that they had all descended from some forty original types. Later he thought that all of them might have come from a single set of parents.

'If the point were once gained', he argued, 'that among animals and vegetables, there had been, I do not say several species, but even a single one, which had been produced in the course of direct descent from other species . . . then no further limit could be set to the power of nature, and we should not be wrong in supposing, with sufficient time, that she could have developed all other organic forms from one primordial type.'

Buffon was not an evolutionist in the modern sense; he did not maintain that the more complex and perfect animals had developed from simple and primitive forms in time. On the contrary he believed for the most part that the various organic species were so many degenerate forms of a more perfect original type, or types. The ass was a degenerate horse, whilst apes and monkeys were degenerate men. The pig possessed side toes on its legs which it did not use and this, Buffon thought, indicated that the pig had degenerated from a more perfect type which had had a use for those toes.

The idea that the organic species might change by a process of degeneration was by no means new: it had been suggested by Plato, and it could find a theological sanction in the doctrine of the Fall of Man. However, during the late seventeenth and early eighteenth centuries it was an unpopular opinion; so too was the idea that any species could become extinct. The species must have remained fixed and constant from the Creation, it was thought, because any evolution, degeneration or extinction of the species would detract from the perfection of the world by leaving gaps, or duplicating grades, in the great chain of creatures inhabiting the world. John Ray said in 1703 that if fossils are the remains of extinct animals,

'It would follow that many species of shell fish are lost out of the world, which philosophers hitherto have been unwilling to admit, esteeming the destruction of any one species a dismembering of the universe, and rendering it imperfect, whereas they think the Divine Providence is especially concerned to secure and preserve the works of Creation.'

The popular dictum of Leibniz that our world is the best of all possible worlds, implied in the same way that none of the organic species could have degenerated or become extinct. Leibniz himself did not deny that fossils were the remains of once living creatures, but he got out of the difficulty by assuming that species extinct on earth flourished on other planets, or other planetary systems.

As we have seen, the French philosophers of the mid-eighteenth century rejected the idea that the world was in any sense perfect. They felt that some improvements could be suggested in the social sphere at least, and that the progress of mankind was a reality. Some of the *philosophes* also rejected the idea that the species formed a graded scale of creatures because they were opposed to the concept of hierarchy which it embodied. Voltaire wrote in his *Philosophical Dictionary*, published in 1764:

'When I first read Plato and came upon this gradation of beings which rises from the lightest atom to the Supreme Being, I was struck with admiration. But when I looked at it more closely, the great phantom vanished. . . . At first the imagination takes a pleasure in seeing the imperceptible transition from inanimate to organic matter, from plants to zoophytes, from these to animals, from these to genii, from these genii endued with a small aerial body to immaterial substances, and finally angels. This hierarchy pleases those good folk who fancy they see it in the Pope and his cardinals, followed by archbishops and bishops, after whom come the rectors, the vicars, the simple curates, the deacons, the subdeacons, then the monks and the line is ended by the Capuchins.'

Voltaire argued that there could not be a continuous chain of beings in the universe because there were no gradations amongst the heavenly bodies. In the medieval world view of course the heavenly spheres and their angelic motors had been hierarchically arranged, but the early modern astronomers on the one hand, and the Calvinists on the other, had done away with the celestial hierarchies, so that Voltaire could now instance the uniformity of the heavenly bodies as an argument against the whole conception of a hierarchically ordered universe.

The French philosophers who questioned the Aristotelian notion that the species could be graded into a linear scale, and who upheld the idea of progress, went back beyond Aristotle to the pre-Socratic Greeks for their theories of evolution. Benoit de Maillet, 1656-1738, put forward a theory similar to that of Anaximander in a work published posthumously in 1749. He suggested that all land animals had come from fishes through the influence of changing habits and differing environments. Birds had come from flying fish, lions from sea-lions and men from mermen. Maupertuis in 1751 revived the view of the Atomists that the various organic species arose from different chance combinations of organic units or atoms endowed with life. Diderot followed in 1754 with a version of Empedocles' theory, according to which various animal organs, heads, limbs and so on, had come together fortuitously, giving some monstrous and grotesque creatures which did not survive, and the present day forms which proved to be more viable. The French philosophers of the mechanical school of thought were not interested primarily in biological questions, however. They did not work out satisfactory mechanisms to account for the development of the species, nor did they elaborate their theories of evolution in any detail.

In spite of the rejection of the idea by some of the French philosophers, the conception that the organic species constituted a graded chain of creatures died hard in Catholic France, and it was incorporated into some of the eighteenth-century theories of evolution. The idea of progress was grafted on to the conception so that the chain of beings came to be seen not as a static hierarchy of species, but as a scale of descent up which the species had evolved in time. The beginnings of such a theory were put forward by Jean Baptiste Robinet, 1735-1820, a one-time Jesuit who turned to natural philosophy, upon which subject he published a large work in five volumes (1761-68).

Robinet held that the organic species formed a linear scale of creatures which was full and complete, without any gaps or duplication of grades. The Creator, he wrote,

'has made all vegetable species which could exist: all minute gradations of animality are filled with as many beings as they can contain.'

But the chain was not a static vertical scale: all creatures, he affirmed, receive

'additions which they are able to give themselves by virtue of an internal energy, or to receive from the action of external objects upon them.'

This internal self-differentiating energy, he thought

'the most essential and the most universal attribute of being . . . a tendency to change for the better.'

Such an immanent force was a spiritual fire, it was a biological expression of the same force that on the human plane expressed itself in the progress of the enlightened opinion that governed the world.

'The human mind must be subject to general law,' Robinet wrote. 'We cannot see what could arrest the progress of its knowledge, or oppose its development, or stifle the activity of this spirit, all of fire that it is.'

In the same way the additions which animals received from external objects were analogous to the conditioning of man by his environment, a doctrine the *philosophes* had derived from their psychological theory.

Like Buffon, Robinet could not restrain himself from looking backwards down the scale of creatures. He did not think of organic change as a degeneration from higher to lower forms as Buffon did, but he looked for the form and qualities of man in the lowest of organisms.

'Envisaging the sequence of individuals as so many steps in the progress of being towards humanity,' he wrote, 'we shall compare each of these with man, first with respect to his higher faculties, that is, his reason.'

Thus Robinet supposed that the characteristics of man were present in a germinal form throughout the scale of creatures: there was life and a soul in the most formless piece of matter. Like Maupertuis, Diderot, and others who thought in terms of evolution, Robinet thought that even the fundamental atoms of the universe were endowed with life and a soul. Such a view simplified their theories of evolution. Inorganic matter could easily generate living beings as it was already alive. The different species were merely different combinations of these living atoms.

Another theory of evolution was developed by Charles Bonnet, 1720-93, a descendant of a Huguenot family which had settled at Geneva. Bonnet was an active naturalist in his younger days, discovering in 1740 that female aphids, or tree-lice, could produce living offspring without fertilization. Later on his sight became defective, and he turned to the theory of biology, publishing his *Philosophical Palingenesis, or Ideas on the Past and Future States of Living Beings* in 1770. His discovery that female aphids would produce living young without fertilization led him to the view that the females of every species contain within themselves all the future generations of those species in miniature. Such a view was widely current at the time. It implied that the species are fixed for all time, as all future animals are already existent in germ. However, Bonnet thought that periodically the world was engulfed by a major catastrophe, the last one being the Mosaic Flood. In these catastrophes, the bodies of all living creatures were destroyed, but the germs of their future generations lived on, and were resurrected after the catastrophe had subsided. Moreover, the new resurrected forms were all higher in the scale of beings than the pre-catastrophic forms, they all moved up a place, so to speak, in the hierarchy. To support his theory, Bonnet pointed to fossil bones and shells which he believed to be the remains of animals killed by past catastrophes. He predicted that another catastrophe was to take place, and that after it stones would have organic structures, plants would be self-moving, animals would develop the power of reasoning, and men would become angels. Then, he wrote,

'there may be found a Leibniz or a Newton amongst the monkeys or the elephants, a Perrault or Vauban among the beavers.'

Robinet, the one-time Jesuit, saw evolution as a continuous ascent of the scale of creatures: Bonnet, the Huguenot, conceived of organic change as the result of vast catastrophic mutations. These men gave expression to the two types of evolutionary theory that were to be current in the next half century, and they illustrate the religious beliefs with which the two types of view were affiliated. Protestant thinkers tended to adopt the catastrophic view, whilst deists, sceptics, agnostics and atheists, particularly, at first, those coming from a Catholic background, tended to assume a continuous theory of evolution. Bonnet's idea that a creature's rise in the scale of perfection was accomplished by a catastrophic death followed by a resurrection, was very much a secularized religious conception. A similar theme ran through the biological speculations of Erasmus Darwin in England, though for him the offspring took the place of the resurrected creature, the embryo taking upon itself the improved qualities which earlier thinkers had ascribed to the soul of the departed parent.

The theories of Robinet and Bonnet, and their differences, were developed in the early years of the nineteenth century by Lamarck and Cuvier respectively. Jean Baptiste Lamarck, 1744-1829, came from a minor noble family of Picardy. He was intended for the Church and was educated at the Jesuit College of Amiens, but he abandoned this career, and after a spell in the army, and in commerce, he was given the post of botanist in the King's Gardens at Paris in 1782. The Gardens were closed 1790-3 during the French Revolution, and Lamarck in this period wrote several pamphlets advocating the reform of the posts in the Gardens. His proposals were adopted for the most part by the National Convention, which reopened the Gardens in 1794, Lamarck being appointed to the chair of invertebrate zoology, a position designed for the study of animals without backbones. Georges Cuvier, 1769-1832, on the other hand, came of a Huguenot family which had settled in Montbeliard near Basle. He acted as tutor to a Protestant family in Normandy, and here studied marine life, his studies obtaining him the chair of comparative anatomy at the Garden of Plants, as the King's Gardens were renamed, in 1795. Lamarck took over the theory of continuous organic evolution from Robinet, whilst Cuvier adopted Bonnet's theory of catastrophes, but the work of both finally ended the conception that the organic species could be arranged in a linear series, which had been embodied in the speculations of both Robinet and Bonnet.

Lamarck passed from the study of plants to the study of the lower animals, his investigations, so to speak, taking him up the old scale of creatures. He was fifty years of age when he changed over to invertebrate zoology, though he made rapid progress in this hitherto neglected field. From the time of Aristotle to that of Linnaeus, animals without backbones had been classified as either insects or worms. Buffon, in attacking the Linnaean system, had drawn attention to the inadequacy of the current invertebrate classification. It was absurd, he suggested, to classify crayfish with the insects. Lamarck studied the problem, and by 1800 had divided the two invertebrate classes of insects and worms into ten different classes, his system being the basis of the modern one.

Lamarck perceived that his ten classes of invertebrate animals exhibited gradations of structure and organization, and in 1802 he arranged them in a linear order, followed by the four classes of vertebrates, the fishes, reptiles, birds and mammals, to exhibit the scale of gradations in the animal world. He then conceived of this scale as an evolutionary order, showing the stages through which the animals had passed from simple single-celled organisms up to man, and in 1809 he published his *Zoological Philosophy*, a work in which he indicated how he thought this evolution had come about. Lamarck today is chiefly remembered for his doctrine of the inheritance of acquired characteristics, the idea that environmentally induced changes in the structure and functions of various animal organs are passed on to their offspring. Such an idea was, however, a secondary element in Lamarck's theory. Like Robinet he thought that within each creature there was an inner force which operated continuously for the improvement of the species. If this force were not impeded in any way, it would lead to a perfectly linear series of creatures, a continuously ascending chain of beings from simple unicellular organisms up to man. Lamarck was reluctant to admit that the scale of creatures as he saw it in nature was at all imperfect. He refused to believe, for example, that any animal species had become extinct, for the dying out of a species would leave missing links in the great chain of living creatures on the earth. Some of the larger animals, such as the Mastodon, might have been wiped out by mankind, Lamarck held, but the smaller animals which appeared to be missing from the scale of creatures existed somewhere on the earth and had yet to be discovered. However, the body of zoological facts known to him forced Lamarck to recognize that the chain of beings was not fully continuous in nature, and he brought in his theory of the inheritance of acquired characteristics to explain the deviations from the linear scale.

'Progress in complexity of organization', Lamarck wrote, 'exhibits anomalies here and there in the general series of animals due to the influence of the environment.'

Characteristics acquired by organisms from environmental causes were of two types. Firstly, there was the direct action of the environment producing mutilations and the like which were not inherited. Secondly, there were environmentally induced changes in animal habits, notably the greater or lesser use of organs, which were inherited and which led to permanent changes in the species. In this way the giraffe had developed a long neck by striving to reach the leaves of high trees, whilst moles had lost the use of their eyes through living underground over many generations.

The idea that the organic species form a scale of creatures distorted early theories of organic evolution in two ways. Firstly, man was placed at the head of the scale, so that the series of animals appeared to be so many degradative steps away from man. Secondly, the evolutionary series was considered to be a rectilinear scale and not a branched genealogical tree as later biologists thought of it. However, the notion that the animal species were products of degradative processes was neutralized by the idea of progress, whilst the pressure of empirical facts led biologists to the view that the scale of creatures was not linear. Lamarck observed that all classifications of animals made before his time were in a degradative order, they started with man and the higher animals, and went down the scale. He himself set out to classify them according to 'Nature's actual order', starting with the simple creatures, and moving up to the higher organisms. But in choosing criteria for his classification, Lamarck could not help but look backwards down the scale, and take characteristics typical of the higher animals for his classification of the animal kingdom as a whole. He thought the nervous system the most important characteristic of animals for classificatory purposes because 'It has produced the most exalted of animal faculties, and is necessary for muscular movement.' As secondary criteria for classification, Lamarck selected the respiratory and circulatory systems of animals, though none of his characteristics existed in the lower four of his ten classes. If he had looked up the series in choosing criteria for classifying purposes, he might have hit upon the digestive system which exists in the higher nine of his ten classes of animals without backbones.

The notion that the chain of creatures was a linear series also left its mark upon Lamarck's theory of evolution, though he got away from the idea more than his predecessors had done. Both Bonnet and Robinet had thought that minerals merge into plants, and plants into animals, to form a linearly ascending scale. Lamarck rejected this conception from the start.

'All known living bodies', wrote Lamarck, 'are sharply divided into two special kingdoms, based upon the essential differences which distinguish animals from plants, and in spite of what has been said, I am convinced that these two kingdoms do not really merge into one another at any point.'

Lamarck thought that minerals, plants, and animals, had all developed from a common source. In the beginning, he held, there were gelatinous and mucilaginous particles, together with the exciting forces of heat and electricity. Such forces had led to the development of animals from the gelatinous particles, and plants from the mucilaginous, whilst the same forces had turned the waste animal and plant products into mineral matter. At first, in the simple organisms, the supply of heat and electricity required to sustain them and drive them towards higher forms derived entirely from the environment. But as the scale of creatures was ascended, organisms began to generate their own heat and electricity, so that they sustained themselves and provided their own evolutionary force. Environmental heat and electricity also brought about the decay of waste organic matter, the degradative series being first blood, then bile, urine, and bone substance, followed by shells, marble, precious stones, metals, and finally rock salt, which Lamarck regarded as the basic substance of the earth.

Lamarck at first arranged his animals into a strictly linear series, but the empirical facts of zoology led him to break down more and more his straight line of evolution into a genealogical tree. In a supplement to his *Zoological Philosophy*, Lamarck postulated that there had been two main lines of evolution from the primordial gelatinous particles. One series led from the unicellular protozóa to the animals with radial symmetry, such as the polyps and the star fish. The other series led to all the animals with bilateral symmetry, branching out from the worms to the insects, arachnids, and crustacea on the one hand, and the annelids, cirrhipedes, and mollusca on the other. From the mollusc branch came the vertebrate animals, first the fishes, then the reptiles. Here the chain branched again, the birds and the amphibia being divergent developments from the reptiles. Finally the amphibia led to the mammals with four limbs and five digits, side branches leading off to the whale-like creatures on the one hand, and the animals with hooves on the other. In his last major work, which was published 1816-22, Lamarck broke down his branched chain even further, admitting that he could find no evolutionary connection between his vertebrate and invertebrate series. In these later schemes the original linear scale of creatures was largely abandoned. Minerals, plants, and animals no longer formed a consecutive series. They now had a common origin, whilst the animal scale branched and divided into a genealogical tree, though it was still more rigid and linear than the subsequent evolutionary trees of the Darwinists.

Lamarck had two main purposes in mind when he wrote his *Zoological Philosophy*. Firstly, he desired to show that the animal species formed an approximately linear evolutionary series, and secondly, he wished to resolve the dilemma which was expressed in its most general form by the opposing doctrines of the French philosophers that 'opinion governs the world' and that 'the world governs opinion'. Lamarck posed the problem in a more specific and psychological sense, enquiring into the degree to which the mind was controlled by physical circumstances and the degree to which the mind in turn governed physical events.

'The influence of the physical on the moral has already been recognized (by the psychologists),' wrote Lamarck, 'but it seems to me that sufficient attention has not yet been given to the influence of the moral on the physical.'

According to Lamarck, the trouble had been that the psychologists and the biologists had studied the body and mind of man in isolation from other creatures, without a historical, evolutionary perspective, so that they had not been able to discover the biological origins of consciousness, and had referred the activity of the mind only to stimuli received from the physical environment.

'After the organization of man had been so well studied, as was the case,' wrote Lamarck, 'it was a mistake to examine that organization for the purposes of an enquiry into the causes of life, of physical and moral sensitiveness. . . . An examination should have been made of the progression which is disclosed in the complexity of organization from the simplest animal up to man, where it is most complex and perfect. The progression should have been noted in the successive acquisition of the different special organs, and consequently of as many new functions as of organs obtained. . . . I may add that if this method had been followed . . . it would never have been said that life is a consequence of movements executed by virtue of sensations received by various organs or otherwise: nor that all vital movements are brought about by impressions received by sensitive parts.'

The view that the actions of animals and human beings were governed by impressions and stimuli received from the environment had arisen from the idea that living organisms were machines. Such an idea was justified, Lamarck held, if only account were taken of the driving force of the organism's machinery. In the case of the lowest organisms the driving forces of heat and electricity derived from the environment, so that such creatures were completely governed by external factors. But as the evolutionary scale was ascended the organisms generated their own driving forces to a greater and greater degree, and so they obtained an ever-increasing measure of self-determination which found its fullest expression in the case of mankind. The environment was a constant external determinant of organic nature, and thus the psychologists who supposed that man was governed by his world were correct to some degree. But so too, perhaps in a more important sense, were the theorists who suggested that man had dominion over the environment through his intellect, as the self-determination of creatures was a progressive factor in the evolutionary scale.

'If nature had confined herself to her original method,' wrote Lamarck, 'that is, to a force purely external and foreign to the animal, her work would have remained very imperfect: animals would have been simply passive machines, and nature would never have produced in such organisms the wonderful phenomena of sensibility, the intimate feeling of existence, the power of acting, and lastly, ideas, by means of which she created the most astonishing of all, viz., thought or intelligence.'

In these ways Lamarck expressed and worked upon the problems preoccupying the late eighteenth century, carrying into the biological realm the idea of progress and the dilemma concerning the relationships between man and his environment. He has been described as the last of the *philosophes,* the French speculative thinkers of the eighteenth century, as he concerned himself with the problems they had discussed and considered those problems in the same speculative way. In keeping with the new trends of the Napoleonic epoch, the work of Lamarck was a little more empirical than that of the *philosophes,* his practical classification of the invertebrate animals finding a permanent place in zoological systematics. His theory of evolution also has had supporters since its inception, though Lamarckism has always been a minority opinion in the world of science.

Chapter 29

German Nature-Philosophy

DURING THE second half of the eighteenth century an interest in the study of nature revived in Germany, where science had been dormant since the days of Kepler. The German naturalists of the late eighteenth century developed philosophies of nature peculiar to themselves for they found the mainstream of scientific thought, which was rather mechanistic and materialist, somewhat uncongenial to their tastes. The German poet and nature-philosopher, Goethe, 1749-1832, reported that when the great French treatise on mechanical materialism, Holbach's *System of Nature*, appeared in 1770 it evoked little response in Germany. The theories in the work were felt to be not so much wrong as irrelevant, corresponding to neither the experience of the Germans nor their ideals.

'Not one of us had read the book through,' wrote Goethe, 'for we found ourselves deceived in the expectations with which we had opened it. A system of nature was announced, and therefore we hoped to learn really something of nature,— our idol. . . . But how hollow and empty did we feel in this melancholy, atheistical half-night, in which the earth vanished with all its images, the heaven with all its stars. There was to be matter in motion from all eternity, and by this motion, right, left, and in every direction, without anything further, it was to produce the infinite phenomena of existence. . . . We indeed confessed that we could not withdraw ourselves from the necessities of day and night, the seasons, the influence of climate, physical and animal condition: we nevertheless felt within us something that appeared like perfect freedom of will, and again something which endeavoured to counterbalance this freedom.'

The German philosophers differed from the French in their method of interpreting natural phenomena. The French *philosophes* made the machine their basic analogy, conceiving of

350 / A History of the Sciences

the universe as a vast mechanical contrivance and the objects within it as smaller mechanical devices. In principle the faculties of the human mind could be analysed in terms of matter in motion, such faculties being determined by external agencies, like the physiology of the body and stimuli received from the external world. The Germans were more introspective. They were interested in the self-activity of the human mind, the inner feeling, as Goethe put it, of what appeared to be the freedom of the will and something which limited and balanced that freedom. The German philosophers were of the view that the universe was permeated by a similar spiritual activity, and so the processes of nature were to be interpreted by analogy with the inner movement of the mind, not in terms of the pure externality of matter in motion.

These two views of nature, the German and the French, derived from the two different kinds of natural philosophy which had arisen in the sixteenth and seventeenth centuries, the vitalistic and the mechanistic. The dominant one was the mechanical school of thought, developed by Descartes and Newton. For all their differences, both the Cartesians and the Newtonians regarded matter as essentially inert, material bodies receiving motion from external forces which were mechanical in character. Separate from the world of matter there was, of course, another world of spirit which in principle at least exercised some guiding influence over the material world. However, it was found difficult to establish any logical or empirical connection between the two worlds, with the result that during the eighteenth century the spiritual world was abandoned by the French materialists and the material world by Bishop Berkeley and his followers. The mechanical school of thought was dominant in France and England during the eighteenth century for even Berkeley's world was, in a sense, a mirror image of the Newtonian world. One of Berkeley's Scottish critics, Thomas Reid, pointed out that the ideas which were supposed by Berkeley to make up the whole furniture of the universe were the idealistic equivalents of the material particles postulated by the atomic philosophy, and another of his Scottish critics, David Hume, suggested that the laws governing the association of those ideas were equivalent to the Newtonian laws of gravitational attraction.

The other philosophy of nature, the iatrochemistry of Paracelsus and Helmont, was the minority report of the naturalists of the sixteenth and seventeenth centuries. For the iatrochemists there was no such thing as inert matter. All substances, even minerals and chemical compounds, were alive, for they were permeated by a vital force which caused growth and determined the forms which that growth assumed. Each natural object was therefore an autonomous being, as it derived its growth and movement from its internal vital force, not from external sources of energy. The iatrochemical school of thought was particularly strong in Germany, for whilst the two most important figures of the school came from the periphery of the Germanic lands, Paracelsus from Basle and Helmont from Brussels, most of the secondary figures were of German origin, as also were the philosophers influenced by the theories of the iatrochemists, namely Boehme and Leibniz.

Jacob Boehme, 1575-1624, a cobbler of Gorlitz, was a mystic who described in the language and the imagery of the iatrochemists the history of the universe as he saw it in his visions. Following Paracelsus, Boehme believed that man was a complete copy of the universe in miniature and that he was sustained by his own inner spiritual life. Man was therefore an autonomous microcosm, a little world unto himself, and in the same way the great world was complete in itself, sustained by God who was the soul of the universe or the spirit of nature. Since man was an epitome of the world, and since life and motion originated in the spirit, Boehme felt that the form of his own spiritual development was analogous to the way that the universe had come into being. Boehme had experienced most acutely what Luther had termed 'the internal warfare of the child of God', the psychological conflict between his self-centred bodily desires and his self-denying spiritual aspirations. The conflict in Boehme's case found its resolution in a mystical vision of images and forms, which revealed to him the workings of God in nature and which gave him spiritual tranquillity, a spiritual rebirth, as he put it. What was true for the development of man, the microcosm, must have been true for the genesis of the universe as a whole, the macrocosm. In the beginning, Boehme suggested, the Spirit of nature, or God, by himself was everything and yet also no-thing, for there was nothing set over against God through which He could manifest himself. Hence arose in the Spirit a centrifugal desire towards self-manifestation which generated in turn its contrary and complement, a centripetal will to conscious self-possession. In the movement deriving from the conflict, the will disciplined and assimilated the desire, producing an image of the Spirit which externalized was nature.

'Without nature God is a mystery,' wrote Boehme, 'for without nature is the nothing, which is the eye of eternity, that stands or sees in the nothing; and this same eye is a will, understand a longing after manifestation, to find the nothing; but now there is nothing before the will, where it might find something, where it might have a place to rest, therefore it enters into itself, and finds itself through nature. And we understand in the mystery without nature in the first will two forms; one to nature, to the manifestation of the wonder eye; and the second form is produced out of the first, which is a desire after virtue and power. And understand us thus; the desire is egressive, and that which proceeds is the spirit of the will and desire, for it is a moving, and the desire makes a form (image or likeness) in the spirit . . . and the outward nature of this visible comprehensible world is a manifestation or external birth of the inward spirit.'

The symbolism and the ideas of the iatrochemists and alchemists abound in the writings of Boehme. His triadic law of development through the opposition and resolution of contraries was expressed in terms of the three iatrochemical principles; Sulphur denoting the expansive desire of the Spirit for self-manifestation; Mercury, the will to self-possession which was born of that desire and which stood in opposition to it; while Salt represented the resultant externalized image, the substance or entity which was generated by their resolution. The alchemists had been of the view that the metals were generated by the interaction and union of two contrary principles, the male and fiery Sulphur, and the female, liquid Mercury. Generalizing the notion, Paracelsus had declared that in alchemy and medicine 'we must set entity against entity, so that each becomes in a sense the wife or husband of the other'.

With Boehme the sexual element in the idea was minimized and the notion of development through the conflict and resolution of contraries in the spiritual realm brought to the fore. Boehme also shared with the alchemists the idea that the process of death and regeneration brought about a mutative advance in the development of a being. The alchemists had believed that base metals could be transmuted into gold by such a process. Death and decay, wrote Paracelsus, 'brings about the birth and rebirth of forms a thousand times improved'. For Boehme, the resolution of contraries was the death of their conflict and the necessary prerequisite for a higher form of development and being, a kind of transformative rebirth.

Boehme, in his mystical, iatrochemical language, gave the primary formulation of the German concept of the dialectic of natural change and development. He was in fact far more interested in the development of nature than in the structure of the world and its creatures, a topic which was discussed more fully by Leibniz, 1646-1716. Leibniz was also influenced by the theories of the iatrochemists, particularly those of Helmont, with which he was acquainted through his friendship with Helmont's son. Van Helmont was of the view that the entities composing the world were a multiplicity of autonomous beings, each deriving its generation, growth, and self-determination, from its own inner vital force. Even the parts of a composite entity, such as the separate organs of the human body, had a life and being of their own which was governed by a specific and individual vital spirit, the archeus as Helmont termed it.

The natural philosophy of Leibniz was very similar to that of van Helmont though it was more abstract and it possessed more of a logical structure. Leibniz supposed that the world was made up of a number of units, or monads as he termed them, which were non-material centres of vital force. The monads of Leibniz, like Helmont's archei, were autonomous beings, having no physical influence one upon the other. However, they did not constitute an anarchy of blindly self-determinate entities for they had all stemmed from a single Creator who had pre-established the harmony of their activities. Similarly there was a 'consensus' of sympathy between the creatures of van Helmont's world and between the individual parts of each of those beings. The action of Helmont's archei was due to a consciousness, however dim, of 'the types or patterns of things to be done by itself', in the same way that the changes of the Leibnizian monads were due to their own 'small perceptions'. Unlike the Newtonian world which was composed of units, or atoms, that were material, inert, and all alike, the world of Leibniz was composed of units, or monads, which were active forces, purely spiritual and infinitely graded one from the other so that no two were alike. The monads differed like 'so many ordinates of a single curve', forming a continuous chain of autonomous beings graded in rank and perfection. Although the gradations between them were continuous, the monads, like the archei, fell into one or other of three main classes which corresponded roughly to the traditional categories of vegetable, animal, and rational souls. Each of the monads, within the limitations set by its degree of perfection, was a microcosm or little world, reflecting within itself the whole of the universe. Since the monads were active centres of force, they must bring about change, but as they were isolated physically one from the other the only change they could effect was their own inner growth. Thus the monads, like the archei, developed within an aggregate of other units towards the perfection of their kind, whereupon the aggregate dissolved and the monads entered fresh assemblies to begin a new cycle of growth.

The views of Boehme and Leibniz did not have a great deal of influence upon scientific thought immediately, perhaps because the German contribution to modern science had already passed its first climax with Kepler early in the seventeenth century. However, towards the end of the eighteenth century there arose in Germany a school of nature philosophers, headed notably by Friedrich Schelling, 1775-1854, and Lorenz Oken, 1779-1851, who in effect endeavoured to show how the dialectic of growth and development postulated by Boehme could generate a world of beings similar in structure and status to those of the philosophy of Leibniz. Some of the themes expressed by Boehme and Leibniz were developed by German philosophers and scientists before the time of the nature philosophers. Immanuel Kant, 1724-1804, for example, in his main works on the physical sciences from the *General History of Nature* (1755) to the *Metaphysical Basis of Physics*

(1786) laid it down that the interaction of the opposite forces of attraction and repulsion was prerequisite to the existence of material objects. If repulsive forces alone existed, Kant argued, matter would be dispersed infinitely over space, while attractive forces alone would gather all matter together at one point. Hence finite objects must be ordered by the interaction and balance of these two forces.

In the biological sciences Kant's pupil, Johann Herder, 1744-1803, gave expression to the death and regeneration theme of the 'Teutonick Philosophy', supposing in 1784 that the scale of organic creatures constituted a historical sequence, and that the large scale destruction of the lower creatures was a necessary prerequisite to the development of the higher species, as they were made up from the materials of which the lower organisms had been composed.

After Herder, Schelling sketched out between 1797 and 1800 the main lines of the German system of nature philosophy, which was expounded notably in the *Elements of Physio-Philosophy* (1810) by Oken, whose ideas had a considerable influence upon the biological sciences during the first half of the nineteenth century. Schelling and Oken brought together the main ideas of Boehme and Leibniz, and infused them with the conception that the universe was the product of a historical development, a notion which had appeared earlier in the works of Herder and to some degree in those of Kant. Adopting the view that man was an epitome of the whole universe, an idea which was of considerable force in German thought, Schelling and Oken suggested that man was a complete microcosm because he was the final product of the development of the world and summed up within himself all aspects of the previous development of nature. He was also a complete microcosm, they suggested, because the World Spirit, which had generated nature by manifesting externally its own internal development, had finally found itself fully manifest in the mind of man. Hence the development of nature was to be understood in terms of the movement of thought within the human mind as the one necessarily reflected the other.

'Man is God wholly manifest,' wrote Oken. 'Man is the summit, the crown of nature's development, and must comprehend everything that has preceded him, even as the fruit includes within itself all the earlier developed parts of the plant. In a word, Man must represent the whole world in miniature. Now since in man are manifested self-consciousness or spirit, Physio-philosophy has to show that the laws of spirit are not different from the laws of nature; but that both are transcripts or likenesses of each other.'

As in Boehme's system, the World Spirit by itself was everything and yet at the same time no-thing, so that there came into being the desire to self-manifestation and, in opposition to the desire, a complementary will to self-possession and self-contemplation. Hence a conflict of opposite polarities emerged which was the archetypal cause of all motion, development, and life in nature, namely an opposition between an individualizing egocentric principle and an exocentric universalizing principle.

'The action or the life of God', wrote Oken, 'consists in eternally manifesting itself, eternally contemplating itself in unity and duality, eternally dividing itself and still remaining one. . . . Polarity is the first force which appears in the world. There is no world, and in general nothing at all without polar force. . . . The law of causality is a law of polarity. Causality is an act of generation. The sex is rooted in the first movement of the world. . . . In everything therefore, are two processes, one individualizing, vitalizing, and one universalizing, destructive.'

Such a process of conflict and resolution within the World Spirit, at every stage of its inner movement, found an externalized manifestation in an individual entity of the natural world. Each individual being summed up the whole of the development preceding it for, wrote Oken, 'the last product of an antecedent stage is always the basis of that which is subsequent'. Thus each being was a microcosm of the whole world within the limits set by its level of development; in fact its grade of perfection could be measured by the degree to which it reflected the whole universe. 'The more a thing has adopted in to itself of the Manifold of the universe,' said Oken, 'by so much the more does it resemble the Eternal.'

The Eternal, the World Spirit, the Absolute Idea—such were the various names given to the driving force behind the development of the universe by the German nature philosophers—possessed the quality that it was completely autonomous and self-determined, forming an individual whole with the world in its entirety. The grade of perfection of a particular entity could be gauged therefore by the degree to which it was a self-determinate individual being. The nature philosophers recognized three main grades or levels of development in nature. Firstly, there were mechanical entities, like the sun and planets, which formed a system, though they possessed the lowest degree of self-determination: secondly, there were chemical substances which had a greater individuality and self-activity, as many chemical processes were spontaneous and highly specific: and thirdly, there were living organisms which were self-developing individuals with self-contained structures. Each grade of entities possessed the qualities and properties of the lower grades as well as those peculiar to itself, so that chemicals possessed mechanical attributes and living organisms both chemical and mechanical qualities, while man at the top of the scale of living creatures subsumed within himself the properties of all entities in the universe and was a complete microcosm.

The entities of the three grades and their subdivisions had not evolved one from the other in time, for they were isolated and had no connection with one another. There was a historical evolution only in the inner self-development of the World Spirit which by the process of resolving self-contradictions generated the series of images, forms, or plans upon which the objects of nature were based. The scale of living organisms was such a succession of separate externalized manifestations of the inner movement of the World Spirit, so that all creatures had a common source in the World Spirit but they had no physical or historical connection one with the other. However, the various living creatures resembled one another, as the higher organisms possessed the properties and qualities of the lower as well as their own particular characteristics, and since all creatures came from a common source it appeared that an individual organism during the course of its embryological develoment should pass through stages bearing some similarity to organisms below it in the scale of beings. Such a view, the law of embryological recapitulation, was put forward in 1793 by the nature-philosopher, Kielmeyer, 1765-1844, professor of biology at Stuttgart. Kielmeyer suggested that each animal developed embryologically up the scale of living organisms to the grade of its species, so that man, for example, first as an embryo had a vegetable life, then an animal life, and finally a rational life.

'It seems', wrote Kielmeyer, 'that the series of individual organic species have arisen separately from our earth, and that the composition and forms of these various kinds of organisms bear real relationships one to the other, even regarding their separate development. The resemblance of the species one to the other, and their diversity, appears in the source, almost as though they came from a common father.'

According to the nature philosophers, the various kinds of living organisms were also related by the fact that they were all composed of the same material units, living cells, or infusorial mucus-vesicles as they were called by Oken who first put forward the idea in 1805. He suggested, rather like Herder, that the species could not have evolved one from the other, as the death and decay of one creature was the prerequisite for the generation of another. The decomposition of one organism into its constituent units provided the material for the formation of another organism based on a different ideal plan or form.

'The process of change in organic individuals is that of their destruction,' wrote Oken; 'death is no annihilation but only a change. Death is only a transition to another life, not unto death. . . . Physically regarded also every individual originates from the Absolute, but not one out of the other. The history of generation is a retrogression into the Absolute of the Organic, or the organic chaos—mucus—and a new evocation from the same.'

Oken was of the view that the inorganic world had originated from a chaos of aether under the organizing influence of polar forces, such as the attractions and repulsions of electricity, magnetism, etc., and that when the inorganic polarities were finally resolved and ordered, a new chaos of infusorial mucus-vesicles came into being, providing the basis for a new and higher phase of organic development.

'As the whole of nature has been a successive fixation of aether,' Oken wrote, 'so is the organic world a successive fixation of infusorial mucus-vesicles. The mucus is the aether, the chaos of the organic world. . . . The infusorial mucus mass originated at the moment when the earth's metamorphosis was at an end; at the moment when the planet succeeded in so bringing together and identifying all the elementary processes that they were all together or at one and the same time in every point.'

The mucus-vesicles were the vital units of which all organic beings were composed. They were, so to speak, the monads of Leibniz in a material form. The mucus-vesicles possessed two lives—a primary one of their own and a secondary one as a constituent part of an organism. Like the monads, the mucus-vesicles did not die with the passing away of the organism but they lived on to form the substance of a new organic being. The mucus-vesicle theory led to the cell theory, and in the 1830's, when the new achromatic microscopes appeared, the living units of organic beings, the cells, were seen and described.

Another relationship between the various organic species perceived by the German nature philosophers was the unity of plan or structure common to whole groups of organic beings. They conceived of the various organic species as so many edifices constructed by a single Architect, the World Spirit, who had developed a number of general ideal plans or designs of which the forms and structures of living creatures were the various modifications. The different organic beings were all constructed from the same bricks, the mucus-vesicles, and in the early stages of their development they were very similar, but they differentiated in their growth as the various ideal plans became more and more manifest. There was an evolution therefore in the ideas and designs of the Architect, but the organic structures produced were separate creations, related only by affinity of plan.

In the 1780's Goethe searched for the ideal archetype of the vegetable world, the general plan common to all plants. In 1787 Goethe wrote to Herder:

'The archetypal plant will be the strangest growth the world has ever seen, and nature herself shall envy me for it. With such a model, and with the key to it in one's hands, one will be able to contrive an infinite variety of plants. They will be strictly logical plants,—in other words, even though they may not actually exist, they could exist.'

The various organic species were regarded as logical modifications of their common forms as the World Spirit was fully manifest in the mind of man and the logical processes of human reason should therefore reflect the development of nature. In 1790 Goethe suggested that the archetype plan of the vegetable world consisted of a number of typical leaf forms, from which all plant structures could be derived, apart from the stem. The first leaves which emerged from the seed were simple in form but successive ones differentiated and became more complex, developing a central rib, indented edges, and so on. Finally flowers and fruit organs appeared, the petals of flowers and the parts of the fruit organs, according to Goethe, being merely modified leaves.

In the field of zoology, Oken suggested in 1807 that the archetype plan of vertebrate animals was a generalized backbone, composed of a number of unit vertebra segments. He noted that the skeleton of the more primitive vertebrate animals, such as the fish, was little more than a simple segmented spinal column, and that the same was true for the skeleton of the early embryonic forms of higher animals. Oken presumed therefore that a number of vertebra segments with rib and limb appendages was the basic structure of the animal body, other skeletal forms in higher animals being modified segment units. He advanced the theory that the skull of mammals was made up of four such vertebra segments, the jaws being modified ribs or limbs. The segments had become enlarged and had changed in shape, though they had preserved their individuality to some degree, as the skull tended to break up into its units along the lines of the sutures.

Such were the main tenets of the German nature-philosophy, and the important lines along which it influenced some of the biological sciences, notably embryology, the study of the development of individual organisms, morphology, dealing with the forms and structures of living creatures, and the cell theory. Embryology was largely a German science during the late eighteenth and early nineteenth centuries, whilst the cell theory was almost entirely a German development. Nature philosophy also had an influence outside of Germany, and morphology was studied notably by Cuvier, 1769-1832, and Hilaire, 1772-1844, in France and Richard Owen, 1804-92, in England. Upon the physical sciences the nature-philosophers had less influence. They were much opposed to the mechanistic point of view of the English and the French which was dominant in those sciences, but here they had little that was positive to offer. The early mystical speculations of their countryman, Kepler, they ranked above the astronomy of Newton, and they considered the Newtonian theory of optics to be quite erroneous. Goethe suggested that white light was not made up of the spectral colours; on the contrary, the colours were produced by the interaction between white light and its opposite, darkness. As proof, Goethe pointed out that if a dark room were entered after looking at the sun, coloured after-images of the sun were seen. Such physiological phenomena produced by the eye were just as valid as objective optical phenomena, for, said Goethe, 'optical illusion is optical truth'.

Elsewhere in the physical sciences the nature-philosophers were interested particularly in electricity and magnetism, the opposite polarities of which seemed to exemplify perfectly their theory of the general conflict and interplay of opposite forces in nature. One nature-philosopher, Oersted, made an important discovery in this connection as we shall see. In the field of chemistry the nature-philosophers postulated that electrical forces were responsible for the combination of chemical substances, a theory which enjoyed a considerable following when experimental evidence for the view was later discovered.

The German nature-philosophers in general were highly speculative. Hegel, who gave a logical structure to the theories of Schelling and Oken, poured scorn on the English term, 'philosophical instruments', used to describe apparatus for scientific experiments. The only instrument one required in natural philosophy, according to Hegel, was the correct set of theoretical concepts. From about the 1820's German science became more empirical and experimental, but the researches pursued in Germany, and the theories deriving from those researches, continued to show the influence of nature-philosophy for several more decades. Even in the late nineteenth century theories of organic evolution based on some of the ideas of the nature-philosophers were formulated in Germany in opposition to Darwinism.

Chapter 30

Embryology: The Development of Individual Organisms

MODERN EMBRYOLOGY may be dated from the work of William Harvey *On the Generation of Animals,* which was published in 1651. Harvey studied the growth of the deer embryo and the development of the chick in incubating eggs. He found that they were somewhat similar, and he came to the conclusion that the mammalian embryo in its sac was equivalent to the egg of birds and other oviparous creatures. Harvey therefore formulated the dictum that all animals came from eggs, though he accepted the possibility that lowly and imperfect animals might be generated spontaneously. Adopting Aristotle's theory of embryological epigenesis, Harvey held that in general the animal foetus gradually developed from amorphous and homogeneous matter into an integrated organism composed of differentiated and heterogeneous parts. In contrast, the scientists of the mechanistic school of thought adopted, for the most part, the preformationist theory according to which organisms were already differentiated and fully formed in their seeds, embryological development being merely an increase in size of the preformed miniature creature without any diffentiation or addition of new parts.

Added weight was given to Harvey's view that all animals came from eggs in 1672 when a Dutch physician, Reinier de Graaf, 1641-73, observed with a lens the close resemblance between the ovaries of a bird and the ovarian follicle of a rabbit. Indeed de Graaf thought that the follicle was the true mammalian egg. Another Dutchman, Jan Swammerdam, 1637-80, observed that the butterfly was fully formed at the chrysalis stage, and from this discovery he presumed that the butterfly was preformed at the earlier caterpillar stage and even in the egg. Swammerdam developed the extreme form of the preformation theory, supposing that an egg contained all the future generations of its kind as preformed miniatures, like a series of boxes one inside the other.

'In nature there is no generation,' wrote Swammerdam, 'but only propagation, the growth of parts. Thus original sin is explained, for all men were contained in the organs of Adam and Eve. When their stock of eggs is finished the human race will cease to be.'

In 1677 a linen merchant of Delft, Antony Leeuwenhoek, 1632-1723, discovered with the microscope the existence of male spermatozoa. Nicolas Hartsoeker, 1656-1725, also of the Dutch school, followed up the discovery and published pictures of the preformed men or Homunculi which he claimed to have seen in spermatozoa with the microscope. Thus by the end of the seventeenth century two versions of the preformationist theory had been formulated; the one supposing that the offspring was present fully formed in the female egg, a view upheld by Swammerdam, Malpighi, and later by Bonnet, von Haller, and Spallanzani; the other supposing that the preformed individual was present in the male sperm, the view held by Leeuwenhoek, Hartsoeker, Leibniz, and Boerhaave.

The preformation theory tended to be associated with the mechanistic school of thought at this period though there were vitalists, such as Leibniz, who were preformationists, and mechanical philosophers, like Maupertuis, who supported epigenesis. The mechanical philosophers held that animals and plants were machines, but they were undoubtedly complex machines and it was difficult to see how they could be formed from the bare components of matter and motion. George Garden wrote in 1691:

'Indeed all the laws of motion which are as yet discovered can give but a very lame account of the forming of a Plant or Animal. We see how wretchedly Des Cartes came off when he began to apply them to this subject (of generation): they are formed by Laws yet unknown to Mankind.'

It was suggested therefore that organisms were fully formed in their seeds and developed by mechanically increasing in size. More completely the problem was solved by assuming that all future generations of the organic species had been preformed in miniature at the Creation inside the first formed adult organisms, the whole collection of preformed germs being handed on from parents to offspring apart from those that became the immediate offspring. As the London physician, George Cheyne, noted in 1715:

'If Animals and Vegetables cannot be produced from these (Matter and Motion), and I have clearly proved that they cannot, they must of necessity have been from all eternity; and consequently that all the Animals and Vegetables that have existed, or shall exist, have actually been included in the first of every species.'

The mechanical philosophers inclined to the view that organisms were materially preformed in their seeds, while the vitalists, as represented first by the iatrochemists and then by the German nature-philosophers, tended towards an idealist preformation theory, the view that a vital force within the seed of an organism caused embryological growth according to a predetermined pattern of development. As van Helmont had put it, 'the seminal efficient cause containeth the types or patterns of things to be done by itself . . . in the business of generation'. Leibniz adopted such a theory, and so too, with some qualifications, did the author of the phlogiston theory, Ernst Stahl. But the idea that an organism was ideally preformed by virtue of its own specific vital force implied that a real physical development, a material differentiation, should be observed empirically in the embryological growth of the organism. Such an implication, the theory of embryological epigenesis, was put forward in 1759 by the German, Caspar Friedrich Wolff, 1738-94, a successor to Stahl at the university of Halle and later a member of the St. Petersburg Academy of Sciences.

The preformation theory was questioned a few years before Wolff by John Turberville Needham, 1713-81, an English Catholic priest, who endeavoured to show in 1748 that single-celled creatures were generated spontaneously in meat broth which had been sterilized and sealed off from the air. Needham argued that if creatures were generated spontaneously they must be formed epigenetically from amorphous matter, not from preformed germs. However, his broth solutions were imperfectly sterilized, as the Italian, Spallanzani, 1729-99, showed in 1767. Wolff, at first, criticized the preformation theory philosophically, his *Theory of Generation* (1759) being almost entirely theoretical. The supporter of preformationism, wrote Wolff, 'does not explain the development of organic bodies but denies the occurrence of it'. If the organism were already preformed in the egg or seed, Wolff argued, then we should be able to see in the embryo limbs and organs of the adult shape and form. He pointed out that such was not the case, the limbs and organs of an embryo passing through a variety of configurations before they reach the adult shapes and forms, an observation he supported by delineating the gradual development of the blood vessels in the chick. Wolff ascribed embryological growth to the operation of a vital force upon homogeneous organic matter. This matter was a clear, viscous, nutritive fluid which at first possessed no organization whatsoever. As development proceeded, cavities developed in the fluid which solidified round the edges, giving cells if the cavities were round or polygonal, and vessels if they were elongated. Thus tissues were formed, and these were further differentiated by the vital force into organs, the whole development being quite independent of external influences.

Wolff's views were criticized by the Swiss, Albrecht von Haller, 1708-77, a professor of medicine at the newly founded university of Göttingen. Haller held that the chick and its blood-vessels were present in the egg from the beginning in the form of a fine invisible network of tissues covering the yolk. The act of fertilization started the process of growth which merely made explicit the invisible preformed structure. The heart was set beating and blood was pumped through the fine network of tissues, which were thus expanded and made visible through the assimilation of food particles carried by the blood. The argument most used by Haller against the epigenesists was that the preformed organisms in the eggs or seeds were at first invisible and that the different parts of them became visible at different times. Haller admitted that the preformation theory as it stood was not complete.

'Although it is not easy to explain everything mechanically', he wrote, 'those difficulties which have been moved (by the epigenesists) cannot overturn such things as have been truly demonstrated, though perhaps some things may remain to which, in so great an infancy of human knowledge, we cannot yet give a full answer'.

Haller adopted the extreme *emboîtement* version of the preformation theory.

'The ovary of an ancestress', wrote Haller, 'will contain not only her daughter, but also her granddaughter, her great-grand-daughter, and her great-great-grand-daughter, and if it is once proved that an ovary can contain many generations, there is no absurdity in saying that it contains them all.'

The *emboîtement* view was supported also by another Swiss, Charles Bonnet of Geneva. Bonnet discovered in 1740 that female tree lice could produce living offspring without being fertilized by the male, and so he concluded that the females of the organic species contained the germs of all future generations.

In 1768 Wolff renewed his criticism of the preformation theory in a work describing his empirical researches on the formation of the intestine in the chick. He showed that the chick intestine could not be a preformed structure in the egg as it developed from a simple sheet of tissue during the course of embryonic development, the sheet first folding along its length to form a gutter and then closing round to form a tube. In the higher animals he discovered an organ, a primitive kidney termed the Wolffian body, which first of all developed and then disappeared during the course of embryological growth. Such discoveries demonstrated that the organs of a creature could not be materially preformed in the egg or seed, they must develop by the differentiation of simple uniform tissue into more complex and heterogeneous structures. However, the hold of the preformation theory proved to be tenacious, and the theory of epigenesis was slow to gain ground in the world of science.

With Wolff some of the characteristic doctrines of the German nature-philosophy made their first appearance. Nature, Wolff suggested, was permeated by a life force which fashioned simple homogeneous materials into complex and differentiated structures, following the same general plan in all of its diverse productions. Plants developed in the same way as animals. Under the impetus of the life force, plants drew up moisture from the soil and passed it to the growing points where the moisture was concentrated and thickened to form new tissue. Plants, like animals, differentiated as they developed. The first leaves of a plant were simple in structure, but later ones were more complex, developing a central rib and crenated margins, and the final leaves were transformed into the petals of the flower. Searching for forms common to plants and animals, Wolff suggested that the sap-vessels of plants were analogous to the blood vessels of animals, and that the leaves of plants were homologous with the limbs of animals. He also compared the leaves of plants with the first simple sheets of tissue formed in animal embryos, terming those sheets leaflets or layers. Within the animal kingdom, Wolff noted that the embryos of different species resembled one another much more closely than did the adult forms and that they developed in very similar ways.

The theories of Wolff were not developed further for some time. They were supported by Johann Blumenbach, 1752-1840, at Göttingen, Friedrich Kielmeyer, 1765-1844, at Stuttgart, and by Johann Meckel, 1781-1833, at Halle. Following Wolff's observation that the embryos of different species resembled one another more closely than the adult forms, Kielmeyer suggested that there was a physiological parallel between the embryological development of higher organisms and the scale of perfection into which the adult forms of organic beings were arranged. Each organism developed up the scale to the grade of its species so that the human embryo, for example, first had only vegetative functions, then those of the lower animals which move but do not feel, and finally those of the higher animals which both move and possess sensation. Autenrieth in 1797 suggested that there was also an anatomical parallelism, the structure of an embryo passing through anatomical forms similar to those of lower creatures on the way to its adult form. Meckel in 1811 supposed that the species might have evolved one from the other, and that the embryos of the higher animals during the course of their growth recapitulated anatomically and physiologically the evolutionary development of their species.

The German romantic philosophy of nature, which was now at the height of its development, had a curious incidental effect upon the arguments used in favour of the theory of epigenesis. The mechanical philosophers had tended to take a standardized and simplified view of nature, regarding the world and its inhabitants as uniformly mechanical in character throughout. Preformationism, for example, emphasized the fixed uniformity of the succeeding generations in each of the organic species. The romantic philosophers on the other hand emphasized the uniqueness of the individual, exalting that which was novel and even grotesque. Following such a train of thought they advanced a telling argument against the preformation theory—the birth of monsters. The argument had been used by Wolff and now it was particularly stressed. Blumenbach in 1789 described monstrous human embryos which resembled frogs, and in 1812 Meckel gave an account of others which had the organs of lower animals, such as the fish. Following Wolff, who had suggested that monsters were due to arrested embryological development, Meckel supposed that the growth of a human embryo might be impeded while it was developing up the scale of organic beings, so that a monstrous child was born with the organs of lower creatures. In order to establish the theory of epigenesis, the Frenchman, Geoffroy St. Hilaire, 1772-1844, endeavoured to produce monstrous chicks artificially in 1826 by tearing the embryos, by coating part of the incubating eggs with wax, and by turning the eggs the wrong way up, or shaking them.

In Germany the rather speculative work of Blumenbach, Kielmeyer, and Meckel, was followed by the important empirical researches of Martin Rathke, 1793-1860, at Dorpat and later Konigsberg, Heinrich Pander, 1794-1865, at Riga, and above all, Ernst von Baer, 1792-1876, who was a professor of physiology at Konigsberg and then an academician at St. Petersburg like Wolff before him. In 1817 Pander showed that the development of the chick embryo proceeded through the formation of three primary tissue layers, or leaflets as he called them following Wolff. From one or other of these layers the several organs of the chick were formed. From one layer, the innermost, the alimentary canal developed, as Wolff had demonstrated, from the next came the muscles, skeleton, and excretory system, and from the outermost layer were formed the skin and nervous system. Rathke discovered in 1829 that the embryos of birds and fishes passed through a stage in which they possessed gill-slits, like adult fishes, the slits disappearing as the lungs developed. He considered this discovery to be good evidence for Meckel's law of biogenesis, the view that during its growth the embryo of an organism recapitulated the development of its species.

Von Baer continued and extended the work of his friends, Pander and Rathke. In 1827 he described the true mammalian egg which he isolated from the ovarian follicle of a dog. De Graaf in the seventeenth century had thought that the ovarian follicle, which he discovered in the rabbit, was the egg itself, but von Baer showed that the mammalian egg originated from within the follicle. In the embryological development of vertebrate animals, von Baer discovered that there appeared a transient cellular rod, a primitive spinal column, termed the notochord, which disappeared with the formation of the true backbone. The notochord was later found in the adult forms of certain lowly creatures, notably the Amphioxus, and so too was the Wolffian body, the primitive kidney discovered by Wolff.

Extending the researches of Pander, von Baer showed that in the embryological growth of a wide variety of animals the first important development was the appearance of four tissue layers, and that the same organs in different animals came from the same particular tissue layer in the embryos. The uppermost layer formed the skin and central nervous system, the second layer the muscular and skeletal systems, the third layer gave the main blood vessels, and the lowest layer formed the alimentary canal and dependencies. Von Baer distinguished three main stages in the development of an embryo. Firstly, primary differentiation, or the formation of the four layers; secondly, histological differentiation, or the formation of different tissues within those layers; and thirdly, morphological differentiation, or the organization of different tissues into organs and organ systems. Generally speaking, the same tissues and the same organs in different animals came from the same layer, so that corresponding structures in different animals could be identified by comparing the embryological developments of those animals. Analogous structures whch had the same function in different animals, like the air-tubes in insects and the lungs in birds and mammals, could be distinguished from homologous structures which had the same origin but might have different functions, like the wings of the birds and the forelimbs of the mammals. Homologous structures came from the same parts of the embryo while analogous structures, unless they were homologous too, came from different parts. In this respect the embryologists had the advantage of the comparative anatomists who studied only the adult forms of organic beings, as the latter based their judgements, said von Baer, 'on an undefined intuition', and confused analogies with homologies. The comparative anatomists, for example, considered that the mammalian skull had been formed from four vertebra segments, and so was serially homologous with the vertebrae, but the embryologists showed that the vertebrae developed from segmented structures round the notochord in the embryo, while most of the skull came from an unsegmented plate beyond the notochord.

Von Baer suggested similarly that the classification of animals and the identification of the general types to which they belonged would be better based on the evidence afforded by comparative embryology than on that provided by comparative anatomy. He pointed out that while all animal embryos began as a single fertilized egg they diverged thereafter, displaying one or other of four main types of development. Firstly, there were the vertebrate embryos which assumed a double symmetry, the four germ layers forming two tubes below the notochord and two above; secondly, the embryos of annulate animals which became symmetrical about the primitive streak in the yolk of their eggs; thirdly, those assuming a spiral form of development as exemplified in the shell of the snail and other molluscs; and fourthly, those developing radially, like the starfish. These four modes of development distinguished the four main groups of animals, the vertebrates, the annulates, the molluscs, and radiata. The structures of all the animals within a given group were based on a common general plan, which, with its variations from animal to animal, became more and more manifest during the course of embryological development. The four main groups or types of animal were independent of each other, though the embryos of different types were similar at very early stages, all embryos starting as single eggs, and passing through a stage where they were a simple sphere of cells. Within a given type, the embryos of different animals continued to resemble one another through succeeding stages of growth and gradually diverged later, but they never resembled the differentiated embryos of the other animal types.

Von Baer therefore rejected Meckel's law of biogenesis, according to which the embryos of higher animals were thought to pass through stages resembling the adult forms of animals lower in the evolutionary scale of organic beings. He denied that the organic species formed a linear scale of creatures and that they had evolved one from the other in time. Under the influence of the nature-philosophers, von Baer held that the species were divergent productions from a common source, there being four main types of divergence, each based on a common general plan. To account for the resemblances observed between the embryos of different animals, von Baer postulated the following four laws:

'I. That the general characters of the big group to which the embryo belongs appear in development earlier than the special characters. In agreement with this is the fact that the vesicular (single-celled) form is the most general form of all; for what is common in a greater degree to all animals than opposition of an internal and an external surface?

'II. The less general structural relations are formed after the more general, and so on, until the most special appear.

'III. The embryo of any given form, instead of passing through the state of other definite forms, on the contrary separates itself from them.

'IV. Fundamentally the embryo of a higher animal form never resembles the adult of another animal, but only its embryo.'

Von Baer himself was much of a nature-philosopher, though his work was less speculative than that of Schelling and Oken. He considered that the main purpose of growth and development in nature was the production of self-contained and self-moving individuals. 'The essential result of development when we consider it as a whole', he wrote, 'is the increasing independence of the developing animal.' Such an aim or creative thought pervaded the whole of the universe.

'It is this same thought', wrote von Baer, 'that in cosmic space gathered the scattered masses into spheres and bound them together in the solar system, the same that from the weathered dust on the surface of the metallic planets brought forth the forms of life. And this thought is nought else but life itself, and the words and syllables in which life expresses itself are the varied forms of the living.'

Von Baer followed up empirically with considerable success some of the ideas suggested to him by the German nature-philosophy. Indeed he all but exhausted the fruitfulness of that philosophy in the field of embryology, for the theory of embryological growth progressed little further until the coming of Darwinism, which suggested fresh avenues for exploration. It was then that Ernst Haeckel revived, in a new Darwinian form, the earlier view that the development of an individual organism recapitulated the history of its race, which gave a further stimulus to the study of embryology.

Chapter 31

The Structure and Function of Living Organisms

DURING THE seventeenth and eighteenth centuries, biologists had been concerned mainly with the classification of animals and plants. The artificial method of classification, based on only a few or even one characteristic, such as the sex organs in the case of plants, had been easier to apply and had achieved greater success than the natural method which required the examination of all of the characteristics of the organisms studied. However, the natural method, which placed animals or plants in their natural families, was felt to throw more light on the relationships of the organisms one to the other than the artificial method with its somewhat forced and arbitrary categories. Linnaeus, the famous botanical classifier of the eighteenth century, wrote in 1764:

'The natural orders teach us the nature of plants, the artificial orders enable us to recognize plants. The natural orders, without a key, do not constitute a method: the method ought to be available without a master.'

For Linnaeus there was no certain key to the natural method, and his understanding of the character and provenance of plants was intuitive.

'A practical botanist', Linnaeus wrote in 1751, 'will distinguish at the first glance the plant of the different quarters of the globe and yet will be at a loss to tell by what marks he detects them.'

A key to the natural orders was provided by the idea that the animals or plants of a given natural family were all variants of a common general structural type. The idea was based in part upon the comparative study of the structures of adult organisms which showed considerable structural resemblances between the creatures of such large groups as the vertebrate animals. The French royal physician, Vic D'Azyr, 1748-94, noted in 1784:

'Nature seems to operate always according to an original general plan, from which she departs with regret and whose traces we come across everywhere.'

The original general plan at first was thought to be the structure of the human body, which had been well studied in connection with medicine compared with the then scanty knowledge of the anatomy of other organisms. The conception that the structure of the human body was a general plan, and the idea that there was a general structural type pervading organic nature, derived also from the philosophical notion which was strong in Germany that man was a microcosm of the universe and so summed up the characteristics of the organic world. The German philosopher, Herder, 1744-1803, wrote in 1784:

'From stones to crystals, from crystals to metals, from these to plants, from plants to brutes, and from brutes to man, we have seen the *form of organization* ascend, and with this the powers and propensities of the creatures have become more various, till at length they have all united in the human frame. . . . Throughout this series of beings we observe . . . a predominant similitude of the principle form which, varying in numberless ways, more and more approaches that of man.'

As we have seen the German nature-philosophers after-Herder thought that man epitomized the characteristics of all organic beings, the structural forms of which had been generated by the self-development of the World Spirit who was completely manifest in man. However, they sought for forms more general than the structure of the human body to serve as the archetype plans of the organic world. Goethe in 1795 suggested that there was one archetype plan for the plant world and another for the animal world which could be discovered by examining the structures of existing animals and plants. Taking as his model the structure of the highest animal class, the vertebrates, rather than that of the highest species, man,

Oken supposed in 1807 that an abstract spinal column was the archetype plan of the animal world. The archetype column consisted of a series of typical units, the vertebra segments with their rib and limb appendages, which were modified in different ways in the various animal genera and species. The mammalian skull, for example, was composed of four vertebra segments.

Oken did not carry his archetype conception beyond the vertebral theory of the skull for, in the style of the nature-philosophers, he concerned himself with a variety of subjects, approaching each in a rather diffuse and speculative manner. However, in France, where science was pursued more empirically, the idea that the general vertebrate pattern was a plan common to all the animal species was worked out in some detail by Geoffroy St. Hilaire, 1772-1844, professor of vertebrate zoology at the Paris Museum of Natural History. In his earliest work of 1796 Hilaire suggested that nature had formed all creatures on the basis of a single plan. Different animals possessed the same organs, which in some were exaggerated, like the trunk, or prolonged nose, of the elephant, and which in others became vestigial, like the rudimentary side toes of hooved creatures. Hilaire then set out to find the organs which were the same, or homologous, in different animals so that he could uncover the general plan behind the diversity of the animal world. He began in 1807 with a comparative study of the fishes and the higher verbebrate animals, and in 1818 he published his *Anatomical Philosophy* in which he gave a general account of his researches.

Hilaire based his search for the homologous parts of different animals upon what he called 'the principle of connections'. He noted that the fore limb in the higher vertebrate animals was adapted to a variety of functions, running, climbing, swimming, or flying, but the arrangement of the bones in the fore limb was always the same. The hand of man and its homologue in the quadrupeds, the fore foot, were both the fourth part of the fore limb outwards from the backbone, and so the homologous parts of different animals could be recognized by their position relative to the other parts of those animals. A given organ, remarked Hilaire, could be enlarged, atrophied, and even annihilated, but not transposed. In different animals it was always placed in the same position relative to the other organs.

As Hilaire pointed out, such a view implied that all animals were made up in principle of the same number of parts, or units of construction as he termed them. Thus there were both a unity of composition and a unity of plan behind the diverse structures of the animal species.

'Nature tends to repeat the same organs in the same number and in the same relations and varies to infinity only their form,' wrote Hilaire. 'From this standpoint there are no different animals. One fact alone dominates: it is as if a single being were appearing. It is an abstract being, residing in Animality, which is tangible to our senses under diverse forms.'

For Hilaire the component material units of animals were individual bones, and those units were arranged in all animals according to a generalized vertebrate plan. Hilaire obtained his abstract archetype plan from a variety of vertebrate animals, basing his archetype model of a particular part upon an animal in which that part had reached its maximum development. In the general archetype, therefore, the potentialities of each organ were fully realized, the archetype indicating what organ transformations were possible in actual animals. All of the units of construction, the individual bones of the archetype, were not perceptible in every animal as the bones had joined and fused or had disappeared, confusing the homologies between one creature and another. Accordingly Hilaire examined the animals in their embryonic forms where the bones had not yet joined and fused, so that the units of construction and the homologies between one animal and another could be more easily seen. In this way he endeavoured to derive such invertebrate animals as the lobster and the crab from the general vertebrate archetype. A vertebral segment of his archetype consisted essentially of four bones forming a ring, and eight bones forming appendages to the ring. Hilaire found that the shell of young crabs and lobsters was made up of segments, each being composed of four elementary pieces which later fused together, and each possessing attachments corresponding to the vertebra appendages. He discovered further that the arrangement of the internal organs of the lobster when turned on its back corresponded to that of the internal organs of vertebrate animals. Thus he came to the conclusion that crabs and lobsters were vertebral animals living inside their backbones and inverted. In Oken's journal, *Isis,* Hilaire wrote in 1820, 'Every animal lives either outside or inside its vertebral column.'

Given a unity of composition and a unity of plan amongst the various animals, Hilaire faced the problem of the origin of their diverse forms. He suggested that the variation of animal forms was mediated by a law of compensation, according to which the development of one part of an animal was balanced by the atrophy of another, as there was only a limited amount of material for the composition of the animal as a whole. Thus in birds the development of a large breastbone was accompanied by the atrophy of thorax bones which were prominent in other vertebrate animals. Following Lamarck, Hilaire supposed that changes in the physical environment had led to the variation of animal forms. The potentialities of the archetype plan during the course of time had been realized more and more fully as suitable environmental conditions arose, so that the various animal species formed a roughly linear evolutionary scale.

'The organization (of animals)', wrote Hilaire, 'only awaits favourable conditions to arise, by the addition of parts, from the simplicity of the first formations to the complication of the creatures at the head of the scale.'

However, according to Hilaire, there was no active evolutionary force within animals, as Lamarck had believed, nor did new habits in animals engendered by environmental changes lead in turn to structural modifications. Hilaire held that the habits and functions of animals were determined by their structures, but that their structures were largely independent of their habits. The possible structural variations of animals were determined by the inherent properties of organic matter. The environment might promote structural changes, but the forms which those changes assumed were determined by the material units of construction and by the archetype plan, just as in the inorganic world the course of a chemical change was determined by the specific properties of the chemical atoms concerned. If the environment changed, the animals species either changed their structural forms, and thus their habits and functions, or they were eliminated. Moreover, such changes were not gradual, as Lamarck had supposed, but sudden and mutative. Hilaire argued that reptiles could not have evolved gradually into birds, a comprehensive change must have taken place within a generation or so. As evidence for sudden mutative changes, Hilaire instanced the birth of monsters, and in 1826 he endeavoured to induce the formation of monstrous birds by interfering with the incubation of their eggs. The normal course of embryological development was considered by Hilaire to be good evidence for the evolution of the species, as he thought that the development of the individual organism recapitulated the evolutionary history of its species.

'An amphibian is at first a fish under the name of a tadpole, and then a reptile under that of a frog,' he wrote, 'in this observed fact is realized what we have above represented as a hypothesis, the transformation of one organic stage into the stage immediately superior.'

The views of Hilaire and Lamarck were strongly controverted by their colleague, Georges Cuvier, 1769-1832, professor of comparative anatomy at the Paris Museum of Natural History. Although he had studied under the nature-philosopher, Kielmeyer, in Germany where an idealist theory of the development of the species was in vogue, and worked in France where a theory of the material evolution of organic beings had some force, Cuvier adhered to the traditional doctrine that the organic species had been fixed in their present forms from the beginning. However, he shared the interest of the biologists of the time in the structural forms of organisms, regarding the forms of creatures as an expression of their fixed characteristics. In his work on the *Animal Kingdom*, Cuvier wrote in 1817:

'It is not in the substance that in plants and animals the identity of the species is manifest; it is in the form. There are probably not two men, two oaks, two rose trees, which have the compound elements of their bodies in the same proportion,—and even these elements change without end, they circulate rather than reside in that abstract figured space which we call the form: in a few years probably there is not left one atom of that which constitutes our body today,—only the form is persistent: the form alone perpetuates in multiplying itself, transmitted by the mysterious operation which we call generation to an endless series of individuals.'

In contrast to Hilaire who supposed that the structure of an animal determined its functions and habits, Cuvier held that the habits and functions which an animal possessed shaped its structural form. Animals possessing the power of self-movement, must have stomachs to carry food with them, and organs to gather or catch food and to divide it and digest it. Plants, which remain stationary, do not require stomachs and possess roots instead. In general, wrote Cuvier, 'plants, having few faculties, have a very simple organization'. In animals the prime function of digestion required a circulatory system to distribute food, though some very primitive animals did not possess a circulatory system. There were different kinds of circulatory systems and these in turn required a breathing apparatus, such as lungs or gills, in all but the simplest animals. In animals, therefore, there was a rational order of functions and organs which performed them. In order to be self-moving, animals required a neuro-muscular mechanism and a digestive apparatus, the latter entailing in general a circulatory system which in turn usually required a respiratory system.

Cuvier suggested that the functional dependence of animal organs one upon the other was of importance for zoological classification. The natural method of classification required the examination of all of the organs of different animals, but it was apparent that not all organs were of equal significance for this purpose, nor were all the variations of each organ equally important. Cuvier suggested therefore that the natural classification of animals should be based primarily upon an examination of the organs performing the basic functions of the animal body, as the organs performing the secondary functions were dependent upon them and they were the most constant in structural form throughout the diversity of the animal species. Cuvier distinguished two basic functions; that of sensation which he termed 'animal', and that of the digestion and distribution of food which he termed 'vegetative'. The heart and organs of circulation', wrote Cuvier, 'are a kind of centre for the vegetative functions, just as the brain and spinal column are for the animal functions.'

Basing himself primarily on the nervous and circulatory systems, Cuvier carried out a natural classification of the animal species, distinguishing in 1817 four main groups or types. Firstly, there was the vertebrate type, covering the four classes, mammals, birds, reptiles, and fishes, all of which had a brain and a spinal chord. Secondly came the mollusc type, consisting of six classes of animals like the octopus, snail, and oyster, which had a nervous system composed of separate neural masses. Thirdly, there were the articulata or jointed animals, a type made up of four classes containing the lobsters, spiders, insects, and some worms, all of which had a nervous system consisting of two ventral chords, and which showed a transition from the blood vessel circulatory system of the two previous types to a tracheal system. Finally, there were the radiata, a miscellaneous type covering the animals with radial symmetry, like the star fish, in contrast to the bilateral symmetry of the first three animal types. The animals of the fourth type had neither a circulatory nor a nervous system, though they possessed both animal and vegetative functions in a rudimentary form. Independently and about the same time von Baer distinguished the same four main types in the animal world. Von Baer based his classification upon a comparative study of the embryological development of different animals, as we have seen, while Cuvier compared the anatomical structures of the adult organisms. Both Cuvier and von Baer held that the animals of different types were quite unrelated, but that the animals belonging to the same type were variants of a common structural plan.

Cuvier's method of selecting the more important and universal characteristics of organisms for the purpose of classification was termed 'the principle of the subordination of characters'. Such a principle had been enunciated in connec-

382 / A History of the Sciences

tion with the natural classification of plants in 1789 by Antoine de Jussieu, 1748-1836, whose work was continued by August de Candolle, 1778-1841. Cuvier and Candolle supposed that all of the organisms belonging to a given natural family were based upon a single archetype, or primitive plan, which could be abstracted by the principle of the subordination of characters. There were only a few possible combinations of the vital and fundamental parts of organisms; according to Cuvier there were only four such combinations in the animal world, but there was a much greater diversity in the less important parts, which, in decreasing order of their significance, could be used to distinguish classes, genera, and species, within a given structural type.

'The natural classification of organized beings', wrote Candolle, 'consists in appreciating the modifying circumstances and abstracting them to discover the real symmetrical type of each group. . . . Symmetry supposes a primitive plan or archetype, and the proofs of symmetry are those of a general order.'

Adopting the natural method of classification, Candolle and Cuvier opposed the conception that the organic species formed a linear chain of beings, an idea associated with the artificial classifications of organisms. Candolle described the conception as 'a facetious image', and Cuvier denied that neither the animals of a given type, nor those of a single class within that type, formed a linear scale of beings, let alone the animal kingdom as a whole. For Cuvier and Candolle the various organic species were divergent modifications of the archetype plan of the group to which they belonged. The archetype plan of a group of organisms was a regulative and conservative principle which prescribed the limits of possible species variations. Thus something else gave rise to the variety of the organic world, but Cuvier and Candolle had no theory as to what this was, for they held that the species were fixed and ascribed the diversity of organisms to an aboriginal creative act. 'We must study the different species as constant things,' wrote Candolle, as this was a more worthy occupation for the scientists 'than the accumulation of doubtful cases in favour of the non-permanence of species'.

Whilst Cuvier had no theory as to the origin of the various modifications of a given archetype, he showed that particular structural animal forms were associated with, and depended upon the particular habits and functions of the creatures in question. Cuvier pointed out that carnivorous animals of their very nature had to possess a digestive tract which could assimilate flesh, while herbivorous animals possessed a different kind of tract for the digestion of vegetable matter. Carnivores of necessity had claws in order to catch their prey and cutting teeth to tear it up. They had the kind of jaw which could accommodate those teeth and a skeleton and muscular system appropriate for such jaws. Thus carnivores never had hooves, or flat molar teeth, like the herbivorous animals. Cuvier suggested that in this way the whole structure, functions, and habits of an animal could be deduced rationally from one of its parts, such as a single bone or an organ. Indeed he considered that such deductions had the same certainty as mathematical demonstrations.

'All the organs of an animal', wrote Cuvier, 'form a single system, the parts of which hang together and act and react upon one another, and no modification can appear in one part without bringing about corresponding changes in all the rest. . . . It is upon this mutual dependence of the functions and the assistance which they lend to one another that are founded the laws which determine the relations of their organs, laws which have a necessity equal to that of metaphysics or mathematics . . . the form of the tooth implies the form of the jaw, that of the shoulder blade that of the claws, just as the equation of a curve implies all its properties.'

However, not all of Cuvier's correlations were rational. Given a hoof, he could deduce the characteristics of the animal from which it came, showing it to be herbivorous, but he could not indicate why herbivores which ruminate possessed cloven hooves and frontal horns. The association of cloven hooves and frontal horns with ruminant animals was purely empirical, unlike the association of hooves in general with herbivorous animals for which reasons could be given.

With the rational and the empirical correlations between

parts of animals which he had discovered, Cuvier endeavoured to reconstruct the structural forms of extinct species from their fossilized bones. With one bone, he searched for the other bones of the animal to which it had belonged, guided by the idea he had formed from his study of living animals of the shapes and sizes which those bones should have had. In this way he built up the skeleton of an extinct animal, and from the skeleton he deduced the forms of the soft parts which the animal had possessed, and thence its functions and habits. Studying first the fossil elephants, Cuvier showed in 1800 that the extinct mammoth had been more closely allied to the Indian elephant than the Indian elephant was to the African. Later, in 1812, Cuvier published his *Researches on Fossil Bones* in which he gave reconstructions of some hundred and fifty extinct mammalian species.

Cuvier found that the extinct species had been built according to the same general structural plans as living animals, all of them belonging to one or other of the four main structural types he had distinguished in the animal kingdom. However, the older they were, the more the extinct animals differed from the present day living forms, becoming simpler in structure the earlier they had occurred in the earth's history. The living species, and the extinct forms stretching backwards in time behind them, thus formed an historical succession of animals made up of four main branches, each branch containing all of the animals, both extinct and living, based on the same general structural plan. Cuvier did not regard his four branches of the animal kingdom as a genealogical tree of evolutionary descent. He adopted from Bonnet the theory that the surface of the earth had been visited now and again by major catastrophes which had largely eliminated the organic species existing at the time. After each castastrophe, a new organic population had appeared upon the earth, the new animals being based on the same general structural plans as the previous species. Cuvier suggested that there had been at least four such catastrophes, the last one being probably the Mosaic Flood about five or six thousand years ago. With the advance of the science of palaeontology, the study of the structures of the extinct species, the number of catastrophes and creations of new species were multiplied. Cuvier's pupil,

Alcide d'Orbigny, 1802-57, the first tenant of the chair of palaeontology at the Paris Museum of Natural History, published a work in 1849 in which he catalogued some eighteen thousand species of extinct animals found in France and postulated twenty-seven catastrophes and creations to account for them.

Between the three colleagues at the Paris Museum of Natural History, Lamarck, Hilaire, and Cuvier, there were considerable differences of opinion. Lamarck and Hilaire were of the view that the animal species formed a roughly linear scale of evolutionary descent, while Cuvier held that they were divergent modifications of four main structural types and that they remained fixed in their first created forms. Lamarck believed that evolution was gradual and continuous, working through the active self-adaptation of the organism to changed environmental conditions, but Hilaire thought that evolution was mutative and that organic variations were not primarily responses to environmental changes. For Hilaire the habits and functions of an animal were determined by its anatomical structure, but for Lamarck and Cuvier the reverse was true, Lamarck emphasizing the structural adaptations of organs caused, through a change of habits, by external conditions, and Cuvier the internal mutual adaptations of organs through their functions within an animal, which depended upon the existence of the animal as a whole in itself.

Such differences of viewpoint led to controversies between the three biologists and their supporters, though Lamarck was not active, as he had few disciples and he had become old and blind with age. Hilaire and Cuvier, however, engaged in an extensive controversy which came to a head in 1830. In that year Hilaire submitted to the Paris Academy of Sciences a paper by two of his pupils, who endeavoured to show that the ink-fish was the missing morphological and evolutionary link between the vertebrate and invertebrate animals, on the grounds that the ink-fish displayed the vertebrate plan if it were regarded as bent in the middle with its head and tail contiguous. In the Academy of Sciences Cuvier made the discussion of this paper the occasion for a general attack upon Hilaire's views. He denied that the animals had evolved one from the other, and that they could be arranged into a linear scale exhibiting the gradual working out of a single structural plan, that of the vertebrate animals. Cuvier laid it down that the animal species were fixed, and that they were divergent modifications of four general plans, the modifications being dependent upon the specific habits and functions of the animals which embodied them, not the reverse as Hilaire suggested. From the debate Cuvier emerged as the victor, and he succeeded in extinguishing the idea of organic evolution in France for several decades.

During the early years of the nineteenth century Cuvier was a man of considerable influence, earning for himself in the sciences the title of 'the dictator of biology'. Under Napoleon he was made Inspector General of the University of France, professor of the Collége de France, and perpetual secretary of the Academy of Sciences. Though of Huguenot stock, Cuvier retained his standing under the Catholic reaction which came with the restoration of the Bourbons in 1814, becoming a baron and Minister for Protestant Affairs. Curiously enough the July 1830 debate in the Academy of Sciences coincided with the second deposition of the Bourbons: indeed Goethe and Eckermann in their *Conversations* confused the two events. A French anthropologist, Topinard, later suggested that the two events were not entirely unconnected, Cuvier associating the doctrine of organic evolution with the cause for social reform.

Chapter 32

The Cell Theory

FROM ANTIQUITY until the nineteenth century it was presumed, following Aristotle, that the materials of which organic beings were composed existed at three main levels of organization. Firstly, there was unorganized material compounded of the four elements; secondly, homogeneous parts or tissues; and thirdly, heterogeneous parts or organs. The new chemistry of the late eighteenth century did not seriously change this classification for the unorganized biological materials were then considered to be compounds of the new chemical elements of Lavoisier instead of earth, water, air, and fire. However, the vitalist current of thought, culminating with the German nature-philosophy of the turn of the eighteenth century, led to the development, and then the change of this classification, organic cells becoming recognized as a level of organization intermediate between that of unorganized biological matter and that of homogeneous tissue.

The Aristotelian classification of biological materials was developed notably at the medical school of Montpellier, which was strongly influenced by the vitalistic views of Stahl and the iatrochemists during the eighteenth century. Basing himself on the work of the Montpellier school, the French physician, Xavier Bichat, 1771-1802, distinguished in 1797 twenty-one different kinds of tissue, such as bone, cartilage, muscle, and so on, in the homogeneous parts of the human body. Bichat indicated that a heterogeneous part, or an organ, was made up of several different kinds of tissue, and that several organs together composed an apparatus or organ system, such as the respiratory or digestive apparatus. The organ systems in turn were linked up to form two main complexes, the one composed of the digestive, circulatory, and respiratory systems, which mediated the growth and nutrition or the vegetative life of the body, and the other composed of the brain, nerves, and muscles which were responsible for the animal life of self-movement and sensation. Bichat supposed that the body was actuated by an inner vital force, as the tissues, organs, and organ systems remained when life departed, so that they could not in themselves constitute a living organism. Developing the views of the iatrochemists, who had supposed that the separate organs had a life of their own deriving from an inner vital force, Bichat supposed that the tissues, the next lower level of organization, had each a *'vie propre'*, a life of its own. The iatrochemists had also held that a disease was a highly specific entity attacking a particular organ, and Bichat developed this doctrine too, supposing that a disease was localized in a specific tissue of the particular organ attacked.

In Germany meanwhile Oken and others had developed their systems of nature-philosophy in which, amongst other things, they delineated what they took to be the typical units composing the diversity of the organic world. Goethe suggested that the leaf was the typical unit structure of the various plant forms and Oken held that the vertebra segment was the basic unit of the general animal archetype structure. Oken supposed further that organic beings were composed of mucus-vesicles or living units which survived the death of the organism to which they temporarily belonged and lived on to form part of another creature. Such a view enjoyed some popularity in Germany during the early years of the nineteenth century and, joined with the microscopic examination of plant and animal structures, it led to the development of the cell theory.

Plant cells had been seen with the microscope during the seventeenth century, notably by Hooke, Malpighi, and Leeuwenhoek, but they were not recognized as the independent, living structural units of the vegetable world. During the seventeenth and eighteenth centuries botanists were more interested in the classification of plants and in the physiological workings of the plant as a whole than in the detail of plant structures, but in the early nineteenth century the study of plant anatomy was revived and the cell was recognized as the structural unit of plants by a number of German botanists, notably Treviranus and von Mohl. About the same time, in the 1820's, Amici in Italy and others developed the improved achromatic microscopes which allowed the detail of the organic cell to be examined. A London doctor, Robert Brown, 1773-1858, observed in 1831 that plant cells in general possessed a nuclear body, though he did not attach much significance to his discovery. In 1835 the Czech, Purkinje, 1787-1869, observed with the microscope a germinal nucleus in the hen's egg, and he pointed out that animal tissues, particularly in embryos, were made up of closely packed masses of cells, similar to those of plants.

Such observations led Mathias Schleiden, 1804-81, professor of botany at Jena, to announce in 1838 the theory that the cell was the basic living unit of all plant structures, and the fundamental entity from which all plants developed. Following the nature-philosophy current of thought, Schleiden suggested that the examination of the development of individual plants would yield more understanding of the nature of plants than the traditional botanical work on the classification of plants and the examination of their adult structures. In botany, wrote Schleiden, plant embryology will be 'at once the sole and richest source of new discoveries and will remain so for many years'. In the development of plants Schleiden supposed that the basic process was the formation of independent living cells, which once formed were arranged in a structural pattern expressive of the unity of the plant as a whole. The cell was a self-contained unit, and as such it had two lives—one of its own that was primary and the other as part of an organized plant structure which was secondary. Both life processes were manifestations of a 'form-building force' which permeated the whole of nature, forming inorganic crystals, fashioning organic cells, and organizing those cells into composite living beings. With regard to the formation of plant cells, Schleiden drew attention to Robert Brown's discovery of the cell nucleus, suggesting that a new cell took its origin from the nucleus of an old cell, forming at first a segment of the sphere of the old cell and then separating off as a complete cell in itself.

In 1839 the cell theory was extended to the animal kingdom by Theodore Schwann, 1810-82, professor of anatomy at Louvain. Like Schleiden, Schwann associated the cell theory with the developmental or embryological approach to organic beings, enunciating the theory in the dictum, 'that there is one universal principle of development for the elementary parts of organisms however different, and that this principle is the formation of cells'. Schwann inclined to the view that the fertilized eggs of all animals were single cells, the germinal vesicle being the nucleus, whether such cells were large like the hen's egg or small like the mammalian egg. Thus all organisms started life as a single cell and developed by the formation of other cells. Following Schleiden, Schwann supposed that new cells in plants and fertilized animal eggs developed inside the old cells, but he held that in later stages of the development of animals new cells were formed from inter-cellular material. According to Schwann, two forces were at work in the formation of cells. One was a metabolic force peculiar to organic cells which transformed intercellular matter into materials suitable for cell formation. The other was an attractive force which formed cells by concentrating and precipitating the prepared inter-cellular material, a force which operated also in the inorganic world, and there led to the formation of crystals. First the nucleolus of the cell crystallized out from the inter-cellular material, and round the nucleolus precipitated a layer of matter forming the nucleus. A further concentric layer of material gave the mucus of the cell, which was completed by the hardening of the outer surface of the mucus into a cell wall. The attractive and metabolic forces gave the cell an autonomy and a life of its own.

'Growth', wrote Schwann, 'is not the result of a force having its ground in the organism as a whole, but each of the elementary parts possesses a force of its own, a life of its own if you will: that is to say, in each elementary part the molecules are so combined as to set free a force whereby the cell is enabled to attract new molecules and so to grow, and the whole organism exists only through the reciprocal action of the single elementary parts. . . . In this eventuality it is the elementary parts that form the active element in nutrition, and the totality of the organism can indeed be a condition, but on this view it cannot be a cause.'

The view of Schleiden and Schwann, that new cells were formed by the crystallization of organic matter inside or outside of the old cells, was corrected in the 1840's by a number of workers, notably the botanists von Mohl, Nageli, and Hofmeister, and the zoologists Kolliker, Leydig, and Remak, who showed that new cells were formed by fission, the nucleus dividing first inside the mother cell, which then split into two daughter cells.

Schwann applied his cell theory to Bichat's classification of the homogeneous parts or organisms, distinguishing five classes of tissues on a cellular basis. He indicated that there were, firstly, tissues in which the cells were independent and separate, as in the case of the blood cells; secondly, there were tissues, such as the skin, in which the cells were independent but pressed together; thirdly, there were bone and teeth tissues where the cells had well-developed rigid walls which coalesced; fourthly, there were tissues in which the cells were elongated into fibres, such as the ligaments and tendons; and fifthly, there were tissues like the nerves and muscles in which, Schwann thought, the cell walls and cavities had joined up. Following Schwann, the medical aspects of Bichat's tissue theory were developed by Rudolph Virchow, 1821-1902, professor of pathological anatomy at Berlin. The locality of a disease had been narrowed down from the body as a whole to a particular organ by the iatrochemists, and from an organ to a specific tissue by Bichat, and now in 1858 Virchow suggested that a disease originated within a single cell, and was propagated by malignant cell formation in a given tissue. Adopting the view of Schwann that the cells were autonomous living entities, Virchow regarded the human body as a 'state in which every cell is a citizen', and a disease as a kind of revolt or civil war.

'Every animal is a sum of vital units,' wrote Virchow, 'each of whch possesses the full characteristics of life. The character and the unity of life cannot be found in one definite point of a higher organization, for example, in the brain of man, but only in the definite, constantly recurring disposition shown individually by each single element. It follows that the composition of the major organism, the so-called individual, must be likened to a kind of social arrangement or society, in which a number of separate existences are dependent upon one another, in such a way, however, that each element possesses its own peculiar activity and carries out its own task by its own powers.'

Virchow thus rejected the idea of the iatrochemical philosophers that the vital units of an organic being were subject to a central power, the 'Archeus Influus' of Helmont, or the 'Central Monad' of Leibniz, and the view of the nature-philosophers and morphologists that the ideal plan of a creature gave it an organic unity. He also turned away from the iatrochemical idea that diseases were vital entities in themselves, invading the human body from outside, an idea which was supported by the discovery of bacteria by Pasteur and Koch. Virchow refused to accept the germ theory of disease, and in the 1880's he turned to other fields, archaeology and anthropology, when more became known concerning the role of bacteria in the causation of disease.

Schleiden, Schwann, and Virchow, carried to an extreme the view that the cells of an organism were largely autonomous vital entities. The romantic movement, of which German nature-philosophy was a part, had accorded pride of place to that which was individual and self-moving, and it was perhaps this general viewpoint which found a particularized biological expression in the doctrine that the fundamental units of organic nature were largely self-contained and self-sustaining. Such a doctrine was criticized by the morphologists of the old school, such as Reichert, who held that an organism in itself formed a whole which was manifest in its structural plan, by the cytologists, such as Strasburger, who pointed out that the cells could not be fully independent of one another as they were connected by protoplasmic bridges, and more fundamentally by the physiologists and psychologists, such as Sherrington and Pavlov, who from the turn of the nineteenth century began to demonstrate the integrative action of the nervous system in the higher animals.

PART FIVE

THE SCIENCE OF THE
NINETEENTH CENTURY:
THE AGENT OF INDUSTRIAL
AND INTELLECTUAL CHANGE

Chapter 33

The Development of Geology

GEOLOGY BEGAN to take shape as a separate science during the eighteenth century, and came to maturity in the early decades of the nineteenth. Before that time geology was a scattered and divided subject. Facts pertaining to the science were known to miners and others concerned with the extraction of metals, clays, coal and salts from the earth, whilst natural philosophers formulated speculative geological theories largely independently of such facts. Generally speaking, the inorganic substances extracted from the earth were thought to be alive, growing under the impulse of an inner plastic force, or the external influence of the stars. It was not an uncommon practice, for example, for mines to be closed down from time to time, so that the ores could grow again and replace what had been mined. Fossils were widely regarded as abortive attempts of nature to produce animals and plants, or as 'sports of nature', fortuitously resembling organic forms. In those periods of history that marked the beginning of a new epoch, there were individuals who suggested that fossils were the remains of once living creatures: amongst the pre-Socratic Greeks there was Xenophanes of Colophon, and during the Renaissance, Leonardo da Vinci, Girolamo Fracastoro and Giordano Bruno.

Such a view was not widely held during the seventeenth century, for it was pointed out, notably by the Cambridge clergyman-naturalist, John Ray, 1627-1705, that if the fossils were the remains of species now extinct, the great chain of living creatures could not be continuous and complete; there would be gaps where species had died out, and thereby the world as a whole would be rendered imperfect. Ray came to accept the view that fossils were the remains of living creatures, though not without reservations. He thought it difficult to conceive how the fossil shells, supposedly of sea creatures, could reach the tops of mountains. Such fossil shells went deeply into the rock strata of the mountains, and thus their presence could not be accounted for by some temporary catastrophe, such as the

Mosaic Flood. Besides, he added, the Flood would have washed the fossil creatures down the mountain as the waters receded.

A reply to Ray was published in 1695 by John Woodward, 1665-1728, a professor of medicine at Gresham College, London, in his *Essay towards a Natural History of the Earth*. Woodward held that the Mosaic Flood was a much more catastrophic event than Ray had assumed.

'The whole terrestrial globe', he wrote, 'was taken to pieces at the Flood, and the strata now visible settled down from the promiscuous mass as any earthy sediment from a fluid.'

The Flood, Woodward suggested, had not only destroyed most of the organic population of the earth, it had also broken up the inorganic surface of the earth and held the constituents of the rocks in suspension. Fresh rock strata had then been formed by a process of sedimentation, the remains of animals and plants being caught up in that process, so that fossils were found in the deepest strata. The heaviest substances, metals, minerals, and the weightier fossil bones, settled out first in the lowest strata, then came the lighter shells of the sea creatures in the chalk above, and finally man, the higher animals and plants, were caught in the sand and marl of the highest strata. Thus Woodward was of the opinion that fossils were the remains of once living creatures, and he regarded them as the surest evidence for the historical authenticity of the Mosaic Flood. Some of Woodward's contemporaries, who accepted his views, were less theologically orthodox, notably Edmund Halley and William Whiston, who suggested that the Mosaic Flood had been nothing more than a huge tidal wave which had been raised by Halley's comet passing too near to the earth. However, it was Woodward's original views that were the more generally accepted during the eighteenth century, and they stimulated the collection of fossils throughout that period. Fossil bones were even hung up in churches, labelled as the 'Bones of giants mentioned in the Scriptures'.

Woodward and his followers laid great stress upon the action of water, notably the Mosaic Flood, in the formation of the rock strata and their fossil content. There was however another school of thought which in opposition emphasized the role of heat and volcanic action. John Ray, in replying to Woodward, had been of the opinion that the mountains and dry land had been raised above the waters of the ocean by the internal fires of the earth at God's command. In countries where volcanoes were active such a view seemed more probable than the Flood theory, and in 1740 the Abbé Anton Moro of Venice put forward an entirely thermal theory of geogony. Moro held that the Mosaic Flood had been an essentially minor and geologically unimportant event. The successive strata of rocks were due, he thought, to a series of volcanic eruptions of liquid rock, each forming a new stratum which entombed the various kinds of animals and plants then living, so that the fossils were deeply embedded in the rocks.

The opposition between the two views, those of Woodward and Moro, marked the beginnings of a controversy that flared up at the end of the eighteenth century between the Neptunists, who stressed the role of water in forming the geological strata, and the Vulcanists, who emphasized the role of heat. Before and after the controversy the two views were seen to be complementary, but there was an acute conflict in the period 1790-1830, when the Vulcanist view became associated with the theory that the rock strata had gradually evolved, and Neptunism with the theory that the strata were formed suddenly and catastrophically, though there were Vulcanists who shared this opinion. During the period of controversy, and prior to it, there was a tendency to adduce purely local evidence in support of a particular geological theory. Thus the examination of active or extinct volcanoes tended to lead geologists to the Vulcanist conclusion, whilst a preoccupation with sedimentary rocks led to the Neptunist view.

Evolutionary theories in geology were put forward before the period of controversy, but they did not attract a great deal of attention as they were mainly speculative. Georges Buffon, the Keeper of the King's Gardens at Paris, published such a theory of the evolution of the earth in 1749 and amplified it in 1778. Buffon was amongst the first to ascribe a much greater period to the age of the earth than the traditional estimate of six thousand years, based upon the genealogy of the persons mentioned in the Old Testament. He supposed that the earth had been in existence for about eighty thousand years, during which there had been seven epochs of development. First, the solar system had been formed out of matter ejected from the sun, following a collision between the sun and a comet. The earth, like the other planets, had been molten or semi-fluid in the beginning, and it had assumed the oblate spheroidal form in which it is found, as a result of axial spin causing the equator to bulge and the poles to flatten. A solid crust had then been formed on the surface of the earth, and the crust had become wrinkled as the earth cooled further, thus giving the mountain ranges and the ocean beds. Next the water vapour in the atmosphere had condensed, and had covered the whole surface of the earth with an ocean. The uppermost parts of the crust of the earth had been eroded away by the universal ocean, and clays were formed by the sedimentation of the detritus. Fossils were enclosed in such sedimentary deposits, being the remains of living creatures with which the ocean had abounded. Later cracks had opened in the crust, and much of the water had entered the interior of the earth, leaving dry land upon which vegetation had appeared. Subsequently land animals arose, and finally man himself.

Buffon's theories were highly speculative, though he attempted to introduce empirical tests into geology, making experiments with globes of iron in a endeavour to estimate the duration of the epochs he delineated. However, the application of laboratory experimentation to geology was limited at this stage: it was primarily a field science, depending upon the collection of observations from a variety of localities. Such observations were made by Buffon's contemporaries in France, Jean Guettard, 1715-86, one-time physician to the Duke of Orleans, and Nicolas Desmarest, 1725-1815, an inspector general and director of manufactures. Their investigations covered much of France, but they were particularly interested in the mountains of Auvergne, which, they saw, were extinct volcanoes. Guettard inclined towards the Neptunist view that the various rocks had an aqueous origin. In 1770 he suggested that basalt, the rock that forms the columns of the Giant's Causeway in Ireland, had been formed by crystallization from water. Desmarest, however, discovered basalt columns in the vicinity of old volcanoes, and he suggested in 1777 that they were formed by the solidification of molten rock. Generally speaking, Guettard and

Desmarest combined the Vulcanist and Neptunist views, holding that heat and volcanic action was important at first in the formation of the rocks, and that later the action of water was more marked.

Guettard was also a pioneer of geological surveying. He noted that bands of minerals and rocks lie side by side on the surface of the earth, and from the run of these bands he concluded that those disappearing into the Channel from the French side should reappear in southern England. Further investigations showed that this was indeed the case. Later, under the auspices of the Paris Academy of Sciences, he traced the layout of these rock bands over the rest of France, with the help of the chemist, Lavoisier, and in 1780 published a large-scale geological map of France. Guettard did not see that rock bands lying side by side on the earth's surface were vertically contiguous below ground, perhaps because France was not then noted for her mines where a vertical cross section of the earth's crust could be explored. As early as 1719 the Englishman, John Strachey, 1671-1743, had traced the succession of the rock strata from the coal to the chalk in the coal mines of the Mendips, and similar investigations were carried out later by the Germans, Johann Lehmann, died 1767, a professor at Berlin, and Georg Fuchsel, 1722-73, a physician, who explored the mining regions of the Harz mountains.

Lehmann and Fuchsel viewed the vertical series of rock strata as a historical succession, each layer gradually building up on top of those beneath it. They distinguished three main types of rock in terms of their age. First came the primary rocks without any fossils, forming the cores of mountains: they were followed by the secondary deposits which contained the fossils of the simple sea creatures; and finally came the tertiary rocks, containing the fossils of land animals and plants. Lehmann and Fuchsel considered that such strata had been formed by the sedimentation of matter from the sea, the rock layers becoming tilted owing to the crinkling of the earth's crust as it cooled. Their views were supported by Peter Pallas, 1741-1811, a German in the service of Catherine II of Russia, who made an extensive survey of the Ural mountains between 1768 and 1784. Pallas observed that the mountain strata of the Urals were markedly tilted, the more recent sedimentary rock being weathered away at the high points, revealing the older primary rocks beneath.

Such evolutionary views were, however, lost upon the next important German geologist, Abraham Werner, 1749-1817, who founded a purely Neptunist school of thought. Werner came of a family that had been associated with the mining industry for three hundred years, and he continued the tradition as director of the Freiberg School of Mines from 1775 to his death in 1817. Werner did not publish a great deal, but he was a popular lecturer and attracted students from all over Europe, in fact it was mainly through his lectures that his teachings became known, his numerous students making Neptunism the more important of the geological theories held in the early decades of the nineteenth century. The geological views of Werner were a secularized and amplified version of the earlier Flood theory of Woodward. In the beginning, he held, the earth was covered by a primeval ocean from which all the rock strata had been deposited by processes of crystallization, chemical precipitation, or mechanical sedimentation.

First of all came the primitive rocks, such as granite, which had crystallized out of the primeval ocean: these were entirely devoid of fossils. Then came the transitional rocks such as the micas and the slates, containing a few fossils, which had been precipitated from the ocean. Next there were the sedimentary rocks, richer in fossils, such as coal and limestone, formed by the deposition of solids from the waters. Finally there were the derivative rocks, such as sand and clay, which were derived from the others by a process of weathering. Werner thought that volcanoes were due to coal catching fire underground, the heat generated melting the neighbouring rocks, and forcing the eruption of volcanic lava from time to time. Thus, for Werner, heat was not an important geological force: volcanic action due to burning coal was a late and subsidiary rock-forming agency, appearing only after the main strata had been laid down.

Werner lived in the period of the German nature-philosophers, and he appears to have been influenced by that school of thought. He was concerned with the problem of the possible origin of the rocks, considering his postulated primeval ocean the common source from which all rocks had been derived. Other geologists contemporary with Werner did not concern themselves with the ultimate origin of the rocks, but with the operation of present-day geological forces, which they supposed to have existed throughout the history of the earth, to account for the formation of the rock strata. Werner too regarded his four classes of rocks as fundamental types: all rocks belonged to one of these types, being formed by the process of crystallization, precipitation, or sedimentation characteristic of that type. In the same way the German biologists of the period regarded the organic species as originating from a common source, all animals and plants being modifications of a few original types. On the practical side Werner was strongly influenced by his interest in mining, in fact his geology in some aspects was subordinate to mineralogy, for he classified the rocks according to the minerals they contained, not according to their fossil content, which later came to be the standard method. His field observations were confined mainly to Saxony and Bohemia, which were mining regions particularly rich in mineral deposits. Werner's classification of the rocks according to their mineral content was therefore very useful, but it did not indicate the historical order of the various rock strata as the fossil classification did. Moreover, it could not be readily applied to regions outside Saxony and Bohemia, where different kinds of rocks and different successions of the strata obtained. However, the most important shortcoming of Werner's geological theory was the lack of an explanation for the disappearance of the primeval ocean after the rock strata had been formed. One of his pupils, Robert Jameson, 1774-1854, a professor of natural history at Edinburgh, published in 1808 an account of Werner's theories in his *Elements of Geognosy*. In this work Jameson wrote concerning the disappearance of the primeval ocean:

'Although we cannot give any very satisfactory answer to this question, it is evident that the theory of the diminution of the water remains equally probable. We may be convinced of its truth, and are so, although we may not be able to explain it. To know from observation that a great phenomenon took place is a very different thing from ascertaining how it happened.'

Such was the assumption upon which Werner's theory rested. He ascribed the origin of the rocks to something that was in principle unobservable, the primeval ocean, and supposed that it disappeared by means unspecified when its role was fulfilled. In opposition to this point of view, James Hutton, 1726-97, an amateur scientist of Edinburgh, put forward the conception that only such geological forces as are seen in operation today should be used to explain the past formation of the rocks. Hutton was trained as a doctor, though he never practised medicine. Instead he participated in the agrarian and industrial enterprise of the time, applying the new agricultural methods which he studied in Norfolk to his estate in Berwickshire, and setting up a chemical factory to manufacture sal-ammoniac, from which he derived sufficient income to pursue freely his scientific and technical studies. In 1785 he read a paper to the Royal Society of Edinburgh which gave the gist of his geological theory, and ten years later he published his main work, *The Theory of the Earth*, giving his views in full. In contrast to Werner, Hutton stressed the geological activity of the internal heat of the earth, though he accepted the formative powers of water. In Norfolk he had found beds of gravel, sand, and mud, fringing the coast and running inland under the fields, which were remains of the weathered detritus of the hills brought down by the rivers. Sedimentary rocks were formed, he thought, from such beds of mud and sand by the combined effect of the internal heat of the earth and the pressure of overlying lands and seas. Such sedimentary rocks were amorphous, whilst the rocks composing some of the Scottish mountains he found to be crystalline, and he presumed that the latter had been formed directly by the solidification of molten rock, not by crystallization from water as Werner believed.

The interior of the earth, Hutton held, was composed of molten lava, the solid surface of the earth serving as a containing vessel, which was closed apart from the volcanoes that served as safety valves. From time to time, he thought, the molten rock escaped through cracks just beneath the earth's surface and tilted up the overlying sedimentary strata. The molten rock then solidified to form the crystalline rocks, such as basalt and granite, thus giving the mountains with their crystalline cores and sedimentary sides. Near the base of some mountains he found horizontal sedimentary strata on top of the tilted sedimentary rocks, from which he concluded that a long time had elapsed between the formation of the tilted rocks and the deposition of the fresh horizontal strata. Hutton indeed saw no beginning to the geological formation of the earth: the age of the earth was indefinitely long, with the same geological forces as are now in operation always at work, forming, breaking down, and reforming the rocks composing the surface of the earth. Like the French mechanical philosophers of the eighteenth century, who derived their idea of progress from the conception that man is always and everywhere much the same, Hutton derived his theory of geological development from the view that the forces of nature are constant. The geological forces of the earth, because they were constant and always the same, produced an historical succession of the rock strata, Hutton believed, just as the French philosophers thought that men, remaining constant in their physical and mental powers, added progressively to the accumulated experience of mankind. Hutton based his view that the rock-forming agencies of the earth were constant on the by now established theory that the solar system was mechanically stable and permanently self-sustaining.

'From seeing the revolutions of the planets', he wrote, 'it is concluded that there is a system by which they are intended to continue those revolutions. But if the succession of worlds is established in the system of nature, it is vain to look for anything higher in the origin of the earth. The result therefore of this physical enquiry is that we find no vestige of a beginning, no prospect of an end.'

Hutton's theories were supported and developed by his friends, John Playfair, 1748-1819, professor of natural philosophy at Edinburgh and Sir James Hall, 1762-1831, an amateur scientist of Edinburgh. Playfair published in 1802 a work entitled *Illustrations of the Huttonian Theory*, which described the theory in a clearer way than Hutton had done, and which contained Playfair's own view that glaciers had been an important geological agency in that they had carried masses of rock from one place to another. Hall from 1790 to 1812 made some important experiments that supported Hutton's theory. Werner's pupils and followers argued against Hutton, firstly, that molten rock would not become crystalline on solidification but would be glassy like lava, and secondly, that some rocks, like limestone, would decompose if subject to heat. Hall observed in a glass factory at Leith that if molten glass were allowed to cool very slowly it became crystalline and opaque, whilst if it were cooled more quickly it became glassy and transparent. He presumed that molten rock would behave in a similar way, and accordingly he obtained some lava from Vesuvius and Etna, and melted it in the blast furnace of an iron works. As he had expected, the molten rock became crystalline, like basalt, when allowed to cool slowly, and glassy, like lava, when cooled rapidly. Hall showed further that if limestone were heated in a closed vessel, it did not decompose as the Neptunists thought, but melted and became marble on cooling as Hutton suggested. In other experiments Hall found that loose sand when heated in an iron pot filled with sea water became hard and compact, like sandstone, which again supported Hutton's views.

In spite of Hall's experiments, Hutton's theories were not widely accepted at first, as they were considered to be subversive to the interests of established religion and indeed to the whole traditional order of things. Hutton was attacked for advocating what were thought to be atheistical views by John Williams of Edinburgh, a geologist who did some important work on the coal strata, in his *Natural History of the Mineralogical Kingdom*, published in 1789. Hutton's

'wild and unnatural notion of the eternity of the earth', he wrote, 'leads first to scepticism and at last to downright infidelity and atheism. If we once entertain a firm persuasion that the world is eternal, and can go on itself in the reproduction and progressive vicissitudes of things, we may then suppose that there is no use for the interposition of a governing power; and because we do not see the Supreme Being with our bodily eyes . . . we commit the care of all things to blind chance.'

All rebellions, he affirmed, 'soon end in anarchy, confusion and misery, and so does our intellectual rebellion'. A similar criticism of Hutton's theory was made by Deluc, reader to Queen Charlotte, in his *Elementary Treatise of Geology*, published in 1809. A knowledge of geology had now become essential for theologians, he wrote, for,

'Certainly no conclusion from the natural sciences can be more important to men than that which concerns Genesis: for to place this book in the class of fables would be to throw into deepest ignorance that which it is most important for them to know; their origin, their duty, and their destination.'

Such an opposition made Hutton's theory most unpopular at the time, though his views were revived in the 1830's, and then came to be generally accepted. Meanwhile a great deal of technical progress took place in geology which provided the empirical basis for the later revival and amplification of Hutton's views. In 1807 the British Geological Society was formed to promote the advance of geological knowledge. Most of the early members of the Geological Society were Neptunists, that is, followers of Werner, though a Scottish member, MacCulloch, supported the Vulcanist aspects of Hutton's teachings. Theoretical controversies were not infrequent at meetings of the Society, though it was the 1830's before they centred round the evolutionary aspects of Hutton's work. Incidentally, the foundation of the British Geological Society illustrates the prominence of Nonconformists in British science during the late eighteenth and early nineteenth centuries: of the thirteen founders, four were Quakers, and one a Unitarian minister.

During the period between 1790 and 1830, known as the 'Heroic Age of Geology', a great deal of field work was carried out on the examination of the succession of the rock strata, and their mineral and fossil contents. An advance in method was the use of fossils to classify the rocks in which they were embedded, a method which had been suggested by Buffon, but first widely used by the surveyor, William Smith, 1769-1839, in England, and by the biologist, Georges Cuvier, 1769-1832, in France. Smith was employed to survey the Somerset coal canal, and he toured England to see how other canals were constructed. During the course of this employment, and other work on drainage and irrigation, he discovered that the different rock strata in England from the coal to the chalk could be characterized by the fossils they contained. Smith suggested that rocks in different places with the same fossil content were of the same age, though he did not advance any theory concerning the formation of the rock strata. He published his method of rock classification in 1799, and in 1815 brought out a geological map of England, showing the horizontal rock bands across the surface of the country. Then, in 1817, he published a chart showing the vertical succession of the strata beneath the surface of England. Smith was concerned mainly with the remains of simple sea animals, notably fossil shells, as he was primarily interested in the classification of the rocks which contained them. Cuvier, on the other hand, was more concerned with the fossil remains of land animals as he was interested in reconstructing extinct animals from their remains, a problem that was solved with less difficulty in the case of vertebrate animals than others. He was also concerned with the geological significance of fossils, and in the introduction to his *Researches on Fossil Bones*, published in 1812, which dealt mainly with the reconstruction of extinct animals, he outlined a theory of the geological development of the earth.

Cuvier was much opposed to theories of evolution in biology, as we have seen in his attitude to Lamarckism, and in the same way he was opposed to the Huttonian theory of geological development. He was of the view that the geological agencies now at work in nature could not account for the development of the rocks as there was no continuity between one rock layer and another. There were, he held, sharp lines of demarcation between the successive strata, and each layer of rock contained its own distinctive fossil remains that were not found elsewhere. Thus each layer must have been the product of a particularized and powerful agency, not of continuously operating minor forces. These agencies were, he thought, a series of catastrophic floods, the last one being about the time of the Mosaic Flood, some five or six thousand years ago. Each flood had largely wiped out the existing organic population of the earth and eroded the surface of the earth, a fresh stratum of rock containing the remains of living creatures being deposited as it subsided. The catastrophes also disturbed and tilted the strata left by previous floods, thus accounting for the fact that the earlier rocks are more distorted and tilted than the later ones. Cuvier had a great influence, and he effectively did away with the idea of geological evolution in France, as he did with the idea of biological evolution, for a number of decades. He himself was a Neptunist, but not all of his followers held that the formative geological catastrophies were due to marine floods. Elie de Beaumont, 1798-1874, a professor at the Paris School of Mines, suggested in 1829 the Vulcanist view that geological catastrophes were caused by the sudden cracking of the earth's solid crust, due to the cooling and contraction of the liquid interior. In Germany Werner's pupil, Leopold von Buch, 1774-1852, adopted a similar theory though neither he nor Beaumont could accept the idea of a slow and gradual geological evolution of the earth which became associated with the Vulcanist view in Britain.

In England the theories of Werner and Cuvier were well received as they conflicted with the theology of the time less than Hutton's theory did. Both of the professors of geology, Adam Sedgwick, 1785-1873, at Cambridge, and William Buckland, 1784-1856, at Oxford were in Holy Orders, and both were ardent Neptunists. Buckland's work, *The Relics of the Flood*, published in 1823, was the last important attempt to combine geology with theology. He postulated the existence of a pre-Adamite period, lasting perhaps some millions of years, covering the time between the original creation of the heavens and the earth and the first day of Genesis. It was during this pre-Adamite period, Buckland thought, that the main geological changes had taken place after the manner that Werner and Cuvier had suggested. Sedgwick, too, was a Neptunist at first. In 1819, shortly after he had been appointed to the chair of geology that Woodward had founded at Cambridge, he wrote that he was 'eaten up with Wernerian notions, ready to sacrifice my senses to that creed,—a Wernerian slave'. However, he and his friend, Roderick Murchison, 1792-1871, a country gentleman, pursued researches which finally ended Werner's geological scheme. William Smith had investigated the later strata that contain fossils, the lower limit of his investigations being the coal series. Sedgwick and Murchison studied the earlier rocks that are found in Wales, primary rocks which contained a few or no fossils, and which were least likely to have been formed from water by chemical or mechanical means. Sedgwick discovered the Cambrian series, and Murchison the Silurian system, whilst both together found the Devonian rocks which lie between the Silurian and the coal series. They then toured the continent to examine similar rocks elsewhere, and by 1829 they had come to the conclusion that the primary rocks were formed by the solidification of molten rock, not by crystallization from water as Werner had supposed. Sedgwick averred that he had wasted two years' work through adhering to Werner's notions, though he now accepted only the Vulcanist aspects of Hutton's work, not his evolutionary views.

However, one of Buckland's pupils, Charles Lyell, 1797-1875, independently arrived at parts of Hutton's geological theory, and later made a study of his work. Lyell himself made no great practical discoveries in geology, his great contribution being to connect up the scattered facts of the subject. He travelled widely, examining the rock strata of various parts of Europe, and read extensively, so that he was able to bring a much greater body of facts to support the theory of geological evolution than Hutton had done. His main work was *The Principles of Geology: being an Attempt to Explain the Former Changes of the Earth's Surface by reference to Causes now in Operation*, published in 1830-33. Here Lyell reiterated the major postulates of Hutton, namely, the premise that only the geological forces now at work should be used to explain the past history of the earth, and that, for this purpose, indefinitely long periods of time may be assumed.

'Confined notions in regard to the quantity of past time have tended more than any other pre-possessions to retard the progress of geology,' he wrote, 'and until we habituate ourselves to contemplate the possibility of an indefinite lapse of ages . . . we shall be in danger of forming most erroneous views in geology.'

Werner had started from a definite if hypothetical origin of the rocks, the primeval ocean, and had argued forwards. Hutton and Lyell started with the present geological forces of nature and argued backwards. Such a point of view was termed Uniformitarianism, as it was assumed that the forces of nature had always been the same as they are now. The earlier mechanical philosophers had assumed that the material systems of nature, the solar system and the organic species, were constant throughout the history of the earth. Now there was a change of emphasis; it was the forces, and not the material systems of nature that were thought to be constant. Thus through the operation of the same unchanging forces the matter of the earth was transformed, and in this way the idea of geological evolution was deduced from an extension of the former unhistorical mechanical view of nature.

Lyell himself was so much of a Uniformitarian at first that he was reluctant to admit that there had been any major changes in the condition of the earth, apart from the successive deposition of rock strata. He admitted that there had been climatic changes, and ascribed them to the varying distribution of land and water, but at first he refused to admit that there had been any changes in the organic population of the earth, and for this reason he rejected Lamarck's theory of biological evolution in the 1820's. However, the succession of fossils in the series of rock strata implied that if there had been a geological evolution, then there must also have been an evolution of the organic species. Thus in the following decade Lyell came to change his opinion on this matter. Writing to John Herschel in 1836, he said:

'In regard to the origination of new species, I am very glad to find that you think it probable that it may be carried on through the intervention of intermediate causes. I left this rather to be inferred, not thinking it worth while to offend a certain class of persons by embodying in words what would only be a speculation.'

Such an implication was obvious also to the older geologists who opposed Lyell's theories. Adam Sedgwick, the most sagacious of the old school, in his presidential address to the Geological Society in 1831, pointed out that one of the major difficulties of Lyell's theory was that it implied the evolution of the organic species.

'I may remind you', he said, 'that in the very first step of our progress, [in tracing geological history backwards], we are surrounded by animal and vegetable forms of which there are now no living types. And I ask, have we not in these things some indication of change and of an adjusting power altogether different from what we commonly understand by the laws of nature. Shall we say with the naturalists of a former century that they are but sports of nature, or shall we adopt the doctrine of the spontaneous generation and transmutation of species with all their train of monstrous consequences?'

When Darwin published his *Origin of the Species* in 1859, Lyell was amongst the first to accept his views. Sedgwick in 1865 remarked again:

'Lyell has swallowed the whole theory, at which I am not surprised, for without it the elements of geology as he expounded them were illogical. . . . They may varnish it as they will,' he continued, 'but the transmutation theory ends, with nine out of ten, in rank materialism.'

For reasons such as these, rather than any serious scientific objections, Lyell's theory was opposed by most of the important members of the British Geological Society, such as Sedgwick, Buckland and Murchison. However, the opposition to Lyell was not so marked as that which had faced the supporters of the Huttonian theory at the beginning of the century. In 1831 Lyell was made professor of geology at King's College, London, then a newly founded Church of England institution. The appointing body consisted of Anglican ecclesiastics, and one of them, the Bishop of Llandaff, expressed concern at Lyell's views, but in spite of this he received the appointment. By the 1830's the situation in geology was very different from what it had been at the beginning of the century. Much more was known about the nature and the historical succession of the rock strata in various parts of the world, particularly in Europe. Moreover, the whole climate of opinion had changed, and was now more deeply imbued with the idea of the historical progress of mankind, which appears to have been conducive to the development of evolutionary theories in science. Lyell in fact pointed out several analogies between geological investigations and the study of history: indeed his Uniformitarianism in geology is said to have been suggested by the gradual growth of the British Constitution, just as the catastrophic theories of the French may have been suggested by the more turbulent recent history of France. Lyell averred that the geological study of the earth was strictly analogous to the archaeological study of human history, and later he went over into this field, producing in 1863 his work on *The Antiquity of Man*.

Lyell's theories were not widely accepted by his contemporaries in the world of science. Notable exceptions were the physicist, John Herschel, and the geologist-politician, Poullet Scrope. Scrope's review of the *Principles of Geology* did much to popularize the work, some six hundred and fifty copies being sold in the three months before the review and fifteen hundred soon after. In the succeeding generation of scientists Lyell's view became generally accepted, the implication of that view, foreseen by Sedgwick and Lyell himself, being developed by the most notable representative of that generation, Charles Darwin. Darwin's early work was in the field of geology, and as he tells us in the *Autobiography*, it was the study of geology that led him to the theory of the evolution of the species, though he obtained the idea of the mechanism for that evolution from another source, the essay of Malthus on population. Whilst the older geologists rejected Lyell's theory, it was immensely popular in middle-class circles, where, incidentally, the belief in progress was most marked. The popular political and historical writer, Harriet Martineau, writing of the 1840's, said, perhaps with some exaggeration, that

'the general middle-class public purchased five copies of an expensive work on geology to one of the most popular novels of the time.'

Disraeli in his novel, *Tancred*, published in 1847, illustrates the influence geology had in upper-class circles, though here it appears that the new theory was received with feelings which were more mixed.

Chapter 34

Theories of the Evolution of the Species During the Nineteenth Century

AT THE turn of the eighteenth century theories of biological evolution, in various forms, appeared in Germany, France, and England. In Germany there was the school of nature-philosophers, who saw the organic species as so many separate and unconnected material realizations of the stages through which the World Spirit had passed during the course of its inner self-movement towards the destined goal of man. In France there was Lamarck, who conceived of the animal species as materially descended from one another, animals progressing by virtue of an inner expansive force, and by inherited additions acquired from the environment. In England there was Erasmus Darwin who put forward views on organic evolution similar to those of Lamarck, but who had in addition one idea that was curiously British and later most fruitful, the notion that organisms progress by competing with one another for sustenance, or for the females of their species. These national differences in biological theory continued on to a considerable degree during the nineteenth century, though there was a certain amount of cross breeding and some fertile hybrid theories appeared. The different theories were each part of their own national current of thought: in Germany the preoccupation with history and the mystical-alchemical tradition, in France the politically orientated psychological and sociological doctrines of progress, and in England the *laissez-faire* theories of economic and social progress, which suggested that men should be left free to pursue their own happiness and their own individual ends in competition with other individuals.

Erasmus Darwin, like other members of the Birmingham Lunar Society, was influenced by the French philosophers of the eighteenth century, though he applied this characteristically British idea of competition between individuals to his biological theory. The elder Darwin believed in progress, as did the earlier *laissez-faire* theorists, Adam Smith (1776) in economics, and Jeremy Bentham (1789) in moral philosophy. They were followed, however, by Robert Malthus who used the idea of competition between individuals to show that human progress was impossible, in opposition to the theories of the French philosophers and their English followers, such as William Godwin. Malthus made known his views in *An Essay on the Principle of Population as it affects the future improvement of society, with remarks on the speculations of Mr. Godwin, M. Condorcet and other writers*, published in 1798.

'I think I may fairly make two postulata,' Malthus wrote in this work. 'First, that food is necessary to the existence of man. Secondly, that the passion between the sexes is necessary and will remain nearly in its present state.'

These things being so, he argued:

'I say that the power of the population is indefinitely greater than the power of the earth to produce subsistence for man . . . (for) population when unchecked increases in geometrical ratio, subsistence only increases in an arithmetic ratio. A slight acquaintance with numbers will show the immensity of the first power in comparison with the second.'

Thus, there could never be enough food for all mankind, as any agricultural advance was immediately neutralized by a greater number of children surviving to maturity, so that the standard of living remained the same.

'Consequently if the premises are just,' Malthus wrote, 'the argument is conclusive against the perfectibility of mankind.'

Malthus was of the view that the life of mankind was of a piece with that of the organic world as a whole.

'Throughout the animal and vegetable kingdom', he wrote, 'nature has scattered the seeds of life abroad with the most profuse and liberal hand. She has been comparatively sparing in the room and nourishment necessary to rear them.

The race of plants and the race of animals shrink under this great restrictive law. And the race of man cannot by any effort of reason escape from it. Among plants and animals its effects are waste of seed, sickness and premature death. Among mankind, misery and vice.'

It was this conception that provided Charles Darwin, 1809-82, with his mechanism of biological evolution: organisms compete for restricted food supplies, and those with favourable variations survive and reproduce their kind. However, Darwin was convinced from his study of geology that there had been an evolution of the species before he had this mechanism to explain how it had happened. Charles Darwin was the son of a Shrewsbury doctor, and the grandson of Erasmus Darwin and Josiah Wedgwood, the potter, both of whom had been connected with the Birmingham Lunar Society. In 1825 Darwin went to study medicine at Edinburgh. Here Robert Jameson, Werner's pupil, was still rampaging against the Huttonian geological theory and the Vulcanists in general. Jameson's lectures he found so 'incredibly dull' that he resolved never 'to read a book on geology or in any way study the science'. Darwin, however, gave up medicine and proceeded to Cambridge with a view to taking Holy Orders. Here Sedgwick and Henslow, the professors of geology and botany respectively, aroused in him the desire to study geology and natural history again, and he accompanied Sedgwick on one of his geological explorations of Wales. Darwin was so highly thought of by his teachers that they recommended him for the post of naturalist on a government exploration voyage to the south Pacific, a post which he accepted. Henslow advised Darwin to take a stack of books with him on the voyage, including the first volume of Lyell's *Principles of Geology* which had just come out, but he advised Darwin 'on no account to accept the views therein'.

The expedition set out in the *Beagle* in December 1831, and after the extensive survey of the coasts of South America and the Pacific archipelagos, returned in October 1836. During the voyage Darwin came not only to accept Lyell's views, but to extend them. Writing home, he said:

'I am become a zealous disciple of Mr. Lyell's views as known in his book. Geologizing in South America I am tempted to carry parts to a greater extent even than he does.'

During the five years of the voyage Darwin made large geological, botanical and zoological collections, but of these the geological was the most important, for he confessed that he had then little biological knowledge and was unable to sketch adequately the organisms he saw. On his return home Darwin's earliest works were upon geological matters, notably on the *Structure and Distribution of Coral Reefs*, published in 1842, in which he put forward the theory that coral reefs and atolls were due to the gradual sinking of land masses or islands, the coral animals building up their reef so that its top was at the surface of the ocean.

However, the biological phenomena he had observed on the voyage of the *Beagle* had already turned his mind towards the possibility of the evolution of the organic species. He had seen how closely related species had succeeded one another as he proceeded southwards down the American continent, and how the species on the Galapagos archipelago resembled those of South America, but differed slightly even from island to island. In his *Autobiography* he wrote:

'It was evident that such facts as these could only be explained on the supposition that species gradually become modified, and the subject haunted me. But it was equally evident that neither the action of the surrounding conditions, nor the will of organisms (especially in the case of plants), could account for the innumerable cases in which organisms of every kind are beautifully adapted to their habits of life. . . . After my return to England it appeared to me that by following the example of Lyell in geology, and by collecting all facts which bore in any way on the variation of animals and plants under domestication or nature, some light might perhaps be thrown on the whole subject. My first notebook was opened in July 1837. I worked on true Baconian principles, and without any theory collected facts on a wholesale scale, more especially with respect to domesticated productions, by printed inquiries, by conversation with skilful breeders and gardeners, and by extensive reading. . . . I soon perceived that selection was the keystone of man's success in making useful races of animals and plants. But how selection could be applied to organisms living in a state of nature remained for some time a mystery to me.'

Thus by extending Lyell's method and viewpoint from geology to biology Darwin arrived at the conclusion that the organic species had evolved in time, but the mechanism whereby this evolution had been accomplished he derived from another source.

'In October 1838,' he wrote, 'that is fifteen months after I had begun my systematic enquiry, I happened to read for amusement Malthus on Population, and being well prepared to appreciate the struggle for existence which everywhere goes on from long continued observation of the habits of animals and plants, it at once struck me that under these circumstances favourable variations would tend to be preserved, and unfavourable ones to be destroyed. The result of this would be the formation of a new species. Here then I had at last got a theory by which to work.'

Darwin spent the next twenty years in collecting information to substantiate this theory of the evolution of the species by natural selection, and in working out its consequences and implications.

Meanwhile another English naturalist, Alfred Russel Wallace, 1823-1913, independently arrived at the theory of natural selection. Wallace visited the Malay archipelago where he observed that neighbouring islands were inhabited by closely related but different species, as Darwin had before him in Galapagos. It was here that Wallace hit upon the theory of natural selection, like Darwin deriving the idea from Malthus. He recorded in his autobiography:

'In February 1858 . . . the problem (of evolution) presented itself to me, and something led me to think of the positive checks described by Malthus in his Essay on Populaton, a work I had read several years before, and which had made

a deep and permanent impression on my mind. These checks,
—war, disease, famine and the like,—must, it occurred to
me, act on animals as well as man. Then I thought of the
enormously rapid multiplication of animals, causing these
checks to be much more effective in them than in the case
of man; and while pondering vaguely on this fact, there suddenly flashed upon me the idea of the survival of the fittest,—
that the individuals removed by these checks must be on the
whole inferior to those that survived. I sketched the draft of
my paper . . . and sent it by the next post to Mr. Darwin.'

Darwin had Wallace's paper published together with one of
his own, and in the following year, 1859, brought out his great
work *On the Origin of the Species by Means of Natural Selection, or the Preservation of Favoured Races in the Struggle for
Life.*

In this work Darwin advanced two main lines of argument
for the theory that the organic species had evolved: firstly,
the distribution of the extinct species in time, which he had
gathered from geology and palaeontology; and secondly, the
geographical distribution of living species in space, which he
had come across during the voyage on the *Beagle*, and amplified from the works of other travellers and geographers,
notably Alexander von Humboldt, 1769-1859. He also relied
to some degree upon the embryological work of von Baer,
which he interpreted as showing that an individual organism
in its growth from a single cell to an adult animal passes
through the evolutionary history of its species. However, Darwin drew very little evidence from French and German sources
on the whole. In the four hundred pages of the *Origin of the
Species*, only ten pages were devoted to the discussion of embryological evidence, and five to the morphological structures
of creatures, whilst the cell theory received scant attention. Unlike the French evolutionists and the German nature philosophers Darwin did not rely upon the classificatory systems of
animals and plants, nor comparisons between the anatomical
structures of adult organisms, to draw up his evolutionary
series. Furthermore, he did not believe that the different organisms formed a linear evolutionary chain of creatures as the
French had thought, nor that they were radial modifications

of a central ideal archetype as the Germans had presumed.

Darwin in fact was the first to develop consistently the idea that the evolutionary series of organisms was a tree of genealogical descent; related forms branching off from common parents, some forms ending in extinction, and others surviving to possess living descendants in various parts of the earth. He drew up his tree of genealogical descent from the geological succession of fossil animals, and showed that the embryological development of individual animals tended to follow roughly the evolutionary development of their races revealed by the fossil remains. Such a tree of evolutionary descent was supported by the facts of the geographical distribution of animals and plants. On islands and other regions cut off by geographical barriers there were found organic species once dominant long ago, like the Kangaroo and other Marsupials in Australia; they were living fossils preserved by their isolation. If we assume organic evolution to have occurred, Darwin wrote,

'we can see why there should be so striking a parallelism in the distribution of organic beings throughout space, and in their geological succession throughout time: for in both cases the beings have been connected by the bond of ordinary generation, and the means of modification have been the same.'

When he came to consider the mechanism of organic evolution, Darwin started out by pointing to variations between the individuals of a particular organic species as an observed fact. A litter of domestic animals contained creatures which varied one from the other. Animal breeders have selected from such litters for breeding purposes those animals which displayed the characteristics they wished to develop in the most pronounced form, and by these means have been produced all the varieties of domestic animals with which we are familiar. In nature the breeder was replaced by the mechanism of natural selection: those creatures with favourable variations survived to reproduce their kind, whilst those with unfavourable variations perished. Darwin suggested that species, which generally are not interfertile, were only developed forms of varieties, which are interfertile. Thus the mechanisms of natural or artificial selection which produced new varieties would, in the long run, produce new species, and ultimately new genera and orders of organisms. As further evidence for the ubiquity of variations in the organic world, Darwin pointed to the fact that the more prolific and widely scattered species produced the greater number of varieties. Such varieties were new species in the making, varieties developing into species when they become widely separated and when the intermediate forms disappeared. Hence in the formation of new species, the emergence of geographical barriers separating varieties, and the gradual divergence of varieties over long periods of time, were particularly important.

Darwin's theory did not demand any mechanism for the production of variations in animals and plants: they could be taken for granted as an empirical fact. However, he speculated on the matter, and suggested that changes in the climate, food, and other environmental causes, particularly those which affected the reproductive organs, brought about variations in animals and plants. He thought that such variations were slight and infinitely variable, so that evolution was gradual and continuous.

'As natural selection acts solely by accumulating slight, successive, favourable variations', he wrote, 'it can produce no great or sudden modification: it can only act by very short and slow steps.'

As his ideas developed, Darwin came to accept Lamarck's view that the use or disuse of organs brought about inherited changes in animals and plants. In the sixth edition of the *Origin of the Species*, Darwin summed up his position in the words that evolution,

'has been effected chiefly through the natural selection of numerous, successive, slight variations, aided in an important manner by the inherited effects of the use and disuse of parts, and in an unimportant manner, that is, in relation to adaptive structures, whether past or present, by the direct action of external conditions, and by variations which seem to us in our ignorance to arise spontaneously.'

Darwin, however, never accepted the theory of Lamarck and his grandfather that there was an inner driving force within each organism which tended to take it towards higher and more perfect forms. The views of Lamarck, he averred, had merely astonished him, whilst the works of his grandfather had been read by him, 'without producing any effect'. On the contrary Charles Darwin stressed the passive character of organic evolution: it occurred by the external mechanism of selection, not by an inner striving for a higher life. The same, he thought, was true of the progress of mankind. In the closing paragraph of *The Descent of Man,* published in 1871, Darwin wrote:

'Man may be excused for feeling some pride at having risen, though not through his own exertions, to the very summit of the organic scale; and the fact of his having risen, instead of being placed there aboriginally, may give him hope for a still higher destiny in the distant future.'

Progress was a reality, therefore, but it occurred in spite of human endeavour. It was an automatic process mediated, as Herbert Spencer put it, by 'the survival of the fittest'. In this belief Darwin, Wallace, and Spencer were fully within the current of early Victorian *laissez-faire* thought. They were men who rose to maturity in the second quarter of the nineteenth century when the theories of the British political economists and utilitarian philosophers gained a firm hold upon the climate of English opinion, as too did the idea of progress and evolution. Malthus expressed the 'competition' idea in a crude and most biological form, and it was perhaps for this reason that he was the specific influence upon Darwin and Wallace, whilst the new feeling of progress and evolution led them to invert the pessimistic conclusion of Malthus, and view the competition between individuals for their livelihood as a progressive rather than a conservative influence in the organic as well as in the human world. Herbert Spencer, 1820-1903, at first was something of a Lamarckist, but like Darwin and Wallace he was influenced by the theories of Malthus and deduced the idea of progress from them. In 1852, before the appearance of Darwin's work, Spencer wrote in his *Theory of Population deduced from the General Law of Animal Fertility* that

'From the beginning, pressure of population has been the proximate cause of progress. All mankind in turn subject themselves more or less to the discipline described: they either may, or may not advance under it, but in the nature of things only those who do advance under it eventually survive.'

When Darwin's *Orgin of the Species* appeared in 1859 Spencer extended the theory of natural selection to human society, viewing the 'survival of the fittest', not only as the mechanism of organic evolution, but also as the mode of progress of mankind. In particular, it exemplified and justified to Spencer's mind the *laissez-faire* policies of the mid-Victorian period: free trade and economic competition were, so to speak, the social forms of natural selection. To tamper with them would interfere with the process of cosmic evolution, and would throw out of gear the vehicle of human progress.

Such an interpretation of Darwinism in terms of the current ethos of liberalism helped to establish the popularity of the theory in middle-class circles. There was much criticism of Darwin's theory on scientific, social, and theological grounds, but it was accepted quite quickly in Britain. The most important scientific opposition came from Richard Owen, 1804-92, director of the Kensington Natural History Museum, and England's foremost student of comparative anatomy and fossil bones. He was a disciple of the German nature-philosopher, Lorenz Oken, and he conceived of the various organic species as the products of the operation of an ideal vital force in nature. This force, he wrote,

'produces the diversity of form belonging to living bodies of the same materials, which diversity cannot be explained by any known properties of matter.'

In an anonymous article written for the *Edinburgh Review* in 1860, Owen was severely critical of Darwin's *Origin of the Species*. He repeated his view that a self-differentiating vital force was responsible for the production of the organic species, as evidence for which he suggested that single-celled creatures were being spontaneously generated all the time, and that this being so, the higher animals could hardly have descended in a single series from them. Owen too, so Darwin believed, provided material for the attacks made by Samuel Wilberforce, Bishop of Oxford, on Darwin's book in the *Quarterly Review*, and at the Oxford meeting of the British Association for the Advancement of Science in 1860. Darwin was ably defended at this meeting by Thomas Henry Huxley, 1825-95, professor of geology at the London School of Mines, who here earned for himself the title of 'Darwin's Bulldog'. After this debate Darwinism became generally accepted by scientifically educated opinion in Britain, though Owen, and the Catholic zoologist, St. George Mivart, 1827-1900, continued steadfast in their opposition, as also did the theologian, Samuel Wilberforce, and the politicians, George Campbell and William Gladstone.

Darwinism was not only widely accepted in Britain, it was also extended to spheres outside biology. The idea of evolution was applied to chemistry, astronomy, linguistics and anthropology, but the full theory of natural selection was applied mainly to social philosophy and ethics, giving rise to a school of Social Darwinists. The doctrines of this school changed with events. Herbert Spencer, the first Social Darwinist, derived the values of the mid-Victorian age from the theory of natural selection, as we have seen. Developments in the late Victorian period, the strife of nations as exemplified in the Boer War, filled him with distaste, for it was the peaceable and industrious competition of individual men that seemed to him to be the main agent of social evolution. However, the new developments could be equally well justified by the Darwinian theory, indeed they were to some degree anticipated by the historian and economist, Walter Bagehot, 1826-77, in his volume of essays, *Physics and Politics, or Thoughts on the Application of the Principles of Natural Selection and Inheritance to Political Society*, published in 1872. In this work Bagehot suggested that, 'the strongest nation has always been conquering the weaker', and it was by these means that, 'the best qualities wanted in elementary civilization are propagated and preserved', since 'the most warlike qualities tend principally to the good'. The evolution of human society had been as gradual and continuous, and indeed as automatic, as Darwin had thought the evolution of the species to have been. 'Judea changed in inward thought, just as Rome changed in external power,' Bagehot wrote. 'Each change was continuous, gradual, and good.' In 1900 Karl Pearson, 1857-1936, at University College, London, wrote an essay *On National Life from the Standpoint of Science* in which he expressed similar views. There always had been, Pearson maintained, 'a struggle of race against race, and nation against nation'.

'The man who tells us', he affirmed, 'that he loves the Kaffir as he loves his brother is probably deceiving himself. If he is not, then all we can say is that a nation of such men . . . will not stand for many generations: it cannot survive in the struggle of nations.'

Such interpretations of Darwinism were quite popular at the turn of the nineteenth century, and as yet they have not entirely lost their appeal.

The biologists themselves, generally speaking, were not given to such interpretations. Darwin in his *Descent of Man* saw in the progress and evolution of mankind the growing dominance of the co-operative over the selfish instincts. 'The more enduring social instincts conquer the less persistent instincts,' he affirmed. Darwin's disciple, Huxley, was much opposed to the conclusions of the Social Darwinists, and he combated them in a series of essays. In his lecture on *Evolution and Ethics*, delivered in 1893, Huxley asserted that human progress does not consist in 'imitating the cosmic process, still less in running away from it, but in combating it'. Alfred Russel Wallace, who had arrived at the theory of natural selection independently of Darwin, deduced the doctrines of the Christian Socialists from the theory in his *Studies Scientific and Social*, published in 1900. In the social struggle for existence, he held, none should have an unfair advantage in wealth or education; we must all start equally to get the full progress of mankind.

'The only mode of natural selection that can act alike on physical, mental and moral qualities,' he wrote, 'will come into play under a social system which gives equal opportunities of culture, training, leisure, and happiness to every individual. This extension of the principle of natural selection as it acts in the animal world generally is, I believe, quite new, and is by far the most important of the new ideas I have given to the world.'

Thus in the end almost any theory of human progress could be deduced from Darwinism, though the more influential interpretations in Britain as elsewhere were those that emphasized the competitive element in human society.

Outside Britain it was in Germany that Darwinism was most widely, and indeed fiercely, debated. In France and America the theory of natural selection did not find a great deal of popular or scientific support. Darwinism, for the most part, was at first opposed by the scientists of those countries, and when theories of evolution appeared in the 1880's, they tended to be Lamarckian in form. In France Cuvier's followers, Elie de Beaumont, 1798-1874, Milne-Edwards, 1800-85, and others, opposed Darwin's theory, so too did the physiologist, Claude Bernard, 1813-78, and the microbiologist, Louis Pasteur, 1822-95. In America, Louis Agassiz, 1807-73, professor of geology at Harvard, was very much opposed to Darwinism, though the professor of botany, Asa Gray, 1810-88, was a friend of Darwin and accepted his views. Agassiz came of a French Hugenot family in Switzerland, was educated under several nature-philosophers in Germany and Cuvier at Paris. He did some important work upon living and fossil fishes, and also upon the geological action of glaciers. In consequence he was a man of some influence, and this he exerted against the Darwinists, holding the view that the species were divinely created and fixed for all time, like other Protestant systematists in biology before him, notably Linnaeus and Cuvier.

However, amongst the next generation of scientists there were theorists of evolution, notably Brown-Sequard, 1817-94, and Alfred Giard, 1846-1908, in France, and Edward Cope, 1840-97, in America, all of whom inclined towards the theories of Lamarck rather than those of Darwin. Brown-Sequard made some experiments in which he injured the brain of a guinea-pig, the injury causing the loss of sensation in the toes and a clouding of the eyes. The animal bit off the insensitive toes, and Brown-Sequard claimed that the clouded eyes and lack of toes was inherited by the offspring. Such experiments were not confirmed, and the belief in the inheritance of mutilations was abandoned, indeed Lamarck himself had rejected the idea that this sort of acquired characteristic was inherited. Brown-Sequard accepted only Lamarck's theory of the inheritance of acquired characteristics: Cope in America accepted also Lamarck's view that there was an inner driving force within each organism which developed that organism towards higher forms. Cope did not identify this force with heat and electricity as Lamarck had done: he held that it was a spiritual force akin to the activity of the human mind. To this degree his theory was a hybrid system of French Lamarckism and German nature-philosophy.

In Germany Darwin's theory aroused considerable controversy, partly because it was opposed to the views of the nature-philosophers and partly because it became involved in the politics of the period. The German Liberals of the middle and late nineteenth century were divided amongst themselves, one group urging collaboration with the Junkers to build national unity, the other group proposing the overthrow of the Junkers as their primary aim. Darwinism was associated by its opponents, and some of its advocates, with the second and more radical group of liberals, whilst nature-philosophy, in its later more materialist and more empirical forms, tended to be associated with the first group, though there were nature-philosophers, Darwinists and scientists who attempted to combine both theories, and who stood outside of this conflict.

When Darwin's *Origin of the Species* reached Germany in 1860 his theory was rejected by most of the older scientists who, generally speaking, were under the influence of the earlier nature-philosophy. Such men as these were the embryologists, von Baer, and Kolliker, Leydig the zoologist, and Braun the botanist, though the cell-theorist, Schleiden, was amongst the first to accept Darwin's theory. The younger biologists were more sympathetic towards Darwinism, and they endeavoured to combine the theory with the sciences of embryology, comparative anatomy and the cell theory which had been much studied in Germany under the influence of nature-philosophy, but not extensively discussed by Darwin. The first important figure in this development was Carl Gegenbaur, 1826-1903, a professor at Jena, where Oken had worked at the beginning of the century. Oken had seen the various organic species as modifications of a few ideal forms or archetypes. Gegenbaur now viewed these ideal archetypes as real ancestral types: they were stages in the descent of the species one from the other, not stages in the thinking of the Architect of nature. He was particularly concerned with the evolution of the hand and feet bones of the vertebrates, maintaining that they had come from the gill-slit apparatus of primitive fishes, which had evolved into the fins of higher fishes and the limbs of land animals. A similar development occurred, he thought, in the embryological growth of the higher land animals which pass through a stage possessing gill-slits like the fishes.

Gegenbaur's most notable pupil was Ernst Haeckel, 1834-1919, who also held a chair at Jena. Haeckel belonged to the radical group of the liberals, and it was he above all who made Darwinism the weapon of philosophical radicalism in Germany. His main empirical work in biology was the investigation of the Radiolaria of which he described some one hundred and fifty species. This work, published in 1862, obtained for him the chair at Jena. Thereafter he devoted himself to the dissemination of a modified form of Darwinism, in a series of works ranging in style from the very popular to the academic. His first main work was the *General Morphology*, published in 1866, in which he combined Darwinism with elements from the theories of Lamarck and the nature-philosophers, stressing more than Darwin the doctrine of the inheritance of characteristics acquired under environmental influence, and, like the nature-philosophers, ascribing the varied productions of nature to the operation of a single cosmic force. Haeckel had a passion for classification and, like Schelling, Hegel, and Oken, he looked everywhere for threefold divisions in nature. All objects had three attributes, he held, matter, form and energy. Thus there was the science of chemistry, dealing with matter; morphology, dealing with forms, and physics dealing with energy. Each of these sciences in turn could be subdivided into three branches: morphology, for example, could be subdivided into the study of animals, of plants, and of protozoa, the simple unicellular creatures. The aim of morphology, he said, was to find causal explanations for the structures organisms possessed which were monistic, that is, which were true for all the grades of nature, covering the inorganic as well as the organic world. Thus he thought that salt crystals and organic cells were strictly comparable in the way that they grew, and in their composition and symmetry of form, for both were the products of the same matter and the same cosmic force. Such a view implied that there was no qualitative distinction between the psychological, biological, and physical grades of nature, and that inorganic nature must possess, at least latently, the qualities of higher organisms and of man himself. Indeed Haeckel affirmed that, 'No matter can be conceived without spirit, and no spirit without matter.' Thus in the end Hackel brought back into biology the World Spirit of the earlier nature-philosophers, though he regarded this spirit as the monistic cosmic force, or energy itself. 'Every atom must have a soul,' he wrote, 'for it possesses some energy.' Such an idea enabled Haeckel to explain the inheritance of acquired characteristics as due simply to the memory of the atoms composing the seed of the offspring, a notion to which others inclined towards Lamarckism had subscribed, notably Herbert Spencer, and of course, Erasmus Darwin.

Haeckel continued Gegenbaur's work on the classification of the organic species into evolutionary series, drawing up several genaealogical trees to illustrate the lines of descent of the various genera and species. He also assimilated the work of the German embryologists into the Darwinian scheme, notably in his book on *The History of Man* (1874). In this work Haeckel revived and amplified the biogenetic principle of Meckel, namely, the view that the individual organism during the course of its embryological development passed through the main stages in the evolution of its species. Haeckel brought together much material to support the principle. Man, he indicated, started life as a single-celled egg; hence the first animal must have been like the single-celled protozoa. The egg developed into a spherical cell group, like the Volvox, which must have come next in the evolutionary series. The sphere of cells then invaginated to give a cup-like double walled gastrula, similar to the adult form of some sponges, which accordingly came after the Volvox. The gastrula elongated, its interior forming the beginnings of the gut cavity, and between the outer layer of cells, the ectoderm, and the inner layer, the endoderm, a fresh cell layer, the mesoderm, developed. These three layers then gave rise to the various organs of the adult body, the endoderm forming the intestinal tract, the mesoderm the muscles, and the ectoderm the connective tissue and nervous system. The biogenetic principle, laying down that the individual organism recapitulates the history of its race, gave a stimulus to embryological research, though it is no longer accepted in the form that Haeckel expressed it. Such a recapitulation, for example, has not been observed in the plant world. However, Haeckel, amongst others, performed a valuable service in assimilating the German work on morphology, embryology, and the cell theory into the Darwinian system, for Darwin himself had only touched upon these subjects, basing his views mainly upon the geological distribution of the extinct species, and the geographical distribution of the living species. .

Of the notable German evolutionists of the nineteenth century, Haeckel perhaps was the one that was nearest to the original views of Darwin. Nature-philosophy was still quite strong in Germany, and it led to other evolutionary theories, one of which was of considerable influence. This theory was put forward by Carl Nageli, 1817-91, who was professor of botany at Freiburg, Zurich, and Munich successively. He studied nature-philosophy under Oken and Hegel, and botany under Candolle at Geneva. Though he was influenced by Darwinism, Nageli never forgot the views of his earlier teachers—he merely put them in a more materialist form. In 1884 he published a work entitled, *A Mechanical-physiological Theory of Evolution*, in which he elaborated ideas he had developed and published earlier in 1844 and 1865. Nageli maintained that the plant or animal cell was not the fundamental unit of organic life, as the cell had a structure that was already differentiated. Cells were composed of smaller units, which he called micelles, that were similar to inorganic crystals. Thus there was no real difference between inorganic and organic matter. Micelles packed together through a physical attraction, and in the presence of water they formed living cells. In this way living creatures were being spontaneously generated all the time, and they evolved into higher forms by virtue of an inner perfecting force of a mechanical character. However, there was no real transition from one species to another: the apes were in no sense man's relatives. Man started as a simple spontaneously-generated unicellular creature first of all, a long time ago. The apes started in the same way a little later, and monkeys later still, whilst the present day protozoa have only just been spontaneously generated. The animals that are monkeys today will be men in the future, and man by that time will have progressed further still. In this way Nageli expressed the view of the nature-philosophers that the organic species have a common origin, but no material connection other than that. All creatures came from micelles, but their character should be judged by the degree of their inner historical development from their origin, not by their external resemblance to other organisms.

Nageli was of the opinion that Darwin had not satisfactorily explained how higher organisms with a wider and superior set of characteristics could have originated from lower creatures. A succession of small favourable variations was not enough, he thought; some inner driving force within the organism was necessary to bring about such marked changes. Nageli did not think of this force as a vital spirit, but as a physico-chemical force analogous to the force of inertia in mechanics. A ball will keep on rolling until it meets an obstacle, and in the same way an organism will evolve until it meets the obstacle of natural selection, which prunes away the forms that do not follow the predominant line of evolution. If there were no struggle for existence, the inner self-differentiating force within organisms would produce an enormous variety of forms and the earth would be overpopulated, but through the mechanism of natural selection only the viable forms are preserved. Nageli supposed that evolution was not a gradual and continuous process: the inner force moved according to the categories of the Hegelian dialectic, it performed leaps. Evolution, therefore, was discontinuous, it consisted of a series of mutations. In fact it was from Nageli that the Dutch botanist, Hugo de Vries, 1848-1935, obtained the idea of biological mutation at the end of the century.

On another point connected with genetic research Nageli made a further important suggestion. He pointed out that both parents contributed equally to their offspring, and yet the female egg was invariably larger than the male sperm. Only part of the egg, therefore, could be the heredity determining substance, which he termed the idioplasm. Nageli held that the idioplasm was made up of micelles linked together in chains, and that it was the sole determinant of the form assumed by the adult organism. Thus evolution consisted primarily in the discontinuous changes brought about in the idioplasm by the operation of the inner force within each organism, natural selection weeding out the nonviable forms. The Austrian plant breeder, Gregor Mendel, 1822-84, found that his genetic researches on pea plants supported Nageli's particle theory of inheritance, and he sent his results to him. However, Nageli wrote that Mendel's formulae seemed 'empirical rather than rational', and he ignored his work. Although Nageli was more materialistic than the earlier nature-philosophers, he was almost as speculative. He claimed that his theory was rational and German, whilst Darwinism was merely an example of English empiricism.

Nageli's theory of a heredity substance or idioplasm distinct from the general body tissues was taken up and developed by August Weismann, 1834-1914, a professor of zoology at Freiberg. In 1892 he published an *Essay on Inheritance and Related Biological Questions* in which he drew a sharp distinction between what he called the germplasm, responsible for the transmission of heredity qualities, that is, Nageli's idioplasm, and the soma or body plasm. He pointed out that the simple unicellular creatures propagate themselves asexually by dividing into two, and thus they are immortal, apart from accidents. In higher animals the body is mortal, and it is only the germplasm, passed on from one generation to another, that is immortal. In Weismann's opinion the germplasm was the important part of an organism: it determined the form and characteristics of the body plasm, which served to nourish the germplasm so that it could reproduce itself. However, the body itself had no effect upon the germplasm, so that characteristics acquired by the body under the influence of the environment could not be passed on to the offspring. Weismann attempted to prove that this was the case by cutting off the tails of rats over a number of generations, and showing that the offspring were always born with tails. He took this experiment as refuting the Lamarckian view, though Lamarck himself had said that mutilations were not inherited.

Weismann rejected Nageli's theory that variations of increasing perfection were brought about in the germplasm by an inner life force within the organism. He thought that variations were produced by the union of two different germplasms, one from the mother and the other from the father. The offspring could not have twice as much germplasm as either of its parents, however, so he suggested, as early as 1887, that the germplasm of each parent splits into two halves when eggs or sperms are formed. Thus the union of an egg and a sperm gives the offspring as much germplasm as either of the parents singly. This prediction of the phenomena of meiosis was made some years before it was fully traced out empirically by microscopic investigation. Weismann suggested further that the germplasm was contained in the thread-like chromosomes of the nuclei in the sexual cells, the germplasm being made up of units, which he called determinants, each of which governed a particular characteristic of the organism. This proposal again was made some years before there was much evidence that the chromosomes were in fact the bearers of hereditary qualities.

The views of Weismann were much opposed by the Neo-Lamarckians, notably Herbert Spencer in England, who held that the various combinations of the male and female germplasm would not give rise to important variations in the offspring, in particular, to variations which were qualitatively new. In Spencer's opinion such variations could only be produced by the Lamarckian mechanism of the inheritance of new characteristics acquired under the influence of environmental changes. Spencer, as we have seen, belonged to the mid-Victorian epoch with its firm belief in progress. Weissman belonged to a later period, and another country, where such a belief was not so marked. The whole emphasis of his theory of 'the continuity of the germplasm' lay upon the preservation of characteristics already possessed by organisms, and not upon the origin of new favourable variations which interested Spencer.

'The saddest thing of all,' wrote Weismann, 'is that in scarcely one case, can we say whether a certain deviation is useful or not: there seems to be no prospect of our ever being able to do so.'

The theories of Weismann were widely accepted in Germany, even before they had much empirical support, a fact ascribed by some authors to the congruence between Weismann's views and the racial theories that were popular in Germany. Summing up the development of the nineteenth century, the biologist, Patrick Geddes, 1854-1932, observed in his book on *Evolution*, which he wrote with another biologist, Arthur Thomson, 1861-1933, that each of the main theories of biological evolution seemed to be part of the general 'social transformations of the age':

'The generation of culminating political revolution in France,' he wrote, 'that of the culmination of the industrial revolution in England, have thus expressed themselves through Lamarck and Darwin more clearly than either thinker ever dreamed, or than their respective exponents and disciples have realized. . . . What are Lamarck's interpretations of the effects of use and disuse, his assured insistence upon the interior freedom of the organism to realize its utmost capacities, but the new step in social progress through abandonment of outworn orders of society, the freedom opening before new ones. 'La carrière ouverte aux talents' is pure Lamarckism; so again the splendid over-assurance of the Napoleonic epic, that 'every French soldier carried a marshal's baton in his knapsack'. But the colder business view so characteristic of English thought came to prevail over such political and military exaggerations; the ideals of mechanical efficiency and of individual and financial success rising above the ruins of liberal aspirations and imperial achievements as they have so often done. . . . 'Competition is the life of Trade': then why not also the trade of Life? Yet with all this freshness and vigour of economic application, there has prevailed in the main, and still prevails, a naïve forgetfulness of the social origins of these naturalists' discoveries. Similarly in neo-Darwinian times. With united and real respect for Weismann, for whose work one of us has once and again acted as translator and editor, the other yet ventures to urge one of the very few criticisms which that wide and fair-minded thinker seems never to have considered: the striking parallelism of his own theory of the germ-plasm, with the thought of contemporary Germany; with the victories and hegemony of Prussia, the renewed claims of its aristocracy also; and above all, with its doctrines of race, political and anthropological combined. The intermediate step between this ruling Prussian world of action and Weismann's ascendency in speculative biology is indicated by the widely diffused doctrine of Count Gobineau, consciously and avowedly bio-social as this has been. All these movements alike have now found eloquent, though hardly scientific expression in Houston Stewart Chamberlain, whose contemporary vogue in Germany is thus earned and explained.'

Count Gobineau was a Frenchman who published an *Essay on the Inequality of the Human Races* in 1853. Houston Stewart Chamberlain was an Englishman, though he was brought up in Germany and wrote in German *The Principles of the Nineteenth Century*, published in 1899. These men were of the opinion that the various human races were fixed types and differed widely one from the other. They believed that the Aryan races were superior, and that these races alone had built civilized society, being the natural rulers of the rest of mankind. The interbreeding of the Aryans with inferior races, they held, would lead to the degeneration of the human species. It was thought that such views were supported by Weismann's theories, as the distinctive features of each race were believed to reside, and perpetuate themselves, in the immortal germ-plasm of its members. Moreover, Weismann stressed the competitive element in Darwinian theory—'The omnipotence of natural selection' as he put it—which seemed to justify the domination of strong nations and races over the weaker, as a case of the survival of the fittest. There was little emphasis on evolution and progress in Weismann's teachings: he held that a species would degenerate unless natural selection con-tinuously weeded out weak combinations of parental germ-plasms. In the last decade of the nineteenth century the belief in progress generally began to wane, and the earlier aphorism of General Pitt-Rivers, that 'history is evolution, and science organized common sense', was received with the scepticism which is still with us today. Darwin had not moved entirely within the confines of the thought of his generation, and, amongst other things, he had pointed out that parasites and degenerate creatures were as much a product of evolution as the higher animals: they were perfectly adapted to their some-what restricted environments. This idea was now stressed. In England Ray Lankester published an essay on *Degeneration, a Chapter in Darwinism* in 1890, while from the Netherlands came the works, *Parasitism, Organic and Social*, by Vander-velde in 1895, and *Evolution by Atrophy in Biology and Sociology* by Demoor and others in 1894.

The social origins of scientific theories are historically of considerable interest and importance, but the value of a scien-tific theory as such depends upon its correspondence with empirical knowledge. It is a measure of the importance of Lamarck that he used the ideas of the French psychologists and sociologists of the eighteenth century to some purpose: he filled the formal analogies between man and the animal world with a real empirical content. The significance of the speculative theories of Nageli and Weismann lies in the fact that some of their views have provided the intellectual framework of the science of genetics. The genius of Darwin led him to interpret a much wider body of facts than those known to Lamarck in terms of the current English thought of his day, or more specifically the ideas of Malthus, and to transcend to some degree the confined notions of his time. In so doing, Darwin produced a theory of more fundamental value; a theory which was capable of assimilating work carried out in other lands and discoveries that were to be made in later times.

Chapter 35

Scientific Institutions in France and Britain During the Nineteenth Century

DURING THE eighteenth century the natural philosophers of France and Britain had been foremost in the world of science. As we have seen, their activities were complementary; the French inclining towards the theoretical interpretation of nature, the British towards empirical investigation. Such methodological divisions in science between the two countries were largely dispelled during the nineteenth century, though faint traces of them persisted. In the early decades of the nineteenth century the French were the leaders of world science, but they did not sustain their efforts, and by the 1850's and the 60's the British were to the front once more. However, the British leadership did not last, for by the end of the nineteenth century Germany had outstripped both England and France in science.

The change in the character of French science, and its rapid growth at the turn of the eighteenth century, had much to do with the events of the French Revolution. The scientists of France found their activities directed towards practical ends, which appears to have given them a greater taste for experimentation than they had previously, whilst scientific institutions were created which trained the French talent that was to be foremost in science during the early years of the nineteenth century. The first practical problem which the revolutionaries posed to the scientists of France was the standardization of weights and measures throughout the country. During the eighteenth century weights and measures in France varied widely from region to region. For example, the metre, measuring 100 centimetres at Paris, was 98 cm. at Marseilles, 102 cm. at Lille and 96 cm. at Bordeaux. At the request of Tallyrand, the Paris Academy of Sciences in 1790 set up a committee composed of Laplace, Lagrange, Lavoisier, Monge and others, to consider the question. In the following year the committee sent in a report to the Constituent Assembly proposing that the metre should be made a natural standard, namely, one ten millionth part of a quadrant of the earth's circumference, and that the gram should be the weight of one cubic centimetre of water at 4° C. The Assembly set up the General Commission of Weights and Measures to carry these proposals into effect, in order to 'Bring to an end the astounding and scandalous diversity in our measures'. The astronomers, Delambre, 1749-1822, and Mechain, 1744-1804, triangulated the distance between Dunkirk and Barcelona to measure the quadrant of the earth's circumference, the new measures being completed by 1799. All countries were invited to adopt the system, and most of the countries on the continent ultimately did so.

With the fall of the Girondists, and the rise to power of the Jacobins in 1793, the French Revolution took a more radical turn, and many of the old institutions were closed down, including the main scientific society, the Paris Academy of Sciences. Moreover, scientists associated with the *ancien régime*, or the Girondists, were executed, notably Lavoisier, who had directed the *Ferme Generale*, the tax organization of the *ancien régime*, and the astronomer, Bailly, who as Mayor of Paris had opposed the Jacobins. A move was made to arrest the secretary of the Academy of Sciences, Condorcet, who had opposed the execution of the King and other measures, but he forestalled his captors by suicide. Coffinhall, the vice-president of the tribunal that tried Lavoisier, declared that, 'The Republic had no need of savants', whilst Durand de Maillane, another jurist, thought that France 'already had too many scholars'. Such a point of view in regard to science was already unrealistic a century and a half ago, for as Maury, the historian of the Academy of Sciences, wrote in 1864:

'Everything was wanting for the defence of the country—powder, cannons, provisions. The arsenals were empty, steel was no longer imported from abroad, saltpetre came not from India. It was exactly those men whose labours had been proscribed who could give to France what she wanted.'

Accordingly the Convention called upon the scientists to meet these technical needs, and founded institutions to train more scientists. Gaspard Monge, 1746-1818, who had developed descriptive geometry dealing with methods of representing solid objects on paper, investigated the casting and boring of cannon, and was made minister of the navy. His friend, Lazare Carnot, 1753-1823, another mathematician, was made minister of war, in which capacity his services earned for him the title, 'Organizer of Victory'. The chemist, Fourcroy, 1755-1809, followed up the researches which Lavoisier had made, in connection with his post of director of gunpowder manufacture, on the extraction of saltpetre from manure. Berthollet, 1748-1822, who had directed the state dyeing industry, experimented with sodium chlorate, a chemical he had discovered, as an alternative to saltpetre, and with Morveau, 1737-1816, another chemist, discovered a method of making saltpetre synthetically by the oxidation of ammonia.

Such contributions indicated that science could solve the technical problems of the period, and to advance the sciences the old institutions were reformed, and new institutions were created. The King's Gardens were reopened in 1794 as the Museum of Natural History, the graded posts of the old institution being transformed into nine professorial chairs of equal status. In 1795 the Academy of Sciences was reconstituted as one of the three sections of the Institute of France, the other sections covering literature, and the political and moral sciences. The old Academy of Sciences had consisted of twelve honorary members, chosen from the nobility, who alone could become presidents or vice-presidents of the academy, and eighteen pensionaries who with the honorary members governed the elections to the society and its business: then there were twelve associates, and twelve adjoints, together with some free associates, superannuated associates, and foreign members, who had very varied responsibilities and rights. The scientific section of the Institute of France, consisted of some sixty members, who like the members of the Royal Society, had an equal voice in the business of their organization. However, like the members of the old Academy of Sciences, they were still salaried officials paid by the state.

The National Convention set up several military and medical schools in 1794, and also the Conservatoire of Arts and Trades, which was a technical college and a museum. At the same time, they founded the Ecole Polytechnique, and the Ecole Normale Supérieure, which were important institutions devoted to scientific education and research in France throughout the nineteenth century. The Supérieure was closed down after four months, and did not become important until 1808 when it was reopened by Napoleon. The Polytechnique, however, flourished from the start. It opened in 1794 with four hundred pupils and a staff composed of the leading scientists of the time. Mathematical physics were taught by Laplace and Lagrange, geometry by Monge, and chemistry by Berthollet. Amongst their pupils and successors were the physicists, Malus, Arago, Poncelet, Poisson, Cauchy, Sadi Carnot, and the chemists, Gay-Lussac, Thenard, Vauquelin, Dulong, and Petit. Under Napoleon several other military, medical and technical colleges were founded by the chemist, Fourcroy, who was made minister of public instruction, though they were of less importance. Napoleon himself encouraged the practical side of science by offering prizes for useful discoveries. He also discouraged the speculative thinkers who continued the tradition of the earlier materialist philosophers, such as the psychologist, Cabanis, by closing down in 1803 the section of the Institute of France devoted to the study of the political and moral sciences, which was their stronghold. Thus French science became more practical and experimental during the Napoleonic period, and at the same time the techniques of the French industries were advanced.

A marked anti-scientific movement arose in the official and fashionable circles of France with the restoration of the Bourbons in 1814. The movement was particularly opposed to the materialist and mathematical tradition of French science, and the Ecole Polytechnique, which was noted for its mathematical physicists and had achieved something of a revolutionary reputation, was disestablished in 1815. Mme de Stael and Chateaubriand voiced their dislike of the 'whole brood of mathematicians'. 'Mathematics were the chains of human thought,' wrote Lamartine; 'I breathe, and they are broken.' The idealist and romantic German nature-philosophy achieved some popularity, though it did not have a great deal of influence upon French science, except perhaps upon biology. The Polytechnique and its mathematical physicists flourished throughout the restoration period, and it was their tradition which was perpetuated in nineteenth-century France.

The scientific institutions set up by the National Convention in 1794 had the effect of concentrating the scientific activity of France in the capital at the Paris schools. During the eighteenth century there had been flourishing scientific academies in the provinces, but during the nineteenth century the Polytechnique and the Supérieure became the Mecca of young French scientists from the provinces as well as from the metropolis. Thus the provinces were impoverished of their scientific talent, and efforts were made to decentralize the concentration of science in Paris, notably through the foundation of the French Association for the Advancement of Science in 1870. In Britain, on the other hand, the growth of scientific activity in the provinces became more and more pronounced, finding an expression in the growth of the provincial literary and philosophical societies in the nineteenth century. As we have seen, the first of these societies in England which was permanent was the Manchester Literary and Philosophical Society, meetings of which were recorded from 1781. The subsequent period of the French Revolution and Napoleonic wars were troubled times, and they saw the demise of the earlier Birmingham Lunar Society. But in 1812 the Liverpool Literary and Philosophical Society was founded, and in 1818 another was set up at Leeds. Four years later a society was formed at Sheffield, and at the same time the large and important Yorkshire Philosophical Society came into being, covering the county as a whole. Thereafter provincial scientific and literary societies began to form at the rate of five, ten, fifteen, and even twenty per decade, so that by the end of the century over a hundred of such societies had been founded, and each main town had its own scientific institution. Most of these societies were associations of amateurs, industrialists, and professional men, who were bent upon advancing the knowledge and the applications of science, and more generally the economy and culture of their region.

The membership of the provincial societies ranged from about one hundred to five hundred persons, a size of the same order as that of the Royal Society in the seventeenth and eighteenth centuries, when it included most Englishmen interested in science, and others besides. We may say, therefore, that the number of persons in England who were actively interested in science increased at least a hundredfold during the nineteenth century. There were also the national specialist societies, such as the Linnaean Society, founded in 1788, the Geological Society (1807), and the Chemical Society (1840), which probably overlapped in membership with the general scientific societies. The officials of the Royal Society frowned upon the formation of these new associations. When the foundation of a metropolitan Chemical Society was mooted in 1806, the suggestion was discouraged by the President of the Royal Society, Sir Joseph Banks. 'I see plainly', he is said to have remarked, 'that all these new-fangled associations will finally dismantle the Royal Society and not leave the old lady a rag to cover her.'

Amateur associations of scientists were not wanting in Britain, but there seems to have been a lack of facilities for training scientists in England during the first half of the nineteenth century. During the eighteenth century the Nonconformist academies had performed a valuable service in this connection, but in the nineteenth century they became, for the most part, narrowly theological in the training they provided. It was not until the 1850's that Oxford and Cambridge were reformed by act of Parliament, following the Royal Commissions of 1850-51, and that the important provincial universities came into being, often fathered by the Literary and Philosophical Society of their locality. During the first half of the nineteenth century colleges were founded at London (1826 and 1828) and Durham (1832), but it would appear that the Mechanics' Institutes were the important establishments in England providing a scientific education in that period. In Scotland the universities were more recent foundations and possessed more modern traditions, science being taught and advanced at Glasgow and Edinburgh at an early date, notably by Joseph Black and his pupils from the 1760's on. The first chemical laboratory for practical instruction was set up at Glasgow in 1817 by Thomas Thomson, the professor of chemistry there, and the first laboratory for the teaching of physics was founded by William Thomson, later Lord Kelvin, when he became professor of natural philosophy at Glasgow in 1846. It was Kelvin in fact who gave shape to the modern structure of science teaching, introducing experimental work as an integral part of the training of the scientist. The Mechanics' Institutes also orig-

inated in Scotland. John Anderson, professor of natural philosophy at Glasgow, lectured upon scientific subjects to artisans
from about 1760, and bequeathed his estate for the foundation
of an institute to teach the sciences upon his death in 1796.
Dr. George Birkbeck was professor of physics at Anderson's
Institution at Glasgow until 1804 when he moved down to
London and gave courses of scientific lectures there, which
led to the foundation of the London Mechanics' Institute in
1823. In the same year the Glasgow Mechanics' Institute was
formed by a number of teachers who had seceded from the
university. Then in 1825 a Mechanics' Institute was founded
in Birmingham, and others soon sprang up in most of the large
towns of the country, so that by 1850 there were six hundred
of these organizations with a membership of over a hundred
thousand. Most of these Institutes had quite a high standard of
education, indeed they are said to have been 'far in advance of
the universities of Oxford and Cambridge in regard to the
physical sciences' at the time. The London Institute ultimately
reached university status as Birkbeck College, but most of
them evolved into, or were supplanted by, the technical colleges.

The French in 1794 had set up a mechanics' institute on a
grand scale in the foundation of the Conservatoire of Arts and
Trades, an institution which achieved the reputation of being
the 'Sorbonne Industrielle'. Such an institution was thought
worthy of imitation by Count Rumford, an American scientist and military man, who had emigrated to England after the
American War of Independence. He formed a 'Society for
Encouraging Industry and Promoting the Welfare of the Poor',
and in 1799 he submitted to the committee of this Society a
proposal to establish a 'Public Institution for diffusing the
knowledge, and facilitating the general introduction, of useful
mechanical inventions and improvements, and for teaching
by courses of philosophical lectures and experiments the application of science to the common purposes of life'. Subscriptions
were raised, and in 1800 the Royal Institution of Great Britain
was set up in London. In 1801 the Cornish pharmacist's apprentice, Humphry Davy, was appointed lecturer in chemistry, and here Davy carried out his celebrated electro-chemical
researches. Unlike the Paris Conservatoire of Arts and Trades,

the Royal Institution was dependent upon private donations for its maintenance, and at first these were not readily forthcoming. However, Davy designed his lectures to appeal to wealthy patrons, and he carried out researches for influential bodies, so that the Royal Institution ultimately paid its way. He lectured and conducted researches upon agricultural chemistry between 1802 and 1812 at the request of Arthur Young, the Secretary to the Board of Agriculture, which was founded in 1793 to meet the shortage in imported foodstuffs consequent upon the French Revolution. In 1816 Davy invented the miner's safety lamp at the request of the Society for the Study and Prevention of Mine Explosions. Thus the character of the Royal Institution changed. It was not the Mechanics' Institute which Rumford had planned, but a professional research institute, providing lectures which were more of a popular than of an educative nature. Rumford wished to implement his original plan, but his views were not shared by the other founders, and after several quarrels, Rumford left England and spent the rest of his days in France.

The Royal Institution was a notable addition to Britain's scientific resources, but it was small, only two scientists and their assistants working there for the first three decades of its existence. Meanwhile science was becoming more complex and less easily apprehended by the untutored mind, whilst experimental research was beginning to involve costly apparatus. Science too was entering more and more intimately into the process of industrial advance, so that a need began to be felt for more educational and research facilities in the sciences. The Scots, as in other matters connected with science, were the first to draw attention to the problem. John Playfair, the professor of natural philosophy at Edinburgh, remarked in a review of Laplace's *Celestial Mechanics*, written in 1808, that there were scarcely a dozen people in Britain who were competent enough mathematicians even to read Laplace's work. He pointed out that hardly anyone in Britain had advanced astronomical theory during the preceding sixty or seventy years, the French almost monopolizing the field.

'Nothing', he wrote, 'prevented the mathematicians of England from engaging in the question of Lunar theory, in which the interests of navigation were deeply involved, but the consciousness that in the knowledge of the higher geometry they were not on a footing with their brethren on the continent.'

The mathematics taught in Britain during the early years of the nineteenth century did not go much beyond the level obtaining in Newton's time. In the calculus Newton's somewhat clumsy notation was adhered to, whilst the more elegant symbolism introduced by Leibniz and the advances made by the French were largely passed by. A move to remedy such a state of affairs began with the formation of the Analytical Society, a Cambridge undergraduate club set up by John Herschel, Charles Babbage, and others, with the aim of introducing continental mathematics into Britain. Babbage proposed to call the club a 'Society for promoting the principles of a pure d-ism (d being the symbol used by Leibniz), in opposition to dotage (a dot being Newton's symbol)'. When these men graduated they took the matter further. John Herschel was one of the first to criticize the state of science as a whole in England, and he was joined by Humphry Davy, who began a book upon the subject but died in 1828 before he could complete it. However, his views and those of Herschel were made known, and were added to, by Charles Babbage, who was now professor of mathematics at Cambridge, in his *Reflections on the Decline of Science in England*, published in 1830. The root of the problem, Babbage thought, was that scientific research in England was still largely an amateur activity: it was neither state aided nor professionalized. 'The pursuit of science', he wrote, 'does not in England constitute a distinct profession as it does in other countries', for 'in England the profession of the law is that which seems to hold out the strongest attraction to talent', and thus it was that, 'by a destructive misapplication of talent we exchange a profound philosopher for but a tolerable lawyer'. The old amateur tradition was inadequate, he thought, for mathematics now, 'require such overwhelming attention that they can only be pursued by those whose leisure is undisturbed by other claims'. Babbage called for an association of interested persons to promote science in Britain, and asked the Government to concern itself with the matter.

Babbage's book created a considerable stir, and was welcomed by the Scottish critics of the state of science in Britain. The subject was discussed in several articles sent in to the *Edinburgh Journal of Science* in 1830. The reform of the universities was urged, for, it was said, that of the nineteen Britons who were foreign members of the Institute of France, not one of them occupied a position in a university. It was pointed out that the French Ministry of the Interior was spending some one and a half million francs a year upon the upkeep of scientific and literary establishments, whilst the British government spent nothing, and had even stopped some small pensions to scientists which had been awarded earlier. The government was taken to task on this point, and the editor of the *Journal*, David Brewster, an amateur scientist, and later vice chancellor of Edinburgh University, reiterated the appeal for an association of interested persons to foster British science.

The Home Secretary, Sir Robert Peel, was somewhat embarrassed by these attacks, and at a meeting in Birmingham to erect a monument to James Watt, denied that science was declining in England, and that the government was indifferent to the fortunes of science in the country. It was being arranged, he said, that the Crown should make grants to scientific amateurs in order to help them defray the cost of their experiments. An award of £300 had already been made to the astronomer, Sir James South, so 'that the country should bear some portion of the enormous expense which Sir James had incurred in pursuing his researches', and in order 'to relieve the country from the charge of perfect indifference to subjects of a scientific matter'.

However, it was felt that such awards merely touched the fringe of the problem, and were not a greatly significant contribution to British science. Accordingly steps were taken to form an organization which would unite scientists throughout the country and promote British science. The prime mover was David Brewster who in 1831 persuaded the Council of the Yorkshire Philosophical Society, one of the largest and most important of the provincial scientific societies, to call a nationwide meeting of the 'Friends of Science.' The meeting was held at York in September 1831, and there the British Association for the Advancement of Science was founded. The aims of the

Association, in the words of its first secretary, Vernon Harcourt, a chemist and canon of York, were

'to give a stronger impulse and more systematic direction to scientific enquiry, to obtain a greater degree of national attention to the objects of science and a removal of those disadvantages which impede its progress, and to promote the intercourse of the cultivators of science with one another and with foreign philosophers.'

The idea of the Association had originated in part from a national congress of German scientists, founded by the nature-philosopher, Lorenz Oken, in 1822, which met annually in the various cities of the German-speaking states to discuss the scientific advances of the year. Babbage had attended the Berlin meeting of 1828 and it was here that he had conceived the idea of the British Association. Other founders appear to have been stimulated by the writings of Francis Bacon, to whom references were made at the first meeting. Bacon in his *New Atlantis*, published in 1626, had suggested the formation of a national academy for the advancement of the sciences and the crafts, the members of which were to make 'Circuits of the divers principal cities of the Kingdom,' as the British Association later did. In the seventeenth century Bacon's project had stimulated the formation of the Royal Society, but the Society gradually lost its early vigour, and in 1831 Vernon Harcourt could say:

'It must be admitted, gentlemen, that the Royal Society no longer performs the part of promoting natural knowledge by any such exertions as those we now propose to revive. As a body it scarcely labours itself, and does not attempt to guide the labours of others.'

The meetings of the British Association were held annually in one of the main cities of the United Kingdom, or occasionally of the Dominions, an average of about two thousand persons attending each meeting. At such gatherings contact was established between the members of the specialist societies and the members of the scattered provincial philosophical societies, many of these organizations being affiliated to the British Association and sending delegates to its meetings. In this way a wide measure of agreement was reached by British scientists concerning many questions germane to the advance of science, questions relating both to the internal development of the subject, and to such externals as the extension of science education, and the financing of scientific research. The discussions concerning the internal development of the sciences served to present an overall picture of the state of science in any given year, and in the nineteenth century they were of considerable importance as they provided the starting point for further investigations and on occasion they indicated promising lines of enquiry. As regards the externals to the advance of science the British Association was active in the movement for the reform of higher education, which was effective from the 1850's on, but it failed to interest the government in financing scientific research to any considerable degree. It was the necessities of the first World War, 1914-18, which brought this matter to the attention of the British government, and led to the foundation of the Department of Scientific and Industrial Research in 1917.

On a small scale the British Association itself financed research, funds being obtained from the subscriptions of the members. Such subscriptions were modest, and as resources were limited, a careful selection of research topics to be financed was made. We find that some sciences were favoured with grants much more than others, and as the British Association was the organization most representative of British science as a whole during the nineteenth century, we may take the various sums allocated to the different sciences as rough indices of the interest displayed in those sciences at the time. Of the £92,000 spent by the British Association upon research during the first century of its existence, £36,000 were devoted to the investigation of problems in physics and mathematics, £18,000 to botany and zoology, £10,000 to anthropology, £7,500 to geology, £4,000 each to chemistry and engineering, whilst smaller sums were awarded for researches in physiology, psychology, economics, geography, education, and agriculture. That the physical sciences received the lion's share is not surprising, for these sciences promised, and gave, the most important applications of the century. Furthermore, they grew out of the scientific investigation of technology, which was made with the view of improving existing machines by determining the principles upon which they worked. Thus the steam engine gave birth to the science of thermodynamics, and in turn the science of electricity led to much of the equipment of the electrical industry. After physics, the biological sciences were allocated the largest amount of money for research by the British Association, the main problems investigated being the discovery and classification of the organic species, and the study of their anatomy, physiology, and habits. Here the interests involved were almost entirely intellectual, centring round the question as to whether the species had been created or had evolved in time. There were few applications of systematic biology in the nineteenth century, and little stimulus to the subject derived from practical questions. Medicine, agriculture, and the fermentation industries stimulated other parts of biology, notably human physiology, biochemistry, and microbiology. The next two sciences which were comparatively well provided for by the British Association, geology and anthropology, derived from interests that were more mixed. Geology in the middle of the nineteenth century was a science of considerable moment, for it was here that the idea of evolution was first established; but it was also of great practical use in the location of coal, metal ores, and other inorganic raw materials used in industry. The Geological Survey of Great Britain was founded for these reasons, and to safeguard the state's interest in mining royalties, as early as 1835. Anthropology too became involved in the evolution question, the communities of primitive peoples being regarded as evolutionary stages in the development of civilized society, such stages being correlated with the archaeological evidence for the development of Stone Age man. This approach was abandoned in the early years of the present century, and primitive communities were studied as static 'social structures' rather than historically developing entities, whilst psychology rather than archaeology was brought in as an ancillary science. On the practical side anthropology was used for the understanding and control of colonial peoples, particularly from the 1880's when investment in the imperial possessions quickened. The British Association set up a separate section to deal with the subject of anthropology in 1881, and financed the study of native tribes in Egypt, India, Australia, and other regions from 1886 on. Compared with physics, biology, and even anthropology, the science of chemistry received comparatively little aid as regards research grants from the British Association. Agriculture too was poorly served compared with engineering. An engineering section of the British Association was set up at its foundation in 1831, a total of £4,000 being spent on engineering research during the course of the century, whilst an agricultural section was not set up until 1912, a total of £5 being spent on the subject before that date. In this connection it is interesting to note that the agriculture of Britain, and the chemical industry, particularly the fine chemical industry where continuous research was essential, were comparatively weak at the beginning of the present century. We were importing nine-tenths of our manufactured dyes from Germany at that time, for example, even though the first synthetic dyes had been discovered in England by Perkin in 1856.

Finally we may note that the Royal Society was reformed by its own efforts in the 1830's and 40's. During the eighteenth century the Society had become more and more of a London club, the proportion of nonscientific members gradually increasing, so that during the first half of the nineteenth century the Society consisted of about equal numbers of scientists and non-scientists. Moreover, the non-scientific members had control of the Society until the 1820's when Humphry Davy was made President and a majority of scientists in the Council of the Society was secured. However, Davy was succeeded by a lawyer, Lord Colchester, and then the Duke of Sussex, one of George III's sons, and it was not until 1847 that admission to the Society was limited largely to men who were scientists. From 1874 peers were no longer allowed privileged entrance to the Royal Society, nor privy councillors from 1902, and finally, in 1945, women were admitted as Fellows to the Society.

Chapter 36

Chemistry and the Atomic Theory of Matter

WITH THE publication in 1789 of the *Elements of Chemistry* by Lavoisier, the science of chemistry severed its remaining connections with the alchemical past and assumed a modern form. Lavoisier stressed the importance of quantitative methods of investigation in chemistry, and in this connection he introduced the principle of the conservation of matter which stated that nothing was lost or gained during the course of a chemical reaction, the weight of the products equalling the weight of the starting materials. He also revived the idea that the chemical elements were just substances that could not be broken down into anything simpler by chemical means—the elements, he said, were 'the actual terms whereat chemical analysis had arrived'—and he listed some twenty-three authentic elements known to him.

Lavoisier's new viewpoint led to the elaboration of several empirical laws in the science of chemistry. The first was the law of equivalent proportions, which was formulated in 1791 by Jeremiah Richter, 1762-1807, a chemist of the Breslau mines and the Berlin porcelain factory. Richter was a pupil of the philosopher, Immanuel Kant, and he thought, with his master, that the physical sciences were all branches of applied mathematics. With this principle in mind, he discovered that the weight of a substance A, that combined with a known amount of a substance B, would also combine exactly with that weight of a substance C, which entered into combination with the same known amount of the substance B. After this discovery tables of equivalent weights were drawn up, showing the relative amounts of the chemical elements that would combine with one another.

A second law, that of constant compositions, was put forward in 1797 by the Frenchman, Proust, 1755-1826, who was a professor of chemistry at Madrid. He found that no matter how a chemical compound was made, the ratio of the weights of the elements which it contained was always the same, the ratio being that of the equivalent weights of the elements. The validity of this law was disputed for a few years by Berthollet, 1748-1822, professor of chemistry at the Ecole Polytechnique, who was of the opinion that the composition of chemical compounds was infinitely variable and not fixed. Berthollet was interested more in the process of chemical change than in the products of that change, and, investigating the subject of his interest, he anticipated some of the discoveries made by the physical chemists during the 1860's. He pointed out that some chemical reactions were reversible, whilst in other reactions the yield of the products was dependent upon the initial quantities of the reactants used, and upon the relative solubilities or volatilities of the reactants and products. From such cases Berthollet concluded that the composition of a compound gradually varied during the course of a reaction. Proust, however, was able to show that it was the amount of a compound, and not its composition, which varied during the course of a reaction, and furthermore, that Berthollet's compounds of indefinite composition were really mixtures. Proust in fact was the first to clearly distinguish between mixtures and compounds, the components of the former being separable by physical means, those of the latter only by chemical means.

Such laws enabled chemists to characterize new compounds and new elements, and they also led to the atomic theory, which provided an explanation as to how it was that these laws of nature were obeyed. As a philosophical speculation the atomic theory had long been current from the time of Democritus on. The theory had not been very popular in later Antiquity and the Middle Ages, but it was revived during the Renaissance, and had been incorporated into the mechanical Newtonian view of the physical world. Before the nineteenth century, not many positive applications of the atomic theory were made, however. Newton had explained Boyle's law, which states that the volume of a gas varies inversely with its pressure, on the assumption that the atoms of a gas were more or less stationary and repelled each other with a force which varied inversely with the distance. The Swiss mathematician, Daniel Bernoulli, in 1738 provided the modern explanation for the same law on the assumption that the atoms of a gas were in random motion, the pressure of a gas being nothing more than impact of the atoms on the wall of the containing vessel. However, the atomic theory was not applied to chemistry before the nineteenth century, as it was generally held that all the diverse substances with which chemists dealt were composed of atoms which were much the same in all respects, a view that could not account for the highly specific character of chemical processes, as Boyle had observed in the seventeenth century.

The atomic theory was modified to meet the needs of chemistry by the Quaker-scientist, John Dalton, 1766-1844, who gave a preliminary outline of his theory in a paper read to the Manchester Literary and Philosophical Society in 1803, and a full account in his *New System of Chemical Philosophy*, published in 1808. Dalton started from Newton's conception that gases were composed of atoms which repelled each other with a force that fell off with distance. Dalton and others thought that this repulsive force was heat, or caloric as it was termed, for in 1801 he had found that the pressure of a gas increased directly with the temperature when it was heated. Gay-Lussac, 1778-1850, in France observed the same phenomenon in 1802, and later found that he and Dalton had been anticipated in 1787 by the Frenchman, Charles, after whom the law of the expansion of gases with temperature is now named. Dalton was much concerned with meteorological problems, particularly the question of the nature of the atmosphere, which by the beginning of the nineteenth century was known to be made up of several constituents, notably oxygen, nitrogen, and water vapour. The atmosphere was homogeneous, but it appeared to Dalton that, if the atoms of gases repel one another, the different constituents of the air should separate out. To overcome this difficulty Dalton suggested that the atoms of different chemical substances were not identical: they form different species such that the atoms of one chemical substance repel one another but not the atoms of a different substance. Thus he wrote in 1802:

'When two elastic fluids denoted by A and B are mixed together, there is no mutual repulsion between their particles; the particles of A do not repel those of B as they do one another.

Consequently the pressure or whole weight upon any one particle arises solely from those of its own kind.'

In this way Dalton arrived at his law of partial pressures, which states that the total pressure of a mixture of gases is the sum of the pressures of each gas taken singly. In other words the different gases in a mixture have no effect upon one another: as Dalton's friend, Henry, put it, 'Every gas is a vacuum to every other gas.'

The importance of Dalton's conception for chemists was that there now appeared to be different species of atoms, the atoms of any one element being all alike with their own specific properties, whilst those of various elements differed one from the other. Dalton held that the atoms of different elements differed in size, weight, and number per unit volume, and that when two elements combined to form a compound each atom of the first element united with one, or a small whole number of atoms of the second element. He made the latter postulate because he found that when two elements united together to form more than one compound, the weights of the element A that combined with a fixed amount of the element B always bore a simple numerical ratio one to the other. In the case of the oxides of nitrogen, which Dalton himself investigated, he found that the amounts of oxygen which combined with a given amount of nitrogen were in the ratio of $1: 2: 4$. This was Dalton's law of multiple proportions, published in 1804, which gave the atomic theory its plausibility. It also indicated that an atom of one element did not always combine with only one atom of another; it must in some cases combine with two, three, four, and so on.

Dalton pointed out that an important property characterizing the atoms of the different elements was their relative weights, and he himself drew up the first table of such weights relative to hydrogen as unity in 1803. The equivalent weights of the elements, the weights that combine together to give definite compounds, could be determined by direct measurement, and from such determinations the atomic weights of the elements could be derived if it were known in all relevant cases how many atoms of one element combined with a single atom of another. At the time there were no means of estimating such combining numbers of the atoms, and so Dalton assumed that, 'When only one combination of two bodies can be obtained, it must be presumed to be a binary one, unless some cause appear to the contrary.' That is, such compounds should be presumed to contain one atom from each element, an assumption which later proved to be untenable.

A discovery that provided an indication of the combining numbers of the atoms was made by Gay-Lussac in 1808. He found that when two gases combine together, the volumes of the gases which unite bear a simple numerical ratio one to the other and also to the volumes of the products if they are gaseous too. Dalton held that the numbers of atoms of two elements which combined together stood in a simple numerical ratio, and thus it seemed not improbable that the volume ratio of two combining gases was the same as the combining ratio of their constituent atoms. Avogadro, 1776-1856, a professor of physics at Turin, went further and suggested in 1811 that the same volumes of different gases contain the same number of particles under the same conditions of temperature and pressure. Ampere, 1775-1836, suggested the same hypothesis in 1814. Avogadro's hypothesis raised the difficulty that when one volume of hydrogen combined with one volume of chlorine, two volumes of hydrogen chloride were produced, implying that the atoms of hydrogen and chlorine were split into halves during the process of combination. Avogadro overcame this difficulty by supposing that the fundamental particles of hydrogen, chlorine, and other gases, were molecules containing two atoms of the element, and that chemical combination between two gases resulted in the splitting up of the elementary molecules and the formation of compound molecules in which there was one atom of each element, as of hydrogen and chlorine in hydrogen chloride.

Avogadro's hypothesis could have provided a general method of determining the combining numbers of the elementary atoms, but it was not widely accepted until the 1860's, as it demanded that the atoms of the same element should combine together to form molecules. Dalton and others rejected such a conception, for they held that like atoms must repel one another and could not combine. Moreover, Dalton himself thought that the various species of atoms differed not only in their atomic weights, but also in their sizes, and the number per unit volume in the gaseous taste. Gay-Lussac's law of combining volumes implied that there were the same number of particles in the same volume of different gases, and Dalton at first questioned the accuracy of the law. The experimental evidence forced him to accept the law, though he denied to the end the validity of Avogadro's hypothesis.

The older atomic theory, which held that the fundamental particles of nature were uniform and all the same, still lived on, and was even merged with the new theory through the postulate that the different atoms of the various chemical elements were all composed of the same primordial matter. Humphry Davy, 1778-1829, at the Royal Institution, London, wrote of

'that sublime idea of the ancient philosophers, which has been sanctioned by the approbation of Newton . . . namely that there is only one species of matter, the different chemical as well as mechanical forms of which are owing to the different arrangement of its particles.'

The atomic weights of a number of elements approximated to whole numbers, relative to hydrogen as unity, and a London physician, William Prout, 1785-1850, suggested in 1815 that the atoms of the other elements were composed of a discrete number of hydrogen atoms. Thomas Thomson, 1773-1852, professor of chemistry at Glasgow, was so taken up by Prout's hypothesis that he rounded off the atomic weights he determined to whole numbers. However, the researches of the Swede, Jakob Berzelius, 1779-1848, and the Belgian, Jean Stas, 1813-91, showed that the atomic weights of the elements were not exact multiples of the weight of an atom of hydrogen, though some were very nearly whole numbers.

From about 1820 to 1860 the atomic theory did not play a prominent part in chemistry. For the most part chemists preferred to use the directly determined equivalent weights of the elements, rather than the atomic weights which involved uncertain estimates as to the combining numbers of the atoms. The rejection of Avogadro's hypothesis left chemists without a general method of ascertaining the combining numbers of the elementary atoms, though a number of specific methods were evolved which were used by the chemists who were still interested in the determination of atomic weights, notably Berzelius and Stas. Such methods were quite effective, the table of atomic weights which Berzelius drew up in the 1830's being similar to that in use today, apart from the cases of silver and the alkali metals. From his law of combining volumes, Gay-Lussac suggested that in the case of the gaseous elements, though not their compounds, identical volumes should contain the same number of atoms. Berzelius accepted Gay-Lussac's principle, and by means of it he was able to assign the modern formulae to a number of compounds. He held that a molecule of water contained two atoms of hydrogen and one of oxygen as two volumes of hydrogen reacted with one of oxygen to give water. Dalton on the other hand thought that a molecule of water contained one atom of hydrogen and one of oxygen, working on the rule that all compounds were to be designated as binary unless there was some good reason to the contrary.

In 1819 one of Berzelius's pupils, Mitscherlich, 1794-1863, noticed that compounds with similar chemical formulae had the same crystalline form. This was his law of isomorphism, and with it Berzelius was able to determine the formulae of many salts and the atomic weights of their constituent elements, though Mitscherlich later showed that his law had many exceptions. In the same year Dulong, 1785-1838, and Petit, 1791-1820, in Paris found that in the case of a number of metals the product of their atomic weight and specific heat, the amount of heat required to raise the temperature of unit weight of the metal through one degree, was constant. Dulong and Petit's rule enabled rough values of the atomic weights of the metals to be determined, but Berzelius did not use it, claiming that it was not universally valid. It was because he rejected the Dulong and Petit rule that the atomic weights Berzelius assigned to silver and the alkali metals were not the modern ones, for he thought that the combining number of their atoms was two, and not one as is held today. Berzelius, like Gay-Lussac and Dalton, rejected Avogadro's hypothesis which would have provided the universal standard for which he searched, as he thought that like atoms repel each other, and thus that they could not combine to form molecules. Dalton had thought that each atom was surrounded by an atmosphere of heat which repelled like atoms, but did not interfere with unlike atoms. Berzelius held a similar, though more developed, electrical theory of atomic attraction and repulsion, for in the meantime the new branch of electrochemistry had developed and had provided another theory of chemical affinity.

During the eighteenth century the effects of electricity upon living creatures, particularly the electrical shock, had aroused much interest; so too had the electrical phenomena exhibited by organisms, such as the sting of the torpedo fish. Working upon such topics, Galvani, 1737-98, professor of anatomy at Bologna, noticed in the 1780's that muscle-nerve preparations of a frog's leg twitched when placed in contact with two dissimilar metals. He thought that this was a biological electrical phenomenon, the frog's leg producing electricity like the torpedo fish. His contemporary, Volta, 1745-1827, a professor of physics at Pavia, suggested that it might be a physical electrical phenomenon, the frog's leg being merely a sensitive detector of the electricity produced by the junction of two dissimilar metals. Volta experimented with various pairs of dissimilar metals, and he found that some combinations were more effective than others, which supported his view. He discovered that with a series of metallic junctions he could produce electrical effects comparable to those exhibited by the frictional electrical machine, particularly if the alternate junctions were moistened with acid. Finally in 1799 Volta found that two dissimilar metals immersed in acid gave a considerable electrical current when they were connected by an external circuit. Thus he discovered the voltaic cell, in which electricity was produced by the chemical action of a metal dissolving in an acid.

Conversely it appeared that electricity, now readily available from the voltaic cell, could bring about chemical action. In 1800 the Englishmen, Nicholson and Carlisle, passed electricity through water by means of two wires dipping into it, and they found that the water was decomposed into its elements, hydrogen and oxygen. In the following year Humphry Davy, newly appointed to the lectureship of chemistry at the Royal Institution, began a series of researches upon similar electrolyses of salt solutions and solid compounds. In 1807 he electrolysed the caustic alkalis, obtaining the alkali metals, sodium and potassium, and later he obtained the alkaline earth metals, calcium, strontium, and barium, all of which were new elements. These, and other researches, led Davy to the theory that the chemical attraction between the elements, which was responsible for compound formation, was essentially of an electrical character, a view that was developed by Berzelius in Sweden from 1811. Incidentally the German romantic nature-philosophers had developed the same idea in a speculative fashion from the consideration that chemical compounds must be a unity of opposite entities, these entities being positive and negative electrically charged bodies. Davy's friend, Coleridge, 1772-1834, toured Germany in 1798-9 and brought back the teachings of Schelling, upon which he lectured at the Royal Institution. Davy himself was something of a romantic, writing a great deal of poetry which had a certain vogue at the time.

However, it was the experimental researches of Davy and Berzelius which established for a time the electrical theory of chemical affinity. Berzelius noted that in the electrical decomposition of chemical compounds one set of elements and chemical groupings—notably, hydrogen, the metals, and the alkalis—went to the negative pole of the electrical circuit, whilst another set, oxygen, the non-metals, and the acids, went to the positive pole. The first set he termed electropositive elements or groups, assuming that they were charged positively and thus attracted towards the negative pole of the electrolytic cell, whilst the second set he termed electronegative as he presumed that they were negatively charged. Berzelius held that chemical combination was due to the mutual electrical attraction of an electropositive and an electronegative element, their union resulting in a partial neutralization of the opposite charges. The remaining charge allowed the group to form a more complex, though looser, compound with a similar group of opposite charge, the complex compound being nearer to electrical neutrality than its constituent groups. From this viewpoint, termed the dualistic theory because it supposed that there were two fundamentally different types of elements and their groupings, Avogadro's hypothesis was inadmissible, as atoms of the same element would possess identical charges and would repel each other, so that they could not form binary molecules in the way that Avogadro had supposed.

The dualistic theory was not inappropriate in the sphere of inorganic chemistry, for the mineral compounds then known were fairly simple ionic compounds which could be understood as unities of oppositely charged elements or groupings. Inorganic chemistry developed rapidly in the period 1790-1830, the 'Heroic Age' of geology, as the geologists discovered numerous minerals for the chemists to analyse. Berzelius himself described the preparation, purification, and analysis of over two thousand inorganic compounds in the decade 1810-20. In the sphere of organic chemistry the position was rather different. Mineral compounds could be characterized by the relative amounts of the elements which were contained in them, but organic compounds from the start were seen to be complex arrangements of few elements, notably of carbon, hydrogen, oxygen, and nitrogen, so that quantitative analysis did not go very far towards the characterization of such compounds. Indeed there were compounds, termed isomers, which had exactly the same elementary compositions but very different properties, the characteristics of isomers appearing to depend upon the arrangement, rather than the numbers, of the elementary atoms contained in them. Such a problem had been foreseen by the English chemist, Wollaston, 1766-1828, at the inception of the atomic theory in 1808, but it did not attract much attention until the rise of organic chemistry in the 1830's.

The development of organic chemistry coincided with the rise of the chemistry in Germany, and it was two Germans, Friedrich Wohler, 1800-82, and Justus von Liebig, 1803-73, who made some of the early important discoveries in the subject. During the early decades of the nineteenth century the foremost continental chemists were Berzelius and Gay-Lussac. Liebig went to Paris in order to study under Gay-Lussac at the Ecole Polytechnique, and Wohler went to Stockholm to study under Berzelius. Both Liebig and Wohler were versed in the mineral chemistry of the day, but they investigated problems bordering upon the field of organic chemistry. In 1824 Liebig made a compound, silver fulminate, which was identical in composition with a compound prepared by Wohler, silver cya-

nate, but it was markedly different in properties. Such a phenomenon was found to be widespread in organic chemistry,
Berzelius coining the term, isomerism, in 1830 to describe a
similar case which he had discovered, the tartrates and the
racemates.

In 1828 Wohler discovered another case of isomerism. He
prepared a compound which was regarded as essentially inorganic, ammonium cyanate, and he found that it rearranged
on warming in aqueous solution to give a well-known organic
compound, urea. Hitherto organic compounds had been obtained only from living organisms, but now it appeared that
they could be prepared from inorganic materials, a discovery
which weakened the current view that organic compounds
were produced only by vital forces in living matter. Berzelius
as late as 1819 had thought that organic compounds did not
obey the law of constant proportions, and that they did not
belong to chemistry proper, as they were the products of vital
forces. Now organic and inorganic chemistry were brought
closer together, and attempts were made to extend the dualistic
theory of chemical combination into the organic sphere. In
1832, Liebig and Wohler showed that there were a whole
series of organic compounds, prepared from the oil of bitter
almonds, which contained a common organic group and
varying inorganic groups, a series formally similar to the set
of inorganic salts formed by a single acid and a variety of bases.
However, it appeared that the organic group or radical could
combine equally well with the electropositive hydrogen, or with
the electronegative oxygen. Furthermore, Dumas, 1800-84, at
the Ecole Polytechnique, found in 1834 that inside organic
radicals themselves electropositive hydrogen could be replaced
by electronegative chlorine without changing fundamentally
the chemical properties of the radical or its compounds.

The further the dualistic theory was pushed into organic
chemistry the more complex and chaotic the situation became.
Wohler wrote to his teacher, Berzelius, in 1835:

'Organic chemistry just now is enough to drive one mad.
It gives me the impression of a primeval tropical forest full of
the most remarkable things, a monstrous and boundless thicket

with no way of escape, and into which one may well dread to enter.'

Wohler himself gave up the study of organic chemistry and returned to the analysis of minerals which he had learned at Stockholm. Liebig, however, adhered to organic chemistry, and trained the next generation of chemists who were to help solve the problems of the subject: men such as Bunsen, Hofmann, Kekule and Wurtz; Hofmann giving up the study of law, and Kekule abandoning architecture, in order to study chemistry. Liebig too rejected the dualistic electrical theory of Berzelius, and accepted the rival theory of structural types put forward by Dumas in 1840, and developed by Gerhardt, 1816-56, and Laurent, 1808-53, who ran a small chemical laboratory for teaching purposes in Paris. Dumas suggested that the chemical properties of organic compounds were due to their particular structural arrangement, or type, and not to the electrical character of the elements which composed them. He held that all compounds of the same structural type should show the same properties: one element could be replaced by another, but, so long as the structural arrangement retained its integrity, the properties of the compound would not be greatly altered. Dumas found that three-quarters of the hydrogen in acetic acid could be replaced by a very different element, chlorine, and yet the resultant compound retained the qualitative properties of acetic acid. Wohler, who still adhered to his master's dualism, ridiculed the theory with the suggestion that it should be possible to replace all the atoms in acetic acid, carbon and oxygen as well as hydrogen, with chlorine, so that it was all chlorine, and still retain the specific properties of acetic acid.

The type theory was generally accepted by the organic chemists, whilst the dualistic theory was accepted by those mineral chemists still concerned with the advance of chemical theories. Generally speaking, the chemists of the second quarter of the nineteenth century were rather untheoretical, particularly the inorganic chemists. Even Dumas used the concept of equivalents, which were directly measurable, rather than the idea of atoms in developing his theory of types. In 1840 he wrote that the compounds which have the same structural type are, 'Those substances which contain the same number of equivalents united in the same manner and have the same fundamental chemical properties.' But the type theory stimulated the development of chemical theory again, for to discover the structure of a chemical molecule it was necessary to know the combining numbers of its constituent atoms. Thus the problem of determining how many atoms of one element could combine with a single atom of another, which had been largely abandoned earlier in the century, came once more to the fore. With it came the revival of the atomic theory, which meanwhile had receded into the background of chemical theory.

The problem of estimating the combining numbers of the atoms, or the valencies as they were called, was investigated by Liebig's pupils, first by Edward Frankland, 1825-99, who in 1851 became professor of chemistry at the newly founded Owen's College, Manchester, and then by August Kekule, 1829-96, who held various posts at Heidelberg, Ghent, and finally Bonn. Frankland investigated the organic compounds of the metal and metalloid elements, discovering in 1852 that each metal atom could only combine with a quite definite number of organic groups, a number he termed the atomicity or valency of the element. He noted that the elements fell into groups which had the same valency: antimony, arsenic, phosphorus, and nitrogen, for example, generally showing the combining number of three or five. The most important element in organic compounds was carbon, and from about 1857 Kekule suggested that the atoms of carbon could each combine with four other atoms, or four groups of atoms. On this basis Kekule designed structural models of organic compounds, and used the models to interpret the reactions of those compounds. However, such ideas of valency and structure were ill-defined for, as yet, there was no general criterion whereby the combining number of an atom could be ascertained. Speaking of this period, Kekule wrote in 1861:

'Besides the laws of fixed and multiple proportions of weight, and in gaseous bodies also of volume, chemistry had as yet discovered no exact laws . . . and all so-called theoretical conceptions were merely points of view which possessed probability or convenience.'

Meanwhile organic chemistry was developing rapidly, and the subject required more and more some theory of molecular structure, so that the problem of determining the valencies of the elements became urgent. The German chemists, in particular Kekule, called a conference of chemists at Karlsruhe in 1860 in an endeavour to settle the problem. Some one hundred and forty chemists attended the conference. From France came Dumas and Wurtz; from England, Frankland and Roscoe; the Germans came *en masse*, Liebig, Wohler, and the more important of their pupils, Kolbe, Bunsen, and Kekule, whilst from Russia came Mendeleef, and from Italy, Cannizzaro. Cannizzaro, 1826-1910, full of the nationalism of the period, averred that his fellow countryman, Avogrado, had solved the problem of valency and atomic weight determination some half a century before. The assembled chemists were not convinced, and the problem remained unsolved at the end of the conference. However, Cannizzaro distributed copies of a pamphlet, setting out his ideas in full, which the delegates took away with them. He showed that, according to Avogadro's hypothesis, the molecular weight of a compound was twice its vapour density, measured relative to hydrogen as unity, since the hydrogen molecule contained two atoms and the same volume of different gases or vapours contained the same number of molecules. Vapour densities were readily ascertained, and so the molecular weights of a number of compounds containing the same elements could be determined. The atomic weight of a particular element, Cannizzaro argued, would then be the lowest weight of that element in a series of its compounds and also the lowest common difference between its weights in the series.

Cannizzaro's pamphlet, and the work which followed it, soon convinced most chemists that Avogadro's hypothesis was generally valid. At first it was thought that there were exceptions to the hypothesis, as to the Dulong and Petit rule of atomic heats and the Mitscherlich law of isomorphism, but the anomalies were found to be due to the breaking up of the compounds concerned in the vapour state. The Avogadro hypothesis gave the definitive atomic weights of the elements, and from such determinations the combining numbers of the elements were readily ascertained by means of the relationship which equated the atomic weight of an element to the product of its equivalent weight and valency. With settled values for the valencies of the elements, structural models of their compounds were constructed. The reactions of those compounds provided tests for the validity of such structures, whilst in turn the structures proposed indicated possible new reactions. With his architect's training, Kekule had a flair for envisaging the possible molecular structures of compounds, suggesting in 1865 the hexagon ring formula for the difficult case of benzene, which was found to be composed of six carbon and six hydrogen atoms. The final addition to the classical theory of molecular structure came in 1874 when Le Bel, 1847-1930, and Van't Hoff, 1852-1911, suggested simultaneously that the four valencies of carbon were directed in space towards the apices of a regular tetrahedron, in order to account for the two isomeric forms of tartaric acid isolated by Pasteur, 1822-95, in 1848, and other cases of optical isomerism discovered later. The two forms of these isomers were identical in chemical properties, but they differed in that one form turned the plane of polarization of a beam of polarized light to the right, and the other to the left. Le Bel and Van't Hoff pointed out that in all such cases four different groups were attached to a central carbon atom, and that two arrangements of those groups would be possible if the four valencies of the carbon atom were orientated tetrahedrally, thus accounting for the isomerism.

The acceptance of Avogadro's hypothesis, followed by the establishment of the definitive valencies and atomic weights of the elements, had its influence upon inorganic as well as organic chemistry. It appeared that elements with the same valencies fell into natural families or groups, a fact which brought attention to bear on the classification of the elements. It had long been known that some of the elements were related to one another, forming family groups. Johann Dobereiner, 1780-1849, a professor of chemistry at Jena, showed in 1817 that the atomic weights of calcium, strontium, and barium, fell roughly into an arithmetic series, and when Balard, 1802-76, at the Sorbonne, discovered bromine in 1826, he predicted from the properties of the element that chlorine, bromine, and iodine, would form another arithmetic series, which Berzelius showed to be approximately true. During the 1830's and 40's, when the atomic theory was under an eclipse, such classifications of the elements according to their atomic weights did not attract much attention, but Dumas attempted to group the elements into natural families according to their properties and reactions, placing boron, carbon, and silicon in one group, and nitrogen, phosphorus, and arsenic in another.

When the atomic weights and the valencies of the elements were finally fixed in the 1860's several new attempts were made to classify the elements into related groups, notably by Chancourtois of France in 1863, Newlands of London in 1864, and more particularly by Lothar Meyer of Germany and Mendeleef of Russia in 1869. Meyer and Mendeleef formulated the periodic law, stating that the properties of the elements varied in a periodic manner with their atomic weights, and they drew up a periodic table of the elements to exemplify the law. Earlier chemists, of whom Newlands was a conspicuous example, attempted to group the known elements into a complete classification, and in so doing they forced some elements into anomalous relationships. Lothar Meyer, 1830-95, and particularly Mendeleef, 1834-1907, emphasized that there were gaps in the periodic table which elements as yet unknown should occupy, and Mendeleef predicted with remarkable accuracy the properties of some of these missing elements, all of which were subsequently discovered.

The periodic classification provided the first theoretical guide to the search for new elements. The twenty-three elements known to Lavoisier had been discovered by the trial and error study of their specific chemical reactions. Practical chemical analysis became more systematized, and, applied to the mineral specimens provided by the geologists, it led to the discovery of thirty-one new elements in the period 1790-1830. Between 1830 and 1860 little was accomplished in regard to the isolation and identification of new elements, save the partial separation of the rare earth elements by Mosander, Berzelius' successor at Upsala in Sweden. However, in 1859 the chemist, Bunsen, 1811-99, and the physicist, Kirchhoff, 1824-87, both at Heidelberg, introduced the spectroscope, by means of which the characteristic colours imparted to flames by chemical substances could be examined and identified. With this instrument Bunsen discovered the new alkali metals, caesium and rubidium, in 1860-1. In London Sir William Crookes, 1832-1919, found the element, thallium, spectroscopically in 1861, and Ferdinand Reich, at the Freiberg School of Mines, discovered indium similarly in 1863. Attempts to find the elements missing from the periodic table provided the next group of discoveries. In 1874 Boisbaudran of France found gallium, the eka-aluminium predicted by Mendeleef, whilst the Scandinavian chemist, Nilson, discovered scandium, or eka-boron, in 1879, and finally Winkler, at the Freiberg School of Mines isolated germanium, or eka-silicon, in 1885.

The orderly arrangement of the elements in the periodic table suggested to some chemists that the various elements might have something in common; they might have evolved from a common origin or be composed of the same fundamental units of matter. The hypothesis of Prout, that the different elementary atoms are built up from a whole number of hydrogen atoms, was revived, and during the last two decades of the nineteenth century it was 'in the air of science', as Sir William Crookes put it. At the 1886 meeting of the British Association Crookes suggested that the elements had evolved from some primordial matter which he called protyle. Prout's hypothesis, he said, contained a truth that was 'masked by some residual or collateral phenomena which we have not yet succeeded in eliminating'. Again at the British Association meeting of 1894, Lord Salisbury declared that, amongst the elements, 'the discovery of co-ordinate families points to some identical origin'. Such views were strengthened in 1901 when Lord Rayleigh, 1842-1919, pointed out that the atomic weights of the elements tend to approximate to whole numbers more closely, by a probability of more than a thousand to one, than would be expected for random distribution of those weights. Thus, he observed,

'we have stronger reasons for believing in the truth of some modification of Prout's law than in that of many historical events which are universally accepted as questionable.'

In order to test the validity of Prout's hypothesis, Lord Rayleigh carried out a series of researches at Cambridge on the density of various gases from 1890 on. In so doing he discovered in 1892 that the density of atmospheric nitrogen was greater than that of nitrogen prepared by chemical means. On removing nitrogen and the other reactive gases from samples of air, Rayleigh and William Ramsay, 1852-1916, of London, obtained a small amount of a new gas which was chemically quite inert and heavier than nitrogen in the ratio of 20 : 14. Crookes examined the spectrum of this gas and showed that it was different from the spectra given by any of the known elements. Thus it was a new element, argon, the first of the inert gases. In 1895 Ramsay obtained from the mineral, Clevite, another inert gas which Crookes again examined spectroscopically. He showed that the spectral lines it gave were identical with those observed in the sun's photosphere during the solar eclipse of 1868 by the astronomers, Janssen in France and Lockyer in England, lines which had been assigned to an element in the sun, helium, then unknown upon the earth. Finally in 1898 Ramsay isolated three other inert gases, neon, krypton, and xenon, from the heavy fractions remaining after the partial evaporation of liquid air, which had been first prepared by Hampson in England and Linde in Germany three years previously.

With the isolation of the inert gases the discovery of the eight main groups or types of chemical elements came to a close, though there were still gaps due to missing elements within those groups. Further advance in the discovery of new elements, and indeed in the development of chemistry in general, became dependent more and more upon the science of physics. Physical methods had already entered chemistry with the advent of spectroscopy, and a development within this field, the use of X-ray spectroscopy by Henry Moseley, 1888-1915, at Manchester, finally cleared up some residual problems concerning the possible number of rare-earth, and other heavy elements. The visible spectra of the elements were found to be a periodic function of their atomic weights, like their chemical properties, but the X-ray spectral lines given by the elements turned out to be related linearly to their atomic weights, or rather their atomic numbers, the ordinal positions of the elements in the periodic table starting with hydrogen as one. In 1913-14 Moseley fixed the absolute number of elements up to uranium as ninety-two, showing that there were fourteen rare-earth metals, and that there were seven elements lighter than uranium as yet undiscovered.

The discovery of radioactivity by Antoine Becquerel, 1852-1909, at Paris in 1896 seemed to confirm what chemists had now suspected for some time, namely, that the elements were genetically connected. The new radioactive elements, such as radium which was isolated in 1900 by Mme Curie, 1867-1934, were found to be breaking up spontaneously into other elements, which in turn decayed to yet lighter elements. Such elements were isolated and characterized by chemical analysis, and in this way three family trees of natural radioactive decay were traced. However, the phenomenon of radioactivity had greater significance for the physicists, indeed it was through atomic physics that the phenomenon made its main impact upon the theories of chemistry. From the second decade of the twentieth century theoretical chemistry became more and more integrated with atomic physics, in regard both to theories of the constitution of the atom and to theories of chemical combination. On the practical side, atomic physics provided the chemist with new materials, firstly, the radioactive versions of the ordinary elements, used to trace the course of chemical reactions, and lately, new elements heavier than uranium, which have amplified the periodic table.

Chapter 37

The Wave Theory of Light

DURING THE eighteenth century there were very few developments in the science of optics. For the most part Newton's view that light rays consisted of a stream of particles in rectilinear motion was adopted, but his idea that the motions of the light particles stimulated, or were accompanied by vibrations in an all pervading ether was abandoned. The validity of the particle theory of light seemed completely assured at the end of the eighteenth century, owing to the development by Lagrange and Laplace of the general Newtonian system, of which the particle theory of light was considered to be an integral part. However, it was at this time that the German nature philosophers began their attack upon the Newtonian philosophy, opposing in particular Newton's theory of light. The nature philosophers held that the various spectral colours did not make up white light. In their view the various colours were a product of the conflict between light and darkness. White light itself did not consist of particles in motion, it was nothing more than a tension in the ether. The ideas of the nature philosophers were highly speculative and they had little direct influence upon the science of optics, but it is possible that they brought about a reorientation from the physical to the physiological study of light, for they held that optical illusions produced by the eye were just as real and as worthy of study as other optical phenomena.

The theories of the nature philosophers were perhaps symptomatic of a wider revolt against the Newtonian philosophy, for Huygens' wave theory of light was revived by the London physician, Thomas Young, 1773-1829, in 1801, although there was then no new evidence to support it. Young himself in fact declared:

'Much as I venerate the name of Newton, I am not therefore obliged to believe that he was infallible. I see . . . with regret that he was liable to err, and that his authority has, perhaps, sometimes even retarded the progress of science.'

Young, like his contemporary, Dalton, came of a Quaker family. He studied medicine and began his researches under the surgeon, John Hunter, 1728-93, at London, investigating problems in physiological optics. Young showed that the accommodation of the eye to objects placed at different distances was due to changes in the curvature of its crystalline lens. He suggested that the retina in the eye had structures sensitive to red, green, and violet lights respectively, in order to explain colour vision and the phenomenon of colour blindness.

Young continued his study of medicine at Edinburgh, Cambridge, and finally at Göttingen in Germany, where he saw reason to revive the wave theory of light. For his doctoral dissertation at Göttingen, Young produced a thesis on sounds and the human voice, a subject which he connected up with his earlier work on optics by suggesting that both sound and light were wave vibrations, colours being analogous to notes of different frequencies. It was generally accepted that sound consisted of wave vibrations in the air along the direction of the sound beam, and Young presumed that light consisted of similar longitudinal vibrations in a luminiferous ether which pervaded all space, as Huygens had done before him. He pointed out that light from a weak source travelled just as fast as that from an intense source, a fact which could be accounted for more easily by the wave theory of light than the particle theory. It was well known that two sets of sound waves or water waves could interfere with one another, and Young performed an experiment in which two light beams were allowed to overlap and interfere, producing alternate light and dark bands where one beam reinforced or cancelled out the other. From the separation of the bands, and the dimensions of the apparatus, he was able to calculate the wavelengths of the light vibrations, showing them to be of the order of a millionth of a metre or so. Since the wavelengths of the light vibrations were very small compared with the size of visible objects, Young pointed out that light would travel in straight lines and could cast sharp shadows. He was aware that light beams did bend round the edges of opaque objects to some degree, producing shadows with coloured edges and other interference effects which had been studied by Grimaldi and others during the seventeenth century, and Young instanced such phenomena as evidence for the wave theory of light. Completing the explanation of the optical phenomena then known in terms of the wave theory of light, Young, together with Wollaston, verified Huygens' analysis of the phenomenon of double refraction observed with crystals of Iceland spar.

The revival of the wave theory of light in England provoked the French Newtonians and gave a stimulus to the study of optical problems in France. Replying to Young, Laplace in 1808 provided an analysis of the phenomenon of double refraction in terms of the particle theory of light. In the same year, Malus, 1775-1812, at the Ecole Polytechnique, discovered the phenomenon of optical polarization by reflection, an effect which occurs when a beam of light meets a transparent medium, such as glass, and is partially reflected and partially transmitted. Malus found that the two images of the sun seen by reflection from glass through a crystal of Iceland spar were of unequal intensities, and Arago, 1786-1853, also at the Polytechnique, discovered that the same phenomenon was observed when the transmitted ray was viewed in a similar way. The matter was investigated further by David Brewster, 1781-1868, at Edinburgh, who showed that, when the reflected and transmitted rays were at right angles to one another, both were fully polarized, that is, only one image could be seen through a crystal of Iceland spar when either the reflected or the transmitted ray were studied. Furthermore he discovered a general empirical law governing the degree of polarization of the two rays when they were not at right angles to each other.

The discovery of the polarization of light by reflection seemed at first to support the particle theory of light. Newton had suggested that light particles possessed 'sides' to account for the splitting of a light beam into two when it passed through a crystal of Iceland spar. It now appeared that the different 'sides' to the light particles caused some of them to be transmitted, and others to be reflected at the surface of transparent media, giving polarized beams. Young regarded the phenomenon as contrary to the wave theory of light for some time, but in 1817 he saw that if the light vibrations occurred transversely across the direction of motion, like water waves or vibrations along a stretched string, instead of along the direction of motion, like sound waves, the problem might be solved. There would be two possible modes of vibration at right angles to the direction of motion of the light beam, and thus the polarization of light could be ascribed to the separation of these two modes at a glass surface, one mode constituting the reflected ray and the other the transmitted ray. Young mentioned this hypothesis in a letter written to Arago in 1817. In the same year the Paris Academy of Sciences offered a prize for the best essay on the subject of optical diffraction, and amongst those who competed for the prize was Fresnel, 1788-1827, a civil engineer, who independently of Young had endeavoured to revive the older longitudinal wave theory of light. Arago mentioned Young's new suggestion to Fresnel, who made it the basis of his prize essay, in which he showed that all of the known phenomena of optics could be explained in terms of the hypothesis that light consists of transverse wave vibrations.

The new wave theory of light raised problems with regard to the luminiferous ether, the medium in which the vibrations of light were presumed to occur. Fresnel pointed out in 1821 that longitudinal vibrations, such as those of sound in air, could be propagated by a gas-like medium, but transverse vibrations, like the tremors of a jelly, could take place only in a medium which possessed characteristics of the solid state of matter. It was difficult to envisage an ether, solid and rigid enough to transmit transverse light waves, which would allow of the free passage of the heavenly bodies on their courses. Moreover Poisson, 1781-1840, at the Sorbonne, showed in 1828 that if the luminiferous ether were a quasi-solid the transverse vibrations of light would always be accompanied by a longitudinal vibration, which added a further difficulty as the longitudinal vibration as well as the transverse vibrations would carry energy away from the light source.

Solids for the most part are resistant to compression, extension, twisting, and bending, and it was seen that these properties did not necessarily go together, so that it was possible to imagine hypothetical solid ethers which could be easily compressed or extended, to permit the resistance-free passage of the heavenly bodies through them, and yet elastic to twisting or bending stresses in order to allow the propagation of wave vibrations.

George Stokes, 1819-1903, at Cambridge, pointed out in 1845 that there were well-known solids, such as pitch and wax, which were rigid enough to transmit tremors or transverse vibrations, and yet yielded to compressions and extensions. The luminiferous ether merely possessed such a combination of properties in a more marked form. Another analogy he suggested was that the ether resembled a highly diluted jelly of glue in water, which permitted the movement of objects through it and at the same time could propagate vibrations. In 1839 James MacCullagh, 1809-47, at Dublin, invented an ether composed of elements which resisted only rotatory twisting stresses, and by means of it he was able to explain a wide variety of optical phenomena in terms of the laws of dynamics. Later, in 1889, Lord Kelvin, 1824-1907, at Glasgow, constructed a mechanical model of an element of MacCullagh's ether. He arranged four rods tetrahedrally, each rod serving as the axis for a pair of contrarotating gyroscope flywheels. This model resisted all rotatory disturbances, but not translatory movements.

Meanwhile Cauchy, 1789-1857, at the Ecole Polytechnique, had put forward two ether theories in an attempt to explain the phenomena of both the reflection and the refraction of light on the assumption that the ether was changed either in density or in elasticity within material bodies. In 1839 he published a third theory in which he suggested that the ether was contractile, or labile, possessing a negative compressibility, in order to overcome the difficulty pointed out by Poisson in 1828 that longitudinal vibrations would accompany the transverse light vibrations. In such an either Cauchy showed that the longitudinal wave would have a zero velocity, and thus it would not be able to carry away any of the energy of the transverse vibrations. George Green, 1793-1841, the founder of the Cambridge school of mathematical physicists, pointed out that this ether would be unstable, tending to contract all the time. Kelvin re-examined Cauchy's ether in 1888, suggesting that it was analogous to a homogeneous foam, free from air, which was prevented from collapsing by adhesion to the walls of a rigid containing vessel. Such an ether would not be unstable, he held, if it stretched throughout infinite space, or if it had a rigid containing vessel as its boundary.

The wave theory of light was well established by the middle of the nineteenth century, and what appeared to be a fundamental verification of that theory was provided by two French amateurs, Fizeau, 1819-96, and Foucault, 1819-68, who measured the velocity of light in various media between 1849 and 1862. In the seventeenth century Descartes had shown that, according to the particle theory, light should travel more rapidly in dense transparent media than in air, whilst the wave theory suggested that light should travel more slowly. Fizeau in 1849 measured the time taken by light to traverse a given distance by means of a rotating cogwheel which, at a certain speed, permitted light to pass through a space between two consecutive teeth and return through the next space. Foucault in 1850 and 1862 used a rotating mirror, which at a measured speed performed a complete revolution in the time taken for light to proceed and return from a stationary mirror. Their results agreed with the value of the velocity of light determined astronomically by Bradley in 1827, and Foucault's work showed that light travelled more slowly in water than in air in the ratio of the refractive indices of water and air, which the wave theory predicted.

Chapter 38

The Development of Electricity and Magnetism

THE SCIENCE of electricity advanced rapidly during the eighteenth century in contrast to optics, which developed slowly during the same period. Stimulated by the discovery of the telescope and microscope, optical problems had been studied intensively during the seventeenth century, but there were few fresh stimuli during the period immediately subsequent. The science of electricity, on the other hand, became very popular, particularly after the discovery of the electric shock in 1745 and the identification of lightning with the electric discharge shortly afterwards. Somewhat extravagant medical claims were made for the vitalizing virtues of the electric shock, and some went so far as to identify electricity with the vital cosmic force of nature. Lamarck, it will be remembered, held that electricity, together with heat, was the driving force of organic evolution. John Wesley, 1703-91, the founder of Methodism, declared that 'electricity is the soul of the universe', a view that the German nature-philosophers came near to sharing, fascinated as they were by the opposite polarities which electricity exhibited.

The study of electricity in modern times, like the study of magnetism, may be said to have begun with the researches of William Gilbert of Colchester during the sixteenth century. The ancient Greeks knew that amber displayed electrical properties, but Gilbert showed that amber was by no means unique, discovering that glass, sealing-wax, sulphur, and precious stones, also attracted bits of paper and straw when rubbed. He noted that magnetic and electrical forces were different in character, magnets acting only on lodestone, or iron objects, and orientating them in a specific direction, whilst electrical forces acted upon a wide variety of materials and were nondirectional. During the seventeenth century Otto von Guericke, who invented the air-pump, made an electrical machine for generating large amounts of electrical charge. He mounted a ball of sulphur so that it could be rotated continuously and rubbed by the hand or a piece of cloth to produce electric charge. Another important electrical instrument was the Leiden jar for concentrating electrical charges, which was discovered together with the electric shock, by Pieter van Musschenbroek, 1692-1761, of Leiden in 1745. He tried to preserve electrical charge from decay in a bottle of water by leading charge from an electrical machine into the bottle through a wire. He held the outside of the bottle in one hand and touched the wire with the other, when, as he put it, 'the arm and the body was affected in a terrible manner which I cannot express; in a word, I thought it was all up with me'.

With these instruments Benjamin Franklin, 1706-90, of Philadelphia, carried out a series of researches to show that lightning was of an electrical character. In 1749 he pointed out that both the lightning flash and the electric spark were practically instantaneous, producing a similar light and sound; both were able to set fire to bodies and melt metals; both flowed through conductors, notably metals, and concentrated on sharp points; and both were able to destroy magnetism, or reverse the polarity of a magnet, whilst both could kill living creatures. In 1752 he carried out his famous kite experiment, collecting charge from a thunder cloud in a Leiden jar, and showing that it was similar in its effects to the charge produced by an electrical machine. To explain the phenomena of electricity known to him, Franklin supposed that there was an imponderable electrical fluid which pervaded all space and all material bodies, such bodies being electrically neutral when the concentration of fluid within them and outside were the same. An excess of the fluid rendered a body positively charged, whilst a deficit rendered it negatively charged. Franklin held that light consisted of vibrations in an ether which filled all space, and like other wave theorists, Leonard Euler before him and Thomas Young after him, he thought that the electrical fluid of space might be the same as the luminiferous ether.

One of the drawbacks to Franklin's theory, pointed out in 1759 by Franz Aepinus, 1724-1802, of the St. Petersburg Academy of Sciences, was that air condensers would automatically discharge themselves if there were an electric fluid in the space between their plates. Aepinus preferred to consider electrical attraction as an action at a distance, like gravity. Another objection was that electrical charge appeared to reside on the surface of bodies, not throughout their bulk as Franklin's theory suggested. Stephen Gray, died 1736, a pensioner of Charterhouse, had shown in 1729 that solid and hollow oak cubes of the same dimensions when charged in the same way exhibited the same electrical effects, indicating that the charge remained entirely on the surface of the cubes. Joseph Priestley made a similar experiment in 1767, showing that a hollow charged body exerted no force on electric charges within its cavity. Newton had shown that if gravitational force decreased with the square of the distance from its source, a spherical shell of matter would exert no gravitational pull on bodies within it, from which Priestley concluded that by analogy electrical force also obeyed the inverse square law. In 1750 John Michell, 1724-93, at Cambridge, had already discovered the inverse square law of repulsion between similar magnetic poles. Suspending a magnet by a thread, Michell brought up another magnet and measured the repulsive force between them by means of the twist imparted to the thread. In France the engineer, Coulomb, 1738-1806, rediscovered Michell's torsion balance, and with it from 1785 to 1789 demonstrated the inverse square variation of force with distance for both electrical and magnetic attractions and repulsions. Such discoveries appeared to show, to the French physicists at least, that electric and magnetic forces were of the same kind as gravity, operating at a distance over empty space, and obeying the inverse square law.

The German nature-philosophers were interested in a different aspect of electricity and magnetism, namely, the phenomenon of polarity, which appeared to exemplify perfectly the dialectical tension they postulated between opposite poles or forces which brought order out of chaos. Since there was only one kind of power behind the development of nature in their philosophy, namely, that of the World Spirit, they held that light, electricity, magnetism, and chemical forces, were all interconnected: all were different aspects of the same thing. One of Schelling's disciples, Hans Christian Oersted, 1777-1851, professor of physics at Copenhagen, announced in 1807 that he was looking for the connection between magnetism and electricity. Franklin had shown in 1751 that iron needles could be magnetized and demagnetized electrically by means of the discharge from a Leiden jar. The Leiden jar gave only transient electrical current, but the voltaic cell, invented in 1799, provided a continuous source of current, and with it Oersted was able to demonstrate in 1820 the magnetical effects of such currents. He showed that a wire carrying an electric current would rotate round a magnetic pole, and conversely that a magnet tended to move round a stationary wire carrying a current. Using the terminology of his school, Oersted wrote:

'To the effect which takes place in the conductor and in the surrounding space, we shall give the name of the conflict of electricity. All non-magnetic bodies appear penetrable by the electric conflict, while magnetic bodies resist the passage of this conflict. Hence they can be moved by the impetus of the contending powers. . . . From the preceding facts we may likewise collect that this conflict performs circles: for without this condition it seems impossible that the one part of the uniting wire, when placed below the magnetic pole, should drive it towards the east, and when placed above it toward the west: for it is the nature of a circle that the motions in opposite parts should have an opposite direction.'

Oersted's discovery aroused considerable interest as the main forces then known were of a linear push-and-pull character, like the gravitational, electrical, and magnetic attractions and repulsions, but here was a case of a rotatory force. The phenomenon puzzled the French school of Newtonian physicists in particular, for they were the staunchest supporters of the view that all actions were the result of push-and-pull forces operating over a distance according to the inverse square law. However, Ampere, 1775-1836, at the Polytechnique had shown by the end of 1820 that a circular coil of wire carrying a current behaved like an ordinary magnet, exhibiting push-and-pull attractions and repulsions, from which he presumed in 1825 that magnetism derived from small resistance-free circular electric currents in the particles of magnetic bodies.

Another member of the German school, Thomas Seebeck, 1770-1831, who assisted Goethe in his scientific work, looked for a connection between heat and electricity. In 1822 he found that a junction of two dissimilar metals when heated gave an electric potential, and a current when the circuit was closed. Unlike the voltaic cells used at the time, Seebeck's thermal source of electricity gave very steady potentials, and it enabled another German, George Ohm, 1787-1854, later a professor of physics at Munich, to work out from 1826 to 1827 the relations between potential, current, and resistance in an electrical circuit. Ohm was influenced by the work of Fourier on the flow of heat through thermal conductors, which had been published in 1822, and he endeavoured to make a similar analysis of the flow of electricity, defining electrical potential by analogy with temperature, and the amount of electric current by analogy with a quantity of heat.

The most important of the researches connecting up electrical effects with other phenomena were those carried out by Michael Faraday, 1791-1867, Davy's assistant and successor at the Royal Institution, London. Faraday at first continued the chemical work of Davy, but he went over more and more into the field of physics. He discovered in 1826 what appeared to be a case of isomerism whilst studying butylene and ethylene, and in 1833 he established that the same amount of electricity brought about the decomposition of the same number of equivalents of different chemical substances. The second discovery indicated that if chemical matter were atomic, then electricity too should be of a particulate character. Faraday, however, rejected both the premise and the conclusion, preferring the conception that 'matter is everywhere present, and there is no intervening space unoccupied by it'.

Faraday's physical researches were more noteworthy. It had long been known that a magnet could induce magnetism in an adjacent piece of iron, and that a static electric charge could induce the appearance of another charge on a neighbouring body. Faraday thought that the same should be true for electric currents, and he began to look for the effect from about 1822, when he first noted down a number of possible connections between natural phenomena, which he subsequently searched for, and in some cases detected. In 1831 Faraday discovered the phenomenon of electromagnetic induction, which showed that one electric current could generate another, and which linked up in general mechanical motion and magnetism with the production of electric current. He found that a current changing in magnitude in one coil of wire would induce a transient electrical current in a neighbouring coil. The same effect could be produced by moving a coil carrying a constant current or, what amounted to the same thing, a permanent magnet, in the vicinity of a second coil of wire. Thus Faraday discovered the basic principle of the dynamo, just as Oersted had discovered the principle of the electric motor.

To explain the phenomena of electricity and magnetism known in his day, Faraday developed a characteristic imagery of his own. As we have seen, he rejected the atomic theory of matter, and with it the conception that forces acted at a distance over empty space. He held that matter was omnipresent in the form of an etherial continuum which acted as a vehicle for the forces of nature. According to Faraday the ether filling all space was composed of lines or tubes of force which connected up opposite electric charges or opposite magnetic poles. The lines composing a magnetic field could be mapped out by means of a small compass needle or by sprinkling iron filings on a sheet of paper in the field, when the lines joining the opposite poles became visible. For Faraday the lines and tubes of force had a real physical significance. Each line of force corresponded to a unit of magnetism or a unit of electric charge. A number of lines made up a tube of force connecting opposite poles or opposite charges, the orientation of a tube at any given point indicating the direction of the magnetic or electric field at that point. The tubes increased, and then decreased, in cross-sectional area along their length as the lines of force diverged from their points of origin, the poles or the charges. The cross-sectional area of a tube of force was a measure of the strength of the magnetic or electric field at that section, as the product of the cross-sectional area and the field strength was constant along the length of the tube, the magnitude of the constant being determined by the number of lines of force composing the tube. Faraday supposed that the tubes of force tended to contract lengthways and expand sideways, so that the tubes joining unlike magnetic poles or opposite electric charges tended to draw those poles or charges together, whilst like poles or like charges repelled each other, as the tubes radiating from them could not join up and pushed against each other owing to their sideways expansion. Furthermore an explanation was provided for the inverse square laws of magnetic and electric attractions and repulsions, since the lines of magnetic and electric force thinned out geometrically with the square of the distance from their origin. In the case of electromagnetic induction, Faraday suggested that the amount of electricity induced in a conductor was dependent upon the number of lines of magnetic force which it crossed, whilst the electro-motive force generated was proportional to the rate at which those lines were cut.

After his discovery of electromagnetic induction, Faraday went on to study the influence of material bodies upon electric fields of force. In 1837 he discovered that an electric condenser, consisting of two conducting plates separated by an insulating material, would take up from a source maintained at a constant potential an amount of electric charge which was dependent upon the particular insulating material used. When the plates were separated by a vacuum, the condenser took up a smaller amount of electric charge than when a material insulator was used, and Faraday designated the ratio of the charges received by the condenser in the two cases the specific inductive capacity of the insulator material. In order to explain his discovery Faraday supposed that the lines of electric force crowded more closely together in a material insulator than in a vacuum in proportion to the specific inductive capacity of the insulator, so that the plates of the condenser could accommodate more electric charge at the ends of the lines of force. In 1845 Faraday discovered a similar kind of interaction between material bodies and magnetic fields of force. He found that bar-shaped samples of many substances, termed diamagnetic, tended to orientate themselves across the magnetic field, assuming a position at right angles to the lines of force, in contrast to bars of iron and a few other substances, termed paramagnetic, which orientated themselves along the magnetic field parallel to the lines of force. To account for such effects,

Faraday supposed that the lines of magnetic force thinned out in diamagnetic substances and crowded together in iron and other paramagnetic bodies.

Faraday came across the phenomenon of diamagnetism while searching for some connection between light, magnetism, and electricity. He placed a piece of glass between the poles of a strong electromagnet, and observed that the glass orientated itself across the magnetic field. On passing a beam of polarized light through the glass along the lines of magnetic force he discovered that the plane of polarization of the light was changed. Such an interaction between magnetism and light led him to suggest in 1846 that light might consist of wave vibrations along the lines of force. Faraday asked:

· 'Whether it was not possible that the vibrations which in a certain theory are assumed to account for radiation and radiant phenomena may not occur in the lines of force which connect particles, and consequently masses of matter together: a notion, which as far as it is admitted, will dispense with the ether which, in another view, is supposed to be a medium in which these vibrations take place.'

Faraday's query was the first suggestion towards the electromagnetic theory of light, which was put forward in 1862 by Clerk Maxwell, 1831-79. A line of enquiry which stimulated the development of the theory was the study of the relation between static and current electricity and, in particular, the estimation of the velocity of current electricity. Charles Wheatstone, 1802-75, a professor of physics at London, measured the velocity of current electricity in 1834 by examining sparks produced at the ends of a lengthy electric circuit with a revolving mirror, and he estimated that electricity travelled at a speed which was one and a half times as great as the velocity of light. In France, Fizeau obtained values for the velocity of electricity in 1850 which ranged from one-third of the speed of light for iron wires to two-thirds for copper wires. Finally Kirchhoff, 1824-87, at Heidelberg, showed in 1857 that static and current electricity were related by a constant which had the dimensions of a velocity, and, by comparing the attractive force of two static charges with the magnetic force produced when they were discharged, he demonstrated that the constant had the same magnitude as the velocity of light.

Clerk Maxwell, who was a professor of natural philosophy first at London and then at Cambridge, endeavoured to put into a quantitative and mathematical form the largely qualitative explanations which Faraday had suggested for electrical and magnetic phenomena. First of all, Maxwell developed the qualitative aspects of Faraday's conception of lines of force, bringing in the ether of the wave theory of light. Lines of force, Maxwell supposed, were tubes of ether rotating on their axes. The centrifugal force of such rotations caused the tubes to expand sideways and contract lengthways, as Faraday had suggested in order to explain attraction and repulsion. But two neighbouring tubes, rotating in the same sense, would move in opposite directions at the points where they touched, a conception which was not feasible mechanically. Maxwell supposed therefore that there were layers of particles between the tubes of ether, the particles rotating in a direction opposite to that of the tubes, like ball bearings or idling wheels. If all of the tubes of ether rotated at the same speed, the particles would not change their positions, but if not, a given particle would move linearly with a speed which would be the mean of the circumferential velocities of the tubes on either side. Thus, if by some means the rotatory speed of one tube were altered, a disturbance would be propagated throughout the system and the particles would be set into a linear motion, rolling from tube to tube. Maxwell considered that the particles were electrical in character, and he thought therefore that such a motion of the particles would constitute an electric current.

Conversely, if a particle were displaced from its normal position, a tangential stress would be exerted on the adjacent tubes and, since such tubes were elastic, they would tend to restore the displaced particle to its normal place. Maxwell suggested that such a state of strain existed in the electrostatic field between two condenser plates, the charges on the plates displacing the electric particles which in turn placed a stress on the ether tubes in the intervening space. Considering the possibility of vibrational stresses and strains in his ether model,

Maxwell deduced from the laws of dynamics governing the mechanics of his model that wave-like disturbances would be propagated through it with the speed of light. It seemed therefore that light was an electromagnetic phenomenon or, as Maxwell put it, 'that light consists in the transverse undulations of the same medium which is the cause of electric and magnetic phenomena'. In a substance other than the etherial medium of empty space Maxwell showed that electromagnetic waves should be propagated with a speed equal to the product of the velocity of light and the square root of the specific inductive capacity of the substance. Since the velocity of light in a transparent substance is related to its refractive index, it appeared that the specific inductive capacity of a substance should equal the square of its refractive index, a prediction which was later verified.

Maxwell did not concern himself a great deal with the experimental verification of the various predictions deriving from his theory, nor did he develop further the qualitative aspects of his model of the electromagnetic ether with its suggestive conception of particles of electricity or electrons. In his later work he abandoned his ether model, and he concentrated upon the mathematical equations he had derived for wave-like disturbances in the ether, applying those equations to optical phenomena. Other scientists, notably Lord Kelvin at Glasgow, who relied upon mechanical models to explain by analogy the natural phenomena they studied, found some difficulty in comprehending Maxwell's mathematical work and they attempted to bring together the phenomena of light, electricity, and magnetism, by developing other models of the ether. Kelvin remarked in 1884:

'I am never content until I have constructed a mechanical model of the object that I am studying. If I succeed in making one, I understand; otherwise I do not. Hence I cannot grasp the electromagnetic theory of light. I wish to understand light as fully as possible, without introducing things that I understand still less. Therefore I hold fast to simple dynamics for there, but not in the electromagnetic theory, can I find a model.'

Accordingly Kelvin in 1890 attempted to explain the phenomena of light, electricity, and magnetism, by means of MacCullagh's optical ether, the elements of which were presumed to resist rotational stresses but not linear displacements. Kelvin suggested that electrical effects were due to translatory motions of the elements of the MacCullagh ether, while magnetic phenomena were due to rotations, and light to wave-like vibrations. The MacCullagh model implied, however, that electrical fields applied to a transparent medium should alter the velocity of light in that medium, but it was shown that such was not the case. Many other ether models were proposed during the latter half of the nineteenth century, models which explained with varying degrees of success the now manifold phenomena of light, electricity, and magnetism. Some attempted to accommodate even the properties of matter into an ether model, Kelvin suggesting in 1867 that the atoms of matter were vortex rings in the ether, like smoke rings in the air, but it was difficult to explain on this basis the weight and density of material substances. Ultimately all models of the ether had to be abandoned, together with the absolute space of which they provided the hypothetical substance, when it was shown that the conception of the absolute velocity of a body, that is, its velocity relative to the ether, was meaningless.

A most important consequence of the electromagnetic theory of light was pointed out in 1883 by Fitzgerald, 1851-1901, professor of natural philosophy at Dublin. He pointed out that, if Maxwell's theory were valid, it should be possible to generate electromagnetic radiations purely electrically by varying an electric current periodically in a circuit. Kelvin had demonstrated in 1853 that the discharge of a Leiden jar and other electrical condensers was of an oscillatory nature, the charge surging to and fro while subsiding to zero. Accordingly Fitzgerald suggested that a discharging condenser would be a good source of the electromagnetic waves predicted by Maxwell's theory, and he showed that the shorter they were in wave length, the greater the amount of energy they would carry, and thus the easier they should be to detect.

A detector for such waves was discovered in 1886 by Heinrich Hertz, 1857-94, later professor of physics at Bonn.

He found that electric sparks would cross a small gap between the two ends of a loop of wire if the loop were held in the vicinity of a discharging Leiden jar, or an induction coil in operation. The loop picked up the electromagnetic radiations given out by the jar or the coil, the radiations being transformed into an electric current which discharged across the spark gap. With this simple apparatus Hertz then went on to show that such radiations had properties similar to those of light. In 1888 he demonstrated that the electromagnetic waves were reflected from the walls of his laboratory and could be refracted through prisms of hard pitch. Furthermore they could be diffracted and polarized like light waves, and they travelled in straight lines with a speed which was of the same order as the velocity of light. In this way Hertz verified the most important of the predictions given by Maxwell's electromagnetic theory of light, and he provided also the fundamental discoveries upon which the later developments of radio broadcasting and radar were based.

Chapter 39

Thermodynamics: The Science of Energy Changes

THE SCIENCE of the relations between the various forms of energy—heat, light, electricity, magnetism, chemical and mechanical energy—arose from the study of the mechanical production of heat through friction, and the thermal generation of mechanical energy by the steam engine. During the seventeenth century some natural philosophers, notably, Bacon, Boyle, Hooke, and Newton, had thought that heat was the mechanical motion of the minute particles of bodies, the rapidity of those motions increasing with temperature. With the development of chemistry in the next century, heat came to be regarded as a weightless material substance which was termed 'caloric', the melting of a solid and the evaporation of a liquid being thought to be a kind of chemical reaction between the material of heat and the matter of the solid or the liquid.

According to the caloric theory, the production of heat by friction was due to the release of the material of heat from its chemical combination or mechanical association with the matter of the two bodies rubbed together, from which it followed that the amount of heat and the quantity of rubbings produced should be proportional to one another. Whilst boring cannon at Munich in 1798, the émigré American scientist, Count Rumford, 1753-1814, observed that the amount of heat produced and the quantity of borings were more or less inversely proportional to each other. Blunt boring instruments gave more heat and performed less cutting work than sharp instruments, contradicting the caloric theory, according to which the sharp borers should have abraded the metal of the cannon more effectively and released more of the material of heat bound up with the metal. Rumford found that a blunt borer, producing little or no abrasion, generated sufficient heat to raise some eighteen pounds of water to the boiling point in two and three-quarter hours. Such an amount of heat was produced by mechanical energy alone, and so Rumford concluded that heat in itself was a form of mechanical motion.

The mechanical theory of heat was not widely accepted at the time, though Rumford found one convert in his future protégé at the Royal Institution, Humphry Davy. In 1799 Davy made an experiment in which two pieces of ice were rubbed together by a clockwork mechanism in a vacuum, the whole of the apparatus being maintained at the freezing point of water. He claimed that some of the ice melted as a result of the mechanical friction, and Davy presumed from the experiment that heat was 'a peculiar motion, probably a vibration of the corpuscles of bodies'. A somewhat different mechanical theory of heat was put forward in 1807 by Thomas Young who supposed, from the study of the radiant heat given out by red-hot bodies and the heating effect of the infra-red region of the spectrum, that heat might be a wave vibration similar to light. However, the mechanical theory of heat found little support at the time and for the most part it was the material theory of heat, the idea of caloric, which was generally received until the 1850's.

Meanwhile in France the factors governing the conversion of heat into mechanical energy by the steam engine were investigated. Such factors had not been studied very intensively in Britain, though by now the steam engine had been in use there for more than a century. Watt had drawn up an indicator diagram showing graphically how the steam pressure varied with the effective volume of the cylinder in a steam engine, but it seems that neither Watt nor any other British scientist deduced anything from such diagrams at the time. The British engineers, such as Watt, were largely self-educated, but the French engineers of the early nineteenth century were trained with the theoretical scientists at the Ecole Polytechnique, and thus it was that they were more able to deal with the theory of the steam engine, and the theory of machines in general.

Both the theoretical scientists and the practical engineers in France studied the problems of heat, and both, generally speaking, adopted the material theory of heat, regarding caloric as a weightless fluid. Fourier, 1768-1830, belonging to the school of theoretical physicists at the Ecole Polytechnique, brought out in 1822 his *Analytical Theory of Heat* in which he dealt with the flow of heat through solids, a new method of mathematical analysis, and the theory of dimensions which had been suggested, but not developed, by Descartes. Fourier was primarily concerned with the phenomena of thermal conduction, and not at all with the mechanical effects of heat. When heated, bodies expand and produce mechanical force, Fourier noted, 'but it is not these dilations which we calculate when we investigate the laws of the propagation of heat'. Fourier in fact was of the view that the study of thermal phenomena was a science distinct from mechanics.

'There exists a very extensive class of phenomena,' wrote Fourier, 'which are not produced by mechanical forces, but which result solely from the presence and accumulation of heat. This part of natural philosophy cannot be brought under dynamical theories: it has principles peculiar to itself and is based upon a method similar to that of the other exact sciences.'

The French engineers on the other hand were concerned primarily with the connection between thermal and mechanical effects. In 1824 a French army engineer, Sadi Carnot, 1796-1832, published his *Reflections on the Motive Force of Fire* in which he attempted to analyse the factors determining the production of mechanical energy from heat in the steam engine and heat engines in general. Carnot drew attention to the fact that in the steam engine heat flowed from a region of high temperature, the boiler, to a region of low temperature, the condenser, mechanical work being generated through the cylinder and piston during the process. In this respect Carnot considered that the steam engine was analogous to another prime mover, the waterwheel.

'We may justly compare the motive power of heat with that of a fall of water,' wrote Carnot. 'The motive force of a water fall depends on the height and the quantity of fluid: the motive force of heat depends upon the quantity of caloric employed and what we may call the height of its fall, that is to say, the difference in temperature of the bodies between which the caloric is exchanged.'

The analogy, and the caloric theory upon which it was based, led Carnot to the incorrect conclusion that no heat was lost, or converted into mechanical energy, during the operation of the steam engine. He thought that the same quantity of heat was given out by the boiler at the higher temperature as was received by the condenser at the lower temperature. However, the analogy led him also to the fruitful conception that the amount of energy produced by a steam engine was solely dependent, in principle, upon the temperature difference between the boiler and the condenser and the amount of heat which passed from the one to the other. Thus it appeared that all steam engines, and all heat engines in general, would have the same efficiency if they worked between the same temperature levels. He substantiated this conclusion, known as Carnot's principle, by pointing out that perpetual motion would be possible if it were not true. If two perfect heat engines operating between the same temperature levels did not have the same efficiency, it would be possible for the more efficient engine to drive the less efficient engine backwards, pumping heat from the lower to the higher temperature, thus leaving the thermal conditions unchanged and yet generating a continuous net excess of mechanical energy. Carnot held that perpetual motion was impossible, and hence he postulated that all heat engines working between the same temperature levels were equally efficient, irrespective of their mode of operation or the material used to transport the heat and do work, that is, they were equally efficient whether they were turbine or cylinder engines, or whether they used steam, air, or any other working substance.

Later, in 1830, Carnot appreciated that his comparison of the steam engine with the waterwheel was not exact, and that some heat was converted into mechanical energy and was lost during the operation of the engine. Thereupon he abandoned the caloric theory and adopted the view that heat was nothing other than the motions of the particles of bodies, heat and mechanical energy being interconvertible and equivalent. However, Carnot died in the cholera epidemic of 1832, and his later views jotted down in his notebooks were not published until 1878. The earlier work of Carnot, based on the caloric theory, was developed by another French engineer,

Clapeyron, 1799-1864, professor at the Paris School of Bridges and Roads. In 1834 Clapeyron revived or rediscovered Watt's indicator diagram which showed how the pressure varied with the volume in the cylinder of the steam engine during a cycle of its operations. He pointed out that the area of the pressure-volume graph provided an estimate of the work done in a cycle of changes, and he suggested that a measure of the efficiency of a heat engine was afforded by the ratio of the work done to the quantity of heat supplied during such a cycle.

The importance of the work of Carnot, which became known through Clapeyron, was not generally appreciated until the 1850's, and meanwhile attention was turned again to the problem of the production of heat from mechanical motion and other sources of energy. In Germany the question was approached from a chemical and a biological point of view. Lavoisier had shown that the ratio of the amount of heat to the quantity of carbon dioxide given out by animals was roughly equal to the ratio of heat to carbon dioxide produced by candle flames. Thus it appeared not unlikely that the heat of warm-blooded creatures was derived from the chemical energy of the combustion of their food. Liebig, who had been trained at Paris, supposed that the mechanical energy of animals, as well as the heat of their bodies, might be derived from the chemical energy of their food. The opinion of the German scientists upon this question was divided, some maintaining that the activities of organisms depended upon a vital force peculiar to living beings. One of Liebig's pupils, Friedrich Mohr, 1806-79, adopted the mechanistic view, from which he derived the conception that all the various forms of energy were manifestations of mechanical force.

'Besides the known fifty-four chemical elements,' Mohr wrote in 1837, 'there exists in Nature only one agent more, and this is called force; it can under suitable conditions appear as motion, cohesion, electricity, light, heat, and magnetism. . . . Heat is thus not a particular kind of matter, but an oscillatory motion of the smallest parts of bodies.'

Such a view was put forward again in 1842 by Robert Mayer, 1814-78, a physician of Heilbronn in Bavaria. While serving on a ship in the tropics, Mayer noted that the venous blood of his patients was redder in colour than he had found it to be in Europe. He ascribed the difference to the greater amount of oxygen in the venous blood under tropical conditions, the excess of oxygen being due to a diminution in the combustion of food providing the body heat. The phenomenon seemed to support the view that the heat of the body came from the chemical energy of the food, and Mayer presumed that the mechanical energy of the muscles came from the same source, mechanical energy, heat, and chemical energy being equivalent and interconvertible. Upon his return to Germany, Mayer pursued the matter further. It had been known since the beginning of the century that gases expanding into a vacuum underwent no thermal change, but that gases expanding against an opposing pressure, and thus performing mechanical work, absorbed heat. Mayer perceived that in the second case the mechanical work produced came from the heat absorbed, the two being equivalent, and from the published data concerning the thermal changes accompanying the expansion of gases he calculated the quantity of heat equivalent to a given amount of mechanical energy.

Mayer's paper, like that of Mohr earlier, was turned down by Poggendorf, the editor of the main German physics journal, on the grounds that it contained no experimental work. This was a fixed point of policy with Poggendorf and other German physicists, as they wished to avoid the speculative tendencies of the nature-philosophy of the period. From his paper, which was ultimately published in a chemical journal edited by Liebig and Mohr in 1842, it appears that Mayer was something of a nature-philosopher, though his speculations resulted in a positive achievement. Forces, he argued, were essentially causes, and since causes were indestructible and convertible into effects, it followed that forces were likewise indestructible and interconvertible.

'Motion in many cases has no other effect than to produce heat,' wrote Mayer, 'and thus the origin of heat has no other cause than motion.'

Another German who arrived at the idea of the conservation and the interconversion of the different forms of energy, again from a biological standpoint, was Hermann Helmholtz, 1821-94, professor of physiology at Konigsberg and then professor of physics at Berlin. Opposing the vitalists, Helmholtz argued that living organisms would be perpetual motion machines if they derived energy from a special vital force over and above the energy obtained from their food. The principle that perpetual motion was impossible indicated therefore that animals obtained their energy from their food alone, the chemical energy of the food being converted into an equivalent amount of heat and mechanical work. Helmholtz argued further that, if heat and other kinds of energy were in themselves forms of mechanical motion, then the principle that the total amount of energy in the universe was constant followed from the law of the conservation of mechanical energy laid down in the seventeenth and eighteenth centuries. Helmholtz's first paper on the principle of the conservation of energy, like those of Mohr and Mayer, was turned down by Poggendorf, though it was published elsewhere.

The experimental work which established the principle of the conservation of energy was carried out in England by James Prescott Joule, 1818-89, a Manchester brewer and amateur scientist. Joule, like Mayer and others, was convinced that energy was indestructible and could manifest itself under various forms but, unlike the Germans, he endeavored to show that this was so by experiment, systematically measuring the amounts of the various forms of energy which could be converted into a given quantity of heat. Joule too had a completely mechanical picture of the material world, believing that heat was the motions of the particles of bodies and thus that heat was fundamentally the same as mechanical energy. He was impatient of the nature-philosopher's viewpoint of Mayer who stressed that the mechanical equivalent of heat was a pure number, expressing the qualitative transformation of one form of energy into another independently of the mechanical theory of heat or any other theoretical model.

Joule studied first of all the subject of electricity, which was then advancing rapidly, but unlike the other great electricians, Davy and Faraday, Joule concentrated on the thermal effects of the electric current. In 1840 he measured the heat given out by an electric current flowing through a wire resistance, and he found that the amount of heat evolved in a given time was proportional to the resistance of the circuit and the square of the current flowing through it, a relation known as Joule's law. Joule presumed from this experiment that electrical energy was converted into heat by the resistance, though he bore in mind the possibility that heat was a material substance, caloric, which was transported from one part of the circuit to another by the current. He disproved the latter possibility in 1843 by measuring the mechanical work expended in operating a dynamo enclosed in a vessel containing water, the increase in the temperature of which provided an estimate of the heat produced. Here the circuit was entirely enclosed, so that the increase in the temperature of the water was due to the conversion of mechanical energy into electricity, and electricity into heat, not to the transport of caloric from one part of the circuit to another. Having convinced himself that the various forms of energy could be quantitatively converted one into the other, Joule measured accurately the amount of heat produced mechanically by a paddle-wheel churning water, finding that 772 foot-pounds of mechanical work produced, and was equivalent to, the heat required to raise one pound of water through 1° F.

Joule's researches did not attract immediate attention. The Royal Society refused to publish two of his papers, at which Joule was not surprised for he was aware of the differences between the interests and values of the gentlemen-scientists of the Royal Society and those of industrial Manchester. However, at the 1847 meeting of the British Association the importance of his work was noted by William Thomson (Lord Kelvin), 1824-1907, who pointed out that Joule's results conflicted with the theory of heat engines elaborated by the French engineers. Joule's experiments showed that mechanical energy was quantitatively converted into heat, and yet the French theory suggested that the reverse change did not occur, that heat was not transformed into mechanical energy in the steam engine, it merely fell from a high to a low temperature.

Kelvin at first adopted the French view as it stood, since it appeared to be more fruitful. In 1848 Kelvin showed that an absolute scale of temperatures could be based on Carnot's theory of perfect heat engines. Up to that time, temperatures had been measured by the expansion of solids, liquids, and gases on heating, equal increments of expansion being taken as estimates of equal increments of temperature. However, temperature scales based on different thermometric substances did not quite agree amongst themselves. The mercury thermometer differed slightly from the gas thermometer, and there appeared to be no reason why the measurements of the one were more fundamental than those of the other. Carnot's theory indicated that all perfect heat engines working between the same temperature differences should be equally efficient, no matter what their working substances, steam, air, and so on. Kelvin suggested therefore that equal increments of temperature on an absolute scale could be defined as ranges of temperature over which a perfect heat engine would operate with equal efficiencies. Later, in 1854 when the caloric theory had been generally abandoned, Kelvin proposed another absolute scale in which equal increments of temperature were taken as temperature ranges over which a heat engine produced the same quantities of work, and he showed that such a scale corresponded quite closely with the scale of the gas thermometer.

The views of Joule, Mayer, and others were assimilated into the theory of heat engines by Kelvin at Glasgow and Rudolph Clausius, 1822-88, at Berlin. They noted that when gases and vapours expanded against an opposing force and performed mechanical work they lost heat, some heat being converted into mechanical energy and lost in this way during the operation of the steam engine. The main obstacle to the law of the conservation and the interconvertibility of the different forms of energy was now overcome, and the law was put forward as a general principle by Clausius and Kelvin in 1851. Whilst the amount of heat decreased during the cycle of operations of the Carnot heat engine, it was seen that there was a quantity which remained constant throughout the cycle. The amount of heat given out was smaller than that taken in by the engine, but the quantity of heat taken in divided by the temperature of the heat source had quantitatively the same value as the amount of heat given out divided by the temperature of the heat sink. Clausius in 1865 termed this quotient, the entropy.

Clausius pointed out that Carnot's perfect heat engine was rather an abstraction, since in everyday experience hot bodies ended to cool down spontaneously and cold bodies to warm up, but if natural objects consisted of pairs of Carnot's heat engines, the one driving the other backwards, hot bodies would remain permanently hot and cold bodies permanently cold. In spontaneous thermal processes, such as the conduction of heat down a bar of metal, the amount of heat remained constant while the temperature decreased. The entropy, the amount of heat divided by the temperature, tended therefore to increase in spontaneous natural processes, not to remain constant as in the perfect heat engine. Such was the second law of thermodynamics. 'The entropy of the world tends to a maximum', as Clausius put it, the first law being the now familiar principle of the conservation of energy, 'The energy of the world is constant'.

The laws of thermodynamics were interpreted dynamically by Clausius and others in terms of the atomic theory of matter. In 1857 Clausius revived the theory that gases consisted of molecules in motion, the pressure of a gas being the result of the impact of the molecules upon the walls of the containing vessel. The heat energy of a gas resided in the kinetic energy of the motions of its molecules, the velocities of those molecules increasing with temperature. Developing the kinetic theory of gases further, Clerk Maxwell at London showed in 1866 that the chance collisions of the molecules in a gas would give a few molecules more energy than the average and leave a few others with less energy. He calculated on probability grounds the fraction of an assembly of such molecules which would possess a given excess of energy above the average, a result that was later important in dealing with situations where a few energized molecules were presumed to surmount an energy barrier and undergo a transformation, as in the case of chemical reactions or the escape of molecules from the surface of a liquid or a solid. Maxwell pointed out that for a being who could handle individual gas molecules the second law of thermodynamics would not hold, as such a 'demon' could separate off the swiftly moving molecules from the slower ones, and so create a temperature difference with no expenditure of energy. Kelvin thought that animals and plants might contain such 'Maxwellian demons', but Maxwell himself held that living creatures obeyed the laws of thermodynamics just as much as inorganic objects.

In terms of the atomic theory, the first law of thermodynamics was seen to be identical with the earlier principle of the conservation of kinetic energy during the impact of bodies, as heat energy was equated with the mechanical energy of the molecules of matter. The second law of thermodynamics was interpreted by the Austrian physicist, Ludwig Boltzmann, 1844 & 1906, to mean that in the spontaneous movements of energy, such as the conversion of mechanical energy into heat or the cooling of hot bodies, the molecules of the system concerned were tending towards a random or Maxwellian distribution of their energies. Such a distribution was the most probable one, being the most random or disordered, whilst other more orderly distributions possessed a smaller probability. Thus the spontaneous increase in the entropy of a system could be correlated with the increase in the probable distribution of the molecular energies of that system, Boltzmann showing in 1877 that the entropy was proportional to the logarithm of the probability.

The second law of thermodynamics and its molecular interpretation gave a physical meaning and a direction to the passage of time which had been lacking hitherto in the Newtonian system of mechanics. In principle the mechanics of the Newtonian world were reversible. A cannon ball theoretically could rebound on striking a target and retrace its path back to the cannon from which it came. According to the second law of thermodynamics, such an eventuality was quite impossible. The orderly, unidirectional motion of the projectile would be continuously transformed by the frictional resistance of the air into heat, that is, into the random, disorderly motions of the molecules of the air and the projectile, and finally all vestiges of the orderly linear motion would be destroyed when the projectile reached its target, the orderly motion being transformed into the random thermal motions of the projectile and its target. Such changes were irreversible: mechanical energy was permanently lost to the world when it was transformed into heat, and when the heat was dispersed.

The spontaneous rate of the dispersion of energy, in such processes as the gradual cooling of the sun through the constant emission of radiation, gave a measure of the flow of time.

Kelvin remarked in 1854, 'as to the sun, we can now go both backwards and forwards in his history upon the principles of Newton and Joule'. The annual output of heat by the sun was measured by Pouillet in France and independently by John Herschel at the Cape of Good Hope in 1837. Their figures showed good agreement, Herschel estimating that enough heat was given out by the sun in a year to melt a crust of ice surrounding the earth some hundred feet thick. Mayer at Heilbronn indicated in 1848 that if the sun were a mass of coal it would burn out in five thousand years at this rate, and he suggested that the kinetic energy of meteors and asteroids falling on the sun might provide sufficient heat for such an annual output. Waterston in England proposed the same hypothesis independently in 1853, but it was shown that if the same density of meteors were to fall upon the earth, their impacts would render the earth permanently red-hot. A more satisfactory hypothesis was advanced in 1854 by Helmholtz, who suggested that the mutual gravitational attractions of the particles composing the sun would cause it to contract, and thereby the potential energy of the particles, the gravitational forces between them, would be converted into kinetic energy, that is, into heat. A shrinkage of a few hundred feet a year would account for the sun's annual output of thermal energy, but a limit was set to the possible age of the sun in both the past and the future. Calculations on this basis showed that the sun had been in existence for some twenty to thirty million years, and that it would last for another ten million years or so.

Similar estimates could be made of the age of the earth from its rate of cooling. Kelvin showed in 1862 that the habitable age of the earth could not be more than two hundred million years, and by 1899 he had shortened the limit to twenty to forty million years. Such estimates stood in opposition to the values of the age of the earth determined by the geologists from the total thickness of the sedimentary strata ranged in their historical order, and the rate of deposition of silt, forming fresh sedimentary rocks, in the deltas of rivers. The geologists estimated in this way that the formation of the sedimentary rocks had occupied a period of at least two hundred million years, some geological estimates including the time taken for the formation of the pre-sedimentary rocks ranging up to four hundred million years. Some of the students of thermodynamics felt that the geologists must be mistaken. At a meeting of the Glasgow Geological Society in 1866, Kelvin remarked:

'A great reform in geological speculation seems now to have become necessary. British popular geology at the present time is in direct opposition to the principles of natural philosophy.'

Huxley at the London School of Mines replied in 1869 that the geological evidence was as valid as the physical, and it might be that the physicists were mistaken. In 1900, the year after Kelvin had made known his short estimate of twenty to forty million years for the age of the earth, the geologist, James Geikie, indicated that the compression of the earth's crust resulting from as much as one hundred million years of cooling was insufficient to account for the thickness of the folded rocks in the Alps. In 1899 another geologist, Chamberlin, suggested that the theory of the physicists might be incomplete since the atoms might well possess complex organizations and enormous energies, which were released under the conditions of the interior of the sun.

The phenomenon of radioactivity, discovered by Becquerel in 1896, led to the development of solar energy theories along the lines which Chamberlin had forecast, and it provided further methods of estimating the age of the earth which substantially verified the estimates made by the geologists earlier. It was found that uranium breaks up at the rate of one per cent in every sixty-six million years, yielding ultimately a light form of lead. The ages of the rock strata could be determined therefore from the relative amounts of uranium and light lead contained within them, the results showing that the estimates of the age of the sedimentary rocks made by the geologists were of the correct order, and that some minerals had been laid down as long as 1,985 million years ago.

Another field upon which the science of thermodynamics had an influence was that of the philosophy of science. Carnot had shown that the operations of a heat engine were independent of the matter composing the particular working substance of the machine, and later physicists stressed that thermodynamics did not entail any presuppositions or hypotheses concerning the nature of matter, as the science dealt only with energy changes. The laws of thermodynamics had been interpreted in terms of the atomic theory of matter, but such an interpretation was not essential to the science. Thermodynamics could proceed without a theoretical model of the nature of matter, indeed it could proceed without the supposition that matter existed objectively. Thus some students of thermodynamics, notably Wilhelm Ostwald, 1853-1932, professor of chemistry at Leipzig, suggested that the phenomena of nature were alone manifestations of energy and its manifold transformations, founding what was termed the 'Energetik' school of thought.

'What we hear', wrote Ostwald, 'originates in work done on the ear drum and the middle ear by the vibrations of the air. What we see is only radiant energy which does chemical work on the retina that is perceived as light. . . . From this standpoint the totality of nature appears as a series of spatially and temporally changing energies, of which we obtain knowledge in proportion as they impinge on the body, and especially upon the sense organs fashioned for the reception of the appropriate energies.'

Once hypotheses as to the nature of matter had been abandoned, there appeared to be no very good reason why hypotheses concerning the nature of energy should be retained. Nature could be regarded as a succession of observed phenomena, and science as an activity which correlated those phenomena. Mayer pointed out that the mechanical equivalent of heat was a pure number, correlating different phenomena, which was independent of the mechanical, the caloric, or any other theory of heat, and as such, in his view, it was of more value than any hypothesis.

'One single number has more real and permanent value than an expensive library of hypotheses,' wrote Mayer; the attempt to penetrate by hypotheses to the inner recesses of the world order is of a piece with the efforts of the alchemists.'

Such a view was developed in 1872 by Ernst Mach, 1838-1916, professor of physics at Prague, and then professor of philosophy at Vienna. Mach attacked the prevailing tendency of scientists to explain the phenomena of nature in terms of theoretical mechanical models, such as the atomic theory of matter employed in chemistry, and the various etherial continuums invented to explain the phenomena of light, electricity, and magnetism. He pointed out that in thermodynamics mechanical models of nature were not employed, observed phenomena being correlated with one another directly. He suggested therefore that thermodynamics, and not mechanics, was the prototype of all the sciences, and that its methodology should be applied to the other sciences in order to rid them of all hypothetical images and theoretical constructs.

Mach was of the view that science consisted essentially of a body of observed facts and phenomena, bound together by a number of laws and rules. The laws of nature were really mental devices developed for the convenient and economical remembering of facts, for the mind of man was too feeble to retain everything that he observed. A scientific law, wrote Mach,

'has not one iota more of factual value than the isolated facts taken together, its worth lying merely in its convenience. It has utilitarian value. . . . If all single facts, all separate phenomena, were as directly accessible to us as we demand that knowledge of them to be, science would never have arisen.'

According to Mach, the task of science was to subsume classes observed facts under general relationships which described all the particular cases of the domain covered without introducing any hypotheses or theoretical models. Such an approach, termed that of phenomenological physics by Mach in 1896, was not unattractive to some investigators in the field of thermodynamics, though most scientists adhered to their theories and their mechanical models of nature.

The severest strictures of Mach and his school were reserved for the followers of the atomic theory, as it was they who had trespassed with success into the field of thermodynamics. The atomic theorists, Mach stated, had attempted 'to make a notion as thoroughly naïve and crude as that which holds matter to be absolutely unchanging the fundamental tenet of physics'. However, Mach himself did not entirely escape from the influence of the atomic theory: indeed, the atomic viewpoint was in a sense fundamental to his system, as he conceived of the world of the scientist as a stream of unit observed facts, or atomic perceptions. One of his followers, Ludwig Wittgenstein, stated explicitly, 'The totality of existent atomic facts is the world.' Such a view rested on the psychological theory that phenomena were perceived as unit elements, a theory that had been a product of the atomic-mechanical model of the universe developed by the English and French natural philosophers of the seventeenth and eighteenth centures.

Mach's phenomenological philosophy of science did not pass unchallenged. Boltzmann protested in 1899 that Mach had obliterated all distinction between metaphysical and scientific theory, and had impoverished the concepts of science by replacing the conceptual framework of space and time with the one-dimensional stream of unit observed facts. Defending the atomic theory of matter, Boltzmann maintained:

'A theory which yields something that is independent, and not to be got in any other way, for which moreover so many physical, chemical, and crystalographic facts speak, must not be combated but further developed.'

The atomic theory at that time was indeed making notable advances. Electrons, units of electricity, were postulated in order to explain the phenomena of the passage of electricity through salt solutions, and through gases at low pressures. The movement of small particles in a liquid, first observed in 1827 by the English botanist, Brown, enabled the French physicist, Perrin, to ascertain Avogadro's number, the number of molecules in two grams of hydrogen or the gram-molecular weight of any other substance. These and other developments led Mach's follower, Ostwald, to recant in 1909 and to accept the atomic theory of matter.

The ether theorists too adhered to their models of the continuum, for, like the atomic theory, the ether models were felt to be useful. They were considered to be useful not so much for the reason that Mach had urged, namely, the 'remembering' of observed phenomena, but because they led to the discovery of new phenomena, as in the case of Maxwell's ether which gave equations leading to the prediction of radio waves. Heinrich Hertz, who had discovered the waves predicted by Maxwell, gave an extensive defence of the use of theoretical models to explain natural processes in 1894.

'It is the first, and in a way the most important task of science to enable us to predict future experience, so that we may direct our present activities accordingly,' wrote Hertz. 'Our procedure in deriving the future from the past, and thus in achieving the desired foresight, is always this:—We set up subjective pictures or symbols of the external objects, and of such a type that their intellectually necessary consequences are invariably symbols again of the necessary consequences in nature of the object depicted. . . . Once we have succeeded in deriving symbols of the desired kind from the totality of past experience, we can develop from them in a short time, as from models, consequences that would appear in the natural world only after a long time, or as a result of our own manipulations.'

Mach had few followers amongst the important scientists contemporary with him, but one aspect of his philosophy has found favour in some quarters during the present century, namely the rejection of the use of theoretical mechanical models to explain natural processes. In theoretical physics mathematical models have come to be used instead of mechanical models, a trend which may be dated perhaps to the 1860's when Maxwell abandoned his model of the ether and confined himself to the study of the equations it had given him. The atomic physicists in particular have rejected mechanical models proposed for the structure of the atom.

'The atom of modern physics can be symbolized only through a partial differential equation in an abstract space of many dimensions,' wrote Heisenberg in 1945. 'All its qualities are inferential; no material properties can be directly attributed to it. That is to say, any picture of the atom that our imagination is able to invent is for that very reason defective. An understanding of the atomic world in that primary sensuous fashion . . . is impossible.'

Chapter 40

Science and Engineering

THE TECHNICAL innovations introduced into engineering and industry generally up to about 1850 were not greatly dependent upon the content of the science then known. On the other hand, science gained considerably from the investigation of engineering problems in certain cases, such as that of thermodynamics which developed in part from the study of the steam engine. After 1850, the application of science to the development of technology became a more and more important factor in the advance of industry, and during the present century most of the outstanding technical discoveries have stemmed in the main from scientific researches. Whilst the content of scientific knowledge did not have much influence upon the development of industry up to 1850, the method of science did. We have seen that the engineers of the eighteenth century, notably Smeaton and Watt, experimented with small-scale models of machines in order to improve the large-scale versions, these men achieving considerable success in the development of the steam engine. During the first half of the nineteenth century, the French engineers, Carnot and Clapeyron, studied the scientific principles upon which the steam engine worked, developing the science of thermodynamics in the course of so doing, whilst the British engineers, Bramah, Maudsley, Clement and Whitworth, devoted themselves to the technical improvement of machines in general, continuing the earlier British tradition of experimental engineering.

The work of the British engineers of the early nineteenth century brought about a change-over from the craft production of individual machines to the industrial mass production of standardized machines. The development of mass production in industry required the manufacture of accurate and interchangeable standard machine components, which focused attention on the problem of precision engineering. The improvement of engines and machine tools, i.e. machines to make machines, was also dependent upon greater technical accuracy in engineering. The Newcomen steam engine of the eighteenth century was built at a level of workmanship little above that of medieval times. In the 1760's Smeaton noted that in one of his engines there was a gap of about half an inch between a cylinder of twenty-eight-inch bore and its piston. Such defects in the Newcomen engine were remedied to some degree by covering the top of the piston with a layer of water. This practice made the engine workable, but lowered its efficiency, as the water cooled down the cylinder, resulting in a loss of steam. James Watt's improvements to the steam engine required that the cylinder should be permanently hot, so that a water seal to the piston could not be used. Accordingly, Watt's invention was held up until a method of shaping cylinders accurately became available, a method which appeared with John Wilkinson's precision cannon borer, patented in 1774.

Wilkinson's cannon borer made possible the commercial development of Watt's improved steam engine. Its original purpose, the fashioning of cannon, exemplifies the other source from which precision engineering arose, namely, the need for the mass production of standard commodities. Such a need became apparent first in the military sphere where large numbers of standard firearms and the like were required. The mass production of muskets from identical, interchangeable parts was started in France towards the end of the eighteenth century. Jefferson, the third president of the United States, recorded that in 1785 he visited the manufacturer, Le Blanc, and assembled several musket locks himself from randomly selected parts. In Britain mass production, and the precision engineering which it entailed, had more of a civil orientation, though the military stimulus was there. The inventor and engineer, Joseph Bramah, 1748-1814, who developed the tumbler lock in 1784 and the hydraulic press in 1795, faced the problems of mass production when he endeavoured to manufacture the tumbler lock in quantity. At first he employed numerous workmen to make the lock components by hand, using the traditional hand tools the hammer, chisel, file, saw, and so on. Later Bramah and his assistant, Henry Maudsley, 1771-1831, introduced mechanical aids to the use of hand tools, which enhanced both the rate and the accuracy of the manufacture of the lock components. Such a development led Maudsley to consider later the possibility of building generalized machine tools for the manufacture of varied types of standardized machine components.

In general the individual parts of a machine consist of metal formed into particular geometrical shapes, or combinations of such shapes, circles, cylinders, rectangles, and so on. A truly accurate shaft would be a perfect cylinder, and the ideal screw would be a perfect spiral impressed on such a cylinder. Thus the problem of mass producing standard machine components resolved itself into the construction of instruments which could confer truly cylindrical or plane surfaces on metal, and cut cylindrical or rectangular apertures in metal work. Maudsley solved the problem of making accurate cylinders and screws by transforming the lathe into a precision instrument between 1794 and 1810. Previously the lathe, like most other machines, had been made largely of wood, only the essential moving parts being constructed of metal. The piece of material to be machined was rotated by a treadle and worked with a cutting tool held in the hand. Maudsley made his lathe entirely out of iron, which was much less susceptible than wood to the distortions that ruined the centring and alignment of the work. Moreover, he introduced the slide rest to maintain the cutting tool at a constant distance from the central axis of the lathe, and in later versions he coupled the slide rest to the rotary motion of the machine, so that it moved linearly parallel to the central axis. Thus any material rotated in the lathe was automatically turned to an accurate cylinder, the dimensions of which were decided by the initial setting of the instrument, and once set the lathe turned out any number of such cylinders, all of the same size. With the coupling device that drove the slide rest parallel to the central axis of the lathe, spiral grooves could be cut on those cylinders, mass producing standard screws. The military connection of such developments is illustrated by the fact that the first major order Maudsley received was from the Admiralty in 1800, who wanted machinery to mass produce standard blocks for ship's tackling.

Besides cylinders and screws, plane work, truly flat surfaces, were needed. Hitherto planar surfaces had been produced by chiselling the surface of a casting or forging until it was roughly flat, and then grinding this surface against another that was judged to be flat. Such a method could lead to gross inaccuracies, as the two surfaces were not necessarily planar even if they touched at all points. The problem was partially solved by John Clement who had worked with both Bramah and Maudsley. In 1825 he invented a metal planing machine, which moved a material to be smoothed in straight lines, so that a fixed cutting tool made parallel planing cuts upon it. However, a material with a truly flat surface was required to standardize the machine and its products. Such standard flat surfaces were produced by Joseph Whitworth, 1803-87, who had worked with both Clement and Maudsley. He saw that two surfaces which touched at all points were not necessarily planar, but that three surfaces which matched, two at a time, must be truly flat. Similarly three bars must be truly rectangular in cross section if they matched perfectly in pairs when resting on a plane surface. With these true cylinders and plane surfaces Whitworth went on to develop in the years 1830-50 his standard screw gauge, measuring instruments sensitive to a millionth of an inch, and the precision lathes, planing, drilling, slotting, and shaping machines that brought him world fame at the Great Exhibition of 1851.

These machine tools speeded up and standardized the output of looms, spinning machinery, steam engines, and other pieces of capital equipment, which themselves could now work at higher speeds owing to the greater accuracy of their component parts and their construction. The Newcomen steam engine of the eighteenth century performed twenty strokes a minute at the most, whilst the engines of the second half of the nineteenth century could make two hundred and fifty strokes a minute and more. Precision engineering and the speeding up of machinery brought a new material into prominence, namely steel. Cast iron was too hard and brittle, whilst wrought iron was too soft for the construction of rapidly moving machine parts; only steel had the requisite strength and toughness. Cast iron with its high carbon content had long been available in quantity directly from the blast furnaces, and wrought iron had been produced on a large scale by burning out nearly all the carbon from cast iron in the reverberatory furnace, invented by Henry Cort in 1784. But steel containing a small amount of carbon was not produced in quantity until 1856, when simultaneously Bessemer invented his converter, and Siemens introduced the open hearth process.

Steel and machine tools led to a new phase in the development of engineering during the second half of the nineteenth century, characterized by the appearance of mass produced and standardized machines which were constructed of accurately turned out steel parts capable of operating at high speeds. At the same time the development of the science of thermodynamics offered a theoretical basis for the improvement of the steam engine and the development of other heat engines. The thermodynamics of the steam engine were developed notably by William Rankine, 1820-72, professor of engineering at Glasgow in his *Manual of the Steam Engine and other Prime Motors*, published in 1859, and also by Zeuner in Germany and Hirn in France. These men popularized thermodynamics amongst engineers, though they could not do a great deal to improve the steam engine. Thermodynamics led to a few developments in this field, as the steam engine was far from being a perfect heat engine, and most of the improvements suggested had already been discovered empirically. Theory indicated that high steam pressures and large expansions would make the engine more efficient, but Richard Trevithick had already developed the high pressure engine in 1802, and Jonathan Hornblower the large expansion engine even earlier.

However, the theory of thermodynamics found applications elsewhere. The science covered the theory of all heat engines, and described their behaviour both when directly acting and when reversed. Kelvin pointed out in the 1850's that if mechanical energy were applied to a heat engine, thus making it work backwards, it would pump heat from a low to a high temperature, acting as a refrigerator at the lower temperature and a warming machine at the higher temperature. Thus refrigerators were an application of the science of thermodynamics, the important modern types being based on the ammonia compression machine developed by Carl Linde of Munich in 1873. The more important applications of thermodynamics were, however, connected with the development of directly acting heat engines, those generating mechanical energy from heat, notably the internal combustion engines and the steam turbine.

The first and somewhat specialized internal combustion engine was the traditional gunpowder-fired cannon in which mechanical energy was developed from heat produced inside a cylinder, instead of outside as in the steam engine. Christian Huygens with his assistant, Denis Papin, attempted to make an internal combustion engine using gunpowder as fuel in the 1680's, though they did not get very far with the project. It was not in fact until the coal gas industry had developed that a suitable fuel became available, whilst the development of machine tools, steel production, and the science of thermodynamics made the engine itself possible. In 1862 Beau de Rochas, continuing the analytical tradition of the earlier French engineers, published a pamphlet in which he set out on thermodynamical grounds a theoretical cycle of operations that would give an efficient internal combustion engine. This was the famous four stroke cycle which was first used in the gas engine patented by the German, Otto, in 1876. Daimler brought out his petrol engine in 1883, and Priestman had developed the heavy oil engine by 1885, though the problems which the latter entailed were only definitively solved by Rudolph Diesel ten years later.

The mechanical principle of the steam turbine similarly had long been known, Hero of Alexandria having constructed a toy based upon that principle in antiquity. Towards the end of the eighteenth century Boulton had feared that a projected steam turbine might compete with the steam engine trade, but his partner, James Watt, quietened his fears with the observation that, 'Without God make it possible for things to move at a thousand feet a second it cannot do much harm.' However, with steel and precision engineering Laval in France had produced by 1889 a turbine in which the periphery of the rotor moved at more than 1,500 feet a second. The velocity achieved by steam expanding from a boiler into a vacuum was found to be of the order of 4,000 feet per second, and to be efficient, the rotor of turbine had to move at about half that speed. Such rotor velocities were still rather dangerous, and moreover they were inconvenient, finding few and limited applications. Laval allowed the steam in his turbine to expand in one step through a single rotor, thus generating these large and undesirable speeds. In England Sir Charles Parsons, 1854-1931, developed a turbine, patented in 1884, in which the steam was allowed to expand in a series of separate stages through several rotors, which moved with more manageable velocities. The shaft speeds of the Laval turbines ranged from 10,000 to 30,000 revolutions per minute, whereas the shaft speeds of the Parsons turbines could be made much lower, ranging from 750 to 18,000 revolutions a minute.

The science of thermodynamics entered much more intimately into the design of steam turbines than into that of steam engines as they approximate much more closely to perfect heat engines. The efficiency of the steam engine is inherently limited by the fact that it is a reciprocating engine. A fresh charge of steam enters a cylinder which has been cooled by the expansion of the previous charge of steam, and thus some of its heat is inevitably lost in warming up the cylinder again. In the turbine, on the other hand, the steam expands continuously from one stage to another, cooling down as it does. Each stage assumes its own temperature, and is maintained at that temperature by the passage of the steam. Hence there are no inevitable losses of steam owing to periodical changes in temperature as in the steam engine, and so in the case of turbines thermodynamical theory may be more closely applied.

One of the more important of the uses which were found for the steam turbine was the driving of the electricity generators that were being developed about the same time, as the shaft speeds of the turbine and the dynamo could be conveniently adjusted to the same value. The dynamo more than the turbine was a product of applied science; in fact most of the equipment of the electrical industry has been dependent at some stage upon the corresponding science. The invention of the voltaic cell led the development of electroplating, patents for which were taken out in 1839 by Karl Jacobi at Konigsberg and Werner Siemens at Berlin. The original electric cell invented by Volta in 1799 was somewhat unreliable, and the early important applications of electricity followed from the development of a cell giving a steady current, which was brought out by John Daniel of King's College, London, in 1836. His colleague, Charles Wheatstone, made a practicable electric telegraph in the following year, using the Daniel cell as a source of electricity, and the electromagnet, invented by Sturgeon in 1825, as a recording device.

Land-line telegraphy raised few novel problems, but when the first submarine telegraph cable was laid between Dover and Calais in 1850, it was found that signals were distorted and came through at a comparatively slow rate. The problem was studied by Kelvin at Glasgow, and he pointed out in 1855 that the essential difference between the conditions of land-line, and submarine cable telegraphy was that sea water acted as a conductor, whereas air was effectively an insulator. Hence a submarine cable, surrounded by an insulated coating, formed an electric condenser with the sea water, so that the cable charged up comparatively slowly at one end and discharged equally slowly at the other when a signal was transmitted. Kelvin indicated that the retardation of signals could be minimized, if a small current were used in a cable of high conductivity and large cross section, protected by a thick layer of insulator. The use of small signal currents required sensitive recording instruments for their detection, and for this purpose Kelvin designed the mirror galvanometer in 1858, and the automatic siphon recorder in 1867. The first Atlantic submarine cable laid in 1858, was ruined after only seven hundred messages had been sent through, as large signal currents were used, but when the second cable was laid in 1866 Kelvin's recommendations were adopted.

Further applications of electricity were developed notably in Germany and America, countries that by-passed to some degree the use of steam engine power and coal gas illumination, which had been characteristic of the early phase of the industrial revolution, and went on more directly than Britain to the use of electricity for illumination and the transmission of power. Moreover, in America, where the density of population was then small and the distances between neighbouring communities large, electrical means of communication were particularly important. The American telegraph was brought out in 1838, only a year after the English invention, by the portrait painter, Morse, who designed the code named after him for the purpose of transmitting signals. The automatic message recorder, the ticker-tape machine, was invented in 1854 by David Hughes, a Kentucky music teacher, whilst the telephone, discovered by Bell and Edison in 1876, was an entirely American invention.

The dynamo was developed in Germany to provide power for the electroplating industry, and in America to provide electric illumination. The small amounts of electricity given by the Daniel cell sufficed for the purposes of telegraphic communication, but not for those of the electroplating industry where large amounts of current were consumed. Faraday had shown in 1831 that electricity could be generated by moving a coil of wire in a magnetic field, and accordingly several electromagnetic machines based on this principle were developed between 1840 and 1865 to provide electricity, primarily for electroplating. Such machines consisted of a coil of insulated wire, which could be rotated mechanically in the field of a permanent steel magnet. They were not very effective, as the best steel magnets gave only a small magnetic field, but in 1866 Werner Siemens of Berlin replaced the steel magnet by a powerful electromagnet, which was energized with some of the electricity produced by the machine itself. All subsequent dynamos were based on Siemens' model, employing electromagnets fed by some of the current they produced, and being more efficient than the earlier electromagnetic machines, they opened up the way for further developments in the field of electrical engineering.

Humphry Davy had discovered that electricity passing between two carbon rods produced a brilliant light, and from the 1850's large candlepower illumination was obtained for use in lighthouses, theatres, and elsewhere, by means of carbon arc-lamps, working first from electromagnetic machines and later from dynamos. Davy had also found that a less intense light was given out when a current was passed through a thin platinum wire, though the wire soon burned away in the air. In 1879 Joseph Swan, 1828-1914, in England and Thomas Edison, 1847-1931, in American simultaneously and independently evolved a lamp based upon this principle, consisting of a carbon filament enclosed in an evacuated glass bulb, that would burn for many hours. Edison made more use of the discovery than Swan, developing the additional equipment required for the widespread adoption of electric illumination. At his Menlo

Park laboratory, near New York, Edison designed a constant voltage dynamo to ensure that the light given out by one lamp would not vary when other lamps in the circuit were switched on or off, and he originated the three wire system for the economical distribution of current. In 1882 Edison set up at New York the first generating station to provide the public with electricity, and he manufactured the lamps required for electric illumination.

In 1883 Edison noted that some of his electric light bulbs gradually became darkened with use, indicating, he thought, that particles of some kind had been emitted from the filament. He sealed a metal plate into such a bulb and found that it became negatively charged when the bulb was in operation, as a positive potential applied to the plate caused a current to flow, whilst a negative potential had no effect. This phenomenon, known as the Edison effect, led to the development of the electronic valve, notably by Fleming in 1904 and Lee de Forest in 1906. In the present century the electronic valve has allowed the electromagnetic waves, predicted by Maxwell and discovered by Hertz, to be utilized, first in radio broadcasting and television, and more recently in the radiolocation of distant objects. Finally the valve has led to the development of complex electronic devices, notably the calculating machines, which possess some attributes of the human mind, such as a memory, an elementary judgement, and the power of computation. It has been suggested that the general adoption of these machines in industry, that is, the process of 'automation,' replacing human beings at tasks requiring the less complex acts of judgement will bring about a second industrial revolution, releasing men from the more mechanical and repetitive exercises of the mind.

Chapter 41

The Applications of Chemistry and Microbiology

THE SCIENCE of chemistry has been applied in the main to the development of a chemical industry and, with microbiology, to the improvement of the older practices of agriculture and medicine. Advances in all these fields were at first largely empirical, and they have remained so, particularly in agriculture and medicine, to a much greater degree than in the case of mechanical and electrical engineering. The technical innovations of the agrarian revolution, notably the new agricultural machinery introduced by Jethro Tull, 1674-1741, and the four-course system of crop rotation practised by Lord Townshend, 1674-1738, and the stock-breeding improvements of Robert Bakewell, 1725-95, were not at all dependent upon the science then known. Nor too were the public health measures based upon the connection between dirt and epidemic disease, established by the *Commission for Enquiring into the State of Large Towns,* which published its findings in 1844. In the same way the early development of the chemical industry was mainly a process of trial and error invention.

Up to the eighteenth century, the main specifically chemical trades were those of the apothecary, who prepared compounds on a small scale for use in medicine, and of the alum makers, who prepared alum on a comparatively large scale for the treatment and colouring of skins, paper, and textiles. The traditional connection between the chemical trade and the textile industry was developed further during the industrial revolution, when the large scale manufacture of chemicals began. The new spinning and weaving machines introduced during the eighteenth century by such men as Kay, Hargreaves, Crompton, Arkwright, and others, gave such an enormously increased output of textiles that the chemical problems of bleaching, and later of dyeing cloth, became considerable. Traditionally textiles had been bleached by dipping them alternately in acid solutions of sour milk and alkaline solutions of plant ashes, and exposing them to the sun on 'bleach fields', a process that occupied all of the summer months in a given year. A shortage was experienced first of all in the supply of natural acid—sour milk—and accordingly attempts were made to bleach with the manufactured acids, of which sulphuric acid was the most accessible. Sulphuric acid had long been prepared in small amounts by the apothecaries, and it was a London apothecary, Joshua Ward, who set up in 1736 the first factory to manufacture acid commercially on a large scale, by burning sulphur with a little saltpetre in large glass globes containing a little water. A Birmingham physician, John Roebuck, replaced the fragile and expensive glass globes with lead chambers in 1746, an innovation which, with that of Ward, brought down the price of sulphuric acid from £2 to 6d. a pound.

The next shortage, that of natural alkali, was not felt for some time in Britain, as soda was prepared in quantity by burning the seaweed which abounded round the coasts, particularly in the north. In France the shortage was more acute, and in 1775 the Paris Academy of Sciences offered a prize of 12,000 francs for a method of making soda from common salt. Such a method was discovered in 1789 by Nicolas Leblanc, 1742-1806, who was physician to the Duke of Orleans. From common salt and sulphuric acid, he obtained sodium sulphate, which he heated with charcoal and limestone, obtaining soda and calcium sulphide. Another French chemist, Berthollet, who was then director of the national dyeing industry, found that the gas chlorine, which had been discovered by Scheele in 1774, would quickly bleach cotton fabrics. He communicated the discovery to James Watt about 1786, and Watt passed it on to his father-in-law, who was connected with the textile trade of Glasgow. There the method was tried out on a large scale, and it was found that bleaching with chlorine took a matter of hours, where formerly it had taken many weeks. The use of the poisonous chlorine gas was at first rather dangerous, but in 1799 John Tennant of Glasgow combined the gas with lime to give a safer and much more convenient agent, known as bleaching powder.

During the French revolution, the government of France asked their chemists to investigate, and improve wherever possible, several of the existing chemical trades. Clement and

Desormes studied the reactions occurring in the manufacture of sulphuric acid, and they found in 1806 that the saltpetre added to the sulphur burning in the lead chambers greatly facilitated the process by forming a gas, nitric oxide. This gas combined with the oxygen of the air to give nitrogen dioxide, which handed on its extra oxygen to the sulphur dioxide by the combustion of sulphur, producing sulphur trioxide which with water formed sulphuric acid. The investigation of Clement and Desormes led to economies in the manufacture of sulphuric acid by reducing the amount of saltpetre consumed. Instead of adding saltpetre to the burning sulphur, it was treated with acid separately to generate directly the nitric oxide gas. Later in 1827, Gay-Lussac showed that the nitric oxide could be recovered from the waste gases of the lead chamber process by absorption in strong sulphuric acid. However, the work of Gay-Lussac did not find an immediate practical application, for it was not until 1860 that a method was found of regenerating the nitric oxide from the sulphuric acid solution. In that year an English acid manufacturer, Glover, passed the hot gases from the burning sulphur, or the pyrites that were then used, through the acid containing the nitric oxide, thus concentrating the acid and removing the nitric oxide for further use in lead chambers. Similarly the French engineer, Fresnel, in 1810 worked out a method of making soda using only lime stone and common salt as the starting materials, with ammonia as an intermediate, but his discovery was not utilized owing to practical difficulties until 1865, when the Solvay brothers in Belgium set up a soda works using the method.

The French scientists also studied the chemistry of plant growth, though again their work was not applied immediately. In 1804 de Saussure, 1767-1845, showed that plants grown in closed vessels derived their entire carbon content from the carbon dioxide of the gas mixture in which they were enclosed, thus demolishing the old theory that plants derived their substance from the so-called humus of the soil. He also found that plants grown in pure water gave the same amount of inorganic ash when burned as their seeds, which indicated that the inorganic material of plants was neither created nor destroyed. In 1817 Pelletier and Caventou isolated chlorophyll, the green colouring matter of plants, and in 1837 Dutrochet showed that carbon dioxide was absorbed only by those parts of a plant which contained chlorophyll, and only when they were exposed to light. In this way the carbon dioxide cycle of nature was discovered: plants build up their substance from the carbon dioxide of the air in the presence of sunlight, and animals, in consuming plants, regenerate the carbon dioxide. In 1841 Boussingault, 1802-87, showed that the amount of carbon, hydrogen, oxygen, and nitrogen, present in various crops was invariably larger than the amount added to the crops in the form of manure, whilst the amount of inorganic salt was invariably less. He found further that the good crop rotations owed their superiority to certain plants, such as clover and peas, which contained an amount of nitrogen greatly in excess of that applied in the form of manure.

The results of the French workers were applied to agriculture notably by the German chemist, Liebig, who had been trained at the Ecole Polytechnique. Liebig argued that, since plants could not create mineral salts, as de Saussure had shown, they must obtain their inorganic constituents from the soil, and what they took from the soil must be restored if fertility were to be maintained. He analysed chemically the mineral content of the ashes of plants, and made up artificial chemical fertilizers identical in composition with plant ashes, consisting mainly of potassium and phosphate salts. His patent manure was not a success, however, as it contained no nitrogen compounds, Liebig believing that all plants obtained their nitrogen from the air. Nevertheless he stimulated a great deal of interest in the subject of agricultural chemistry, and his lecture on *Chemistry and its Applications to Agriculture and Physiology* was very well received at the Liverpool meeting of the British Association in 1837.

Liebig visited England again in 1842 when he met the prime minister, Peel, together with several large landowners, and the foundation of a chemistry college was proposed. Sir James Clark, Queen Victoria's physician, collected subscriptions for the foundation, and in 1845 the Royal College of Chemistry was set up under the presidency of the Prince Consort. Liebig was asked to nominate a professor for the institution, and he sent over one of the more able of his pupils, August von Hofmann. From the start Hofmann's work was orientated towards the industrial rather than the agricultural side of chemistry, for he investigated the chemistry of the coal-gas industry, first the inorganic side, the gases produced, and then the organic side, the constituents of coal tar. Although some important chemical researches were carried out at the Royal College of Chemistry, the interest of the landlords in the institution rapidly waned, as it turned out little that was useful to them, and the College was only saved from dissolution by being merged with the Royal School of Mines in 1853.

One of the landlords, Sir John Lawes, 1814-1900, carried out researches in the field of agricultural chemistry on his own estate at Rothamsted, together with Joseph Gilbert who had studied under Liebig. Together they investigated the use of artificial fertilizers in agriculture, discovering by 1855 most of the basic facts of agricultural chemistry. In opposition to the views of Liebig, they showed that plants in general do not require the same proportion of mineral salts as is found in their ash for optimum growth, and that most plants require fertilizers containing nitrogen compounds, such as ammonium salts or nitrates, only leguminous plants, like peas and clover, thriving without them. They found further that if ground was left fallow the nitrogen content of the soil gradually increased, and that the fertility of the soil was not impaired if continuously cropped when artificial fertilizers alone were used. The work of Gilbert and Lawes drew attention to the singular position of nitrogen in the economy of nature, some plants requiring nitrogen compounds, whilst others, and the soil itself, appeared to prepare their own. These facts were elucidated by the development of microbiology, which brought to light the hitherto unknown stages in the nitrogen cycle of nature.

The founder of microbiology was Louis Pasteur, 1822-95, who was a professor of chemistry at Strasbourg and then at the Sorbonne. Pasteur first of all studied the brewing industry, investigating the long known fact that the fermentation of two samples of the same wash would sometimes result in different products. With the microscope he demonstrated the presence of small yeast organisms in fermenting liquors, and he discovered that different breeds of yeast gave different products. In 1863 he found that the souring of wine was caused by a micro-organism, and he showed that the micro-organism could

518 / A History of the Sciences

be killed by warming the wine to 55° C. In the following year he was asked by the French Ministry of Agriculture to investigate the diseases of silkworms. Within a few months he had isolated the micro-organisms responsible for two of the diseases of silkworms, and had shown how the disease-free eggs, worms, and moths could be recognized, so that they could be separated and used for breeding purposes. A decade later he studied cattle anthrax and chicken cholera, and finally in the 1880's he investigated some of the diseases which affected human beings.

The medical implications of Pasteur's work were appreciated in England by the Quaker-surgeon, Lord Lister, 1827-1912, some time before Pasteur himself studied the problem of human disease. Chemistry had already provided the surgeons with anaesthetics, which reduced the suffering involved in surgical operations, but not the high post-operative mortality. Humphry Davy in 1799 had discovered that nitrous oxide, or laughing gas as it was called, induced intoxication and then insensibility. He suggested that nitrous oxide might be used in surgical operations to render the patient unconscious, a suggestion which was first adopted in 1844 when Horace Wells in America utilized the anaesthetic properties of the gas for dental surgery. A friend of Wells, William Morton, found that ether was a better anaesthetic, and he showed in 1846 that it could be used in major operations. In the following year Sir James Simpson at Edinburgh discovered that chloroform was a superior anaesthetic in certain cases, notably childbirth.

However, the number of surgical patients that recovered remained small, owing to the fact that infections were not infrequently contracted during the course of the operation. Lister's statistics of 1864 show that 45 per cent of his patients died after the operation, and other surgeons of the time could record success only in one case out of every five. Pasteur's work on fermentation and putrefaction suggested to Lister that septic operation wounds were a kind of putrefaction caused by micro-organisms. He looked for chemical methods of killing micro-organisms and, after trying out various compounds, he found that phenol, a substance obtained from coal tar, acted as a good antiseptic. Lister sprayed his operating theatre and the operation wounds of his patients with a solution of phenol in water, and he discovered that blood poisoning after operations was thereby greatly reduced. His first operation conducted with the new antiseptic technique was performed in 1865, and by 1868 he had reduced the surgical death rate from 45 to 15 per cent.

The medical, as opposed to the surgical applications of microbiology were made by Robert Koch, 1843-1910, in Germany, and by Pasteur himself in France. In 1876 Koch discovered that the micro-organisms responsible for cattle anthrax could be grown outside of the animal body in a culture medium consisting of meat broth jelly. By these means he discovered the bacillus of tuberculosis in 1882 and isolated the cholera microorganism in the following year. Pasteur repeated and extended the work of Koch. He discovered that some bacteria became inactive when cultivated outside of the animal body, a culture of chicken cholera left for some time producing no disease when injected into chickens. Furthermore the same chickens remained healthy when later injected with the virulent cholera bacteria, indicating that the inactive organisms had immunized the animals against the normal active strains. In 1881 Pasteur prepared an inactive strain of anthrax which protected cattle against the active forms of the disease, establishing another case of the principle of preventive inoculation.

A specific example of this general principle had been known long before the germ theory of disease appeared. The deliberate infection of children with mild forms of small-pox to protect them from the deadly forms had been practised since the 1720's when Lady Mary Whortley Montague had brought the method from the Middle East. Later, in 1798 Edward Jenner, a country doctor of Gloucestershire, showed that the much milder disease, cow-pox, immunized human beings against small-pox, a discovery he derived from the observation that dairymaids rarely caught small-pox. Now, in the 1880's, the practice of inoculation was generalized, finding a rational basis in the germ theory of disease. It was suggested that bacteria produce chemical poisons, or toxins, which are primarily responsible for the symptoms of the disease, whilst the body defences produce antitoxins to counteract the effects of the bacteria and their toxins. Thus it appeared that dead bacteria injected into the body should produce the mild symptoms of their disease, and stimulate the production of antitoxins, which would counteract future infections. Such was found to be the case, and it was discovered further that the antitoxin produced by one animal body was effective in counteracting the bacteria to which it corresponded in the body of another animal.

In agriculture the discovery of micro-organisms helped to clear up the problem of the nitrogen cycle in nature. Warrington, one of Lawes' assistants at Rothamsted, showed in 1878 that micro-organisms in the soil converted nitrogenous fertilizers consisting of ammonium compounds first into nitrites, and then into nitrates. He found that the micro-organisms were killed with chloroform, and that in these circumstances plants would not grow, even if they were provided with plenty of nitrogen in the form of ammonium compounds, indicating that plants could only take up nitrogen in the form of nitrates. In 1885 the French chemist, Berthelot, discovered other types of micro-organisms which could utilize the nitrogen of the atmosphere directly, converting it into ammonia. Some of these micro-organisms lived freely in the soil, but others were found only in the root nodules of leguminous plants. If the latter kind of micro-organism were killed, the plant with which it was normally associated would not form nodules on its roots, and required nitrogenous fertilizers. With these micro-organisms leguminous plants were independent of fertilizer nitrogen, atmospheric nitrogen being converted into ammonia by the root nodule organisms, and thence into nitrates by other soil micro-organisms. Some soils, such as the virgin soils of Canada and America, were lacking in nitrifying micro-organisms of certain kinds, and there a rotation of crops depending upon leguminous plants proved to be a failure. However, by the end of the century cultures of the nitrogen fixing organisms associated with clover, peas, and other leguminous plants, were generally available, and when inoculated into barren soils permitted the practice of crop rotation.

The applications of agricultural chemistry stimulated the growth of an artificial fertilizer industry. Guano, the dehydrated droppings and carcases of sea birds, was imported from Peru for agricultural purposes as early as 1839. Sir John Lawes set up a factory at Deptford in 1843 to manufacture superphosphate fertilizer, by treating insoluble phosphates with sulphuric acid to render them more soluble. First he used animal bones as a source of phosphate, and then from 1847 he exploited the deposits of mineral phosphate found in Suffolk, Bedfordshire, and elsewhere. Ammonia had been removed from coal gas with sulphuric acid from about 1815, as it was an undesirable impurity, and the resulting ammonium sulphate was used extensively as an artificial fertilizer after 1850. Completing the first stage in the development of chemical manures, the nitrate deposits in Chile, and the potassium sulphate deposits at Strassfurt in Germany, were first exploited in 1852, the crude salts being used directly as fertilizers.

The Royal College of Chemistry at London, which had been founded by the large landowners with the hope that chemical researches would lead to the improvement of their estates, produced little that was of importance to agricultural chemistry, but work carried out there led to the foundation of the fine chemical industry. The professor at the College, Hofmann, like his master, Liebig, was much concerned with the applications of chemistry, particularly those in the medical field, and he hoped that naturally occurring drugs might be manufactured artificially. Hofmann suggested that quinine might be made from coal-tar products, and in 1856 one of his pupils, William Perkin, 1838-1907, endeavoured to make the drug by oxidizing some aniline derivatives with which he happened to be working at the time. He obtained no quinine, but a mauve colouring matter was formed which proved to be an excellent dye. The organic chemists had not yet evolved the theory of molecular structure, and the nature of organic compounds and their reactions was then unknown. Thus syntheses which would be ambitious today were not uncommonly attempted, as in the case of Perkin's venture to make quinine, which was synthesized only in 1945.

For Perkin the industrial significance of his discovery was important, and though only a youth of eighteen he set up a factory to make the colouring matter in quantity, founding a fine chemical industry. In France Perkin's work was extended by Girard and de Laire, who treated aniline derivatives with different ozidizing agents and produced another dye, magenta. They then treated magenta with more aniline, and obtained a whole range of dyes known as the aniline blues. Hofmann in

London investigated further the compounds prepared by Perkin and the French chemists, producing another range of dyes termed the Hofmann violets in 1863. Two years later Hofmann left the Royal College of Chemistry to take up the chair of organic chemistry at Berlin, and at the same time the German chemist, Caro, who had been working at a Manchester chemical factory, returned to Germany as director of a newly founded and large chemical works, the Badische Soda und Anilin Fabrik. From then on the Germans assumed a more and more prominent place in chemical science and chemical industry, particularly in the fine chemical trade. Hofmann helped to design the large new laboratories of Bonn and Berlin universities, which were completed by 1869, and from them came the chemists who gave Germany her scientific and industrial strength.

Two of the more important natural dyes used in the nineteenth century were alizarin, obtained from the madder plant, and indigo blue, derived from the indigo plant. By the end of the century the Germans had synthesized both, and were producing them in quantity. The important figure on the scientific side was Adolf von Baeyer, 1835-1917, who became assistant professor in chemistry at Berlin in 1860. He and his pupils, Graebe and Liebermann, in 1866 showed alizarin to be a derivative of anthracene, one of the common constituents of coal-tar, and shortly afterwards they synthesized alizarin in the laboratory. Their method was impracticable for the production of alizarin on a large scale, but by 1869 Graebe and Liebermann, with Caro of the Badische Soda und Anilin Fabrik, had evolved another method that was commercially feasible. In the same year Perkin in England discovered two different methods by which alizarin could be made, but the Germans had the industrial strength and by 1873, the year that Perkin retired, the Badische Soda und Anilin Fabrik were turning out one thousand tons of alizarin a year, as opposed to Perkin's four hundred and thirty-five tons. Finally Baeyer, who had now succeeded Liebig to the chair of chemistry at Munich, synthesized indigo in 1878, though again there were technical difficulties, and the dye was not manufactured on a large scale until 1897. By that time the Germans were well ahead, the six largest German chemical firms taking out nine hundred and forty-eight dye patents in the period 1886-1900, whilst the six largest British firms took out only eighty-six.

The Germans were supreme only in the fine chemical industry where the development and application of organic chemistry had been essential from the start. The British chemical industrialists were slow to appreciate the importance of chemical research in the development of their trade, and thus they fell behind in the fine chemical industry, though they retained their standing in the field of heavy chemicals where continuous research did not become necessary until the present century. In 1909, for example, ninety per cent of the dyes used in Britain were manufactured in Germany, though British chemical exports, mainly heavy chemicals, were £644,000 in excess of chemical imports. The important practical innovations introduced into the heavy chemical industry during the nineteenth century were made in fact mainly by the British chemical manufacturers. As we have seen, the acid manufacturer, Glover, in 1860 made practicable the method suggested by Gay-Lussac for recovering the nitric oxide used in the lead chamber process of making sulphuric acid.

The Leblanc soda process, discovered in France, was adopted in Britain when the government abolished the duty on common salt in 1823, and thereafter it was considerably improved. The process gave two important by-products, hydrogen chloride and calcium sulphide, and it was with these materials that the improvements were concerned. William Gossage, an alkali manufacturer of Stoke Prior, invented a tower in 1835 for the absorption of the hydrogen chloride gas in water, which was extensively adopted after 1863 when the Alkali Act was passed forbidding the liberation of the gas into the atmosphere. A method of generating chlorine from the waste hydrogen chloride of the soda works was discovered by Henry Deacon, a manager of a glass factory at St. Helens, in 1868. The hydrogen chloride and air were passed over heated cupric chloride to give chlorine and steam, the chlorine then being used to manufacture bleaching powder. In the same year the chemist, Walter Weldon, improved the old method of making chlorine from manganese dioxide and hydrochloric acid, using lime and a stream of air to regenerate the manganese dioxide. Finally an alkali manufacturer of Oldbury, Alexander Chance, developed a method of recovering sulphur from the waste calcium sulphide of the soda works in 1887. He passed flue gases containing carbon dioxide through a suspension of the calcium sulphide in water, thus liberating hydrogen sulphide, which was passed with air over a heated metallic oxide to give sulphur.

Such developments made the Leblanc process for the manufacture of soda reasonably efficient. Meanwhile the alternative method of making soda, worked out by Fresnel in 1810, was made practicable by the Solvay brothers of Belgium in 1865. The Solvay process, as it was called, gave a purer and cheaper product than the Leblanc process, and it was adopted in Britain by Brunner and Mond in 1873. The Brunner and Mond works went rapidly ahead, and to compete with them the other alkali works of Britain formed the United Alkali Company in 1890. It is of interest to note that the important figures of the alkali industry, Brunner and Mond, were the first notable chemical industrialists in Britain to finance scientific research. Brunner gave donations to the University of Liverpool in the 1890's, and Mond presented the Davy-Faraday Laboratory to the Royal Institution in 1896. In Germany, where the chemical industry had been more large scale from the start, the industrialists had financed scientific research much earlier.

Towards the end of the nineteenth century the German chemists began to introduce new methods into the heavy chemical industry, applying in particular the new physical chemistry that indicated the optimum conditions under which a chemical reaction would proceed. An alternative to the lead chamber process of manufacturing sulphuric acid was evolved, termed the contact process, whereby sulphur dioxide and the oxygen of the atmosphere were combined directly by means of a catalyst, such as platinum. The contact process gave a much more concentrated acid than the lead chamber process, and it was developed notably from 1897 when a supply of the concentrated acid became necessary for the manufacture of dyes. An even more important problem facing the German chemists was the manufacture of nitrogen compounds for fertilizers and explosives, as Germany was largely dependent upon imported supplies of nitrates and ammonium compounds, which would be cut off in the event of hostilities. Fritz Haber studied physicochemically the direct combination between hydrogen and nitrogen to give ammonia, discovering that the reaction was favoured by high pressures and moderate temperatures. Simultaneously Ostwald investigated the conversion of ammonia into the oxides of nitrogen, and thence into nitric acid. By 1912 the investigations were completed, and they were applied on an industrial scale by the Badische Soda und Anilin Fabrik, providing Germany with ample fertilizers and explosives during the First World War.

The use of such physico-chemical methods to determine the optimum conditions under which chemical reactions proceed has become a characteristic feature of industrial practice during the present century. The chemical industry now has numerous ramifications. A turning point in the explosives industry was marked by the development of nitro-glycerine, dynamite, and the gelignites, by Nobel in Sweden from 1862. Artificial fibres date from 1883 when Joseph Swan produced nitro-cellulose filaments by extrusion, a process adopted commercially by the French chemist, Count Chardonnet. The first thermosetting plastic, Bakelite, was made by Leo Baekeland of Columbia University in 1907, whilst the first important thermoplastic substance, Celluloid, was discovered by Alexander Parkes of Birmingham in 1865. The Germans in their search for substitutes gradually evolved a serviceable synthetic rubber from the end of World War I, whilst Fischer and Tropsch made a substitute petrol from water-gas in 1925, and Bergius produced another motor fuel by hydrogenating coal in 1935. In the fine chemicals industry attention had turned from dyes to drugs and perfumes in the present century. William Perkin, who synthesized the first dye, was the first chemist to prepare a naturally occurring perfume, *Coumarin,* which he made in 1868 from coal-tar derivatives. The synthesis of drugs too was associated with the manufacture of dyes. It was found that some dyes were highly selective in their action, colouring wool but not cotton, and when applied to organic tissues, staining some parts but not others. The founder of chemotherapy, Ehrlich, 1854-1915, suggested that since organic dyes were absorbed by some organic cells but not by others, it should be possible to make toxic compounds which would be taken up by a parasitic microorganism but not by the host it had infected. Thus the microorganism would be killed and the host cured of the disease which it had caused. Ehrlich prepared and tested numerous compounds, achieving success with *Salvarsan* which specifically counteracted syphilis, yaws, and other spirochaetal infections. The chemists of the German dye industry later prepared *Pamaquin* (1926), and *Mepacrine* (1930), which were toxic to the Malaria parasite, and in 1935 produced a red dye, *Prontosil*, which was the first sulphonamide drug. A further line of chemical research, now of considerable medical importance, is the synthesis of naturally occurring biologically active compounds, such as the vitamins, hormones, and natural antibiotics produced by living organisms, such as *Penicillin*.

PART SIX

TWENTIETH-CENTURY SCIENCE:
NEW FIELDS AND NEW POWERS

case difficulties appeared, their calculations indicating that there

Chapter 42

Some Aspects of Modern Biology

FROM THE 1870's a number of technical advances were made in the field of experimental biology which allowed the processes occurring in the asexual reproduction of cells and in the union of sexual cells to be observed more closely. The achromatic microscope was further improved by the introduction of the high-power immersion lens and substage illumination, while the newly discovered aniline dyes, together with natural dyes and some inorganic salts, were found to stain selectively certain parts of the organic cell, particularly the nucleus. It was known from the 1840's that the organic cell reproduced asexually by fission, the nucleus dividing first. In the 1870's it was shown by Hertwig, 1849-1922, at Berlin, and Fol, 1845-92, at Geneva, working on animals, and Strasburger, 1844-1912, at Bonn, working on plants, that sexual reproduction involved the union of the nuclei of the male and female cells, from which Hertwig and Strasburger suggested in 1884 that the nucleus of the cell was the physical basis of heredity.

The new techniques showed that in the ordinary resting cell nucleus there was a fine network of material, termed the chromatin by Flemming, 1843-1915, at Kiel, in 1879, because it stained deeply with basic aniline dyes. As the time of cell division approached in asexual reproduction, it was found that the chromatin arranged itself into a number of separate filaments, termed the chromosomes, each of which consisted of two contiguous threads similar along their lengths. The chromosomes were generally paired, the number of chromosomes in the cells of a given animal or plant species being constant. When the chromosomes became distinguishable as separate filaments, it was discovered that the nucleus membrane disappeared and that two point bodies, the centrosomes, arranged themselves on either side of nucleus. From the centrosomes radiated lines, termed the asters, towards the chromosomes at the equator of the nucleus and the chromosomes then split, the halves of each chromosome moving in opposite directions along the aster lines to the centrosomes, where they formed the nuclei of the two daughter cells. The chromosomes then disappeared into the chromatin network, reappearing only at the next cell division. Although the chromosomes could not be seen in the chromatin, Rabl, 1853-1917, at Leipzig, suggested in 1895 that the chromosomes must retain their individuality within the network, a hypothesis that became important when it was accepted that the chromosomes were the carriers of heredity.

In the case of a union between two sexual cells it was found that the chromosomes behaved rather differently. Van Beneden, 1845-1910, at Liége, observed in 1887 that in the primary cell division leading to the formation of an egg, the chromosomes did not split into two along their lengths as in asexual cell division, but each chromosome pair separated, forming two cells each with only half of the usual number of chromosomes. Both cells then divided again by the normal asexual process but only one of the four products became a mature egg, the other three forming small bodies which usually degenerated. The same process was found to occur in the formation of spermatozoa, though in this case all four products of the primary and secondary divisions became mature sperms. The eggs and the sperms therefore had only half of the number of chromosomes generally found in the cells of their species, but in the union of the sexual cells the chromosome number of the species was restored, one half of the chromosomes coming from the mother and the other half from the father. In 1894 Strasburger showed that in some plants the cells with half the usual number of chromosomes formed a separate generation, a discovery which explained the alternation of generations found in non-flowering plants by Hofmeister in 1851.

Meanwhile the speculative biologists had developed theories of inheritance which postulated that the genetic materials of organisms should show the phenomena exhibited by the chromosomes during the formation of the sexual cells. Following Nageli, Weismann suggested that each organism possessed a heredity substance or germplasm, distinct from its body substance, which was made up of separate units, each determining a particular characteristic of the creature. In order to avoid the doubling up of the germplasm units at each sexual generation, Weismann postulated in 1887 that prior to sexual union the germplasm of both the male and the female divided into two, a half from each parent uniting to form the germplasm of the offspring. When the behaviour of the chromosomes during the formation of the egg and the sperm was elucidated, Weismann proceeded to identify the germplasm with the chromosomes; suggesting that the latter were divided into units along their lengths.

Another of Nageli's speculations, his idea that there was an inner force within the germplasm of organisms which gave rise to sudden and large mutations, was taken up by de Vries, 1848-1935, at Amsterdam in order to accommodate the history of organic evolution to the short estimates for the age of the earth made by the physicists, such as Kelvin. From about 1885 de Vries began to look for such mutative changes in organisms, and he found them in a wild colony of the American evening primrose. Some of the individual plants differed markedly from the normal type, and from them de Vries obtained further variants over a few generations. Bateson, 1861-1926, in England and Johannsen, 1857-1927, in Denmark, also looked for mutations. Johannsen bred self-fertilized beans, obtaining pure lines which always produced seeds of the same average weight, whether the heaviest or the lightest of a crop were sown. In one case however he found a mutation; the average weight of bean shifted spontaneously, and the change was preserved in the succeeding generations. De Vries and others now searched for past work on the subjects of inheritance and mutation, and they found it in the papers published by Mendel in 1866 and 1869.

Mendel, 1822-84, a monk of Brno, experimented upon the crossing of sweet peas in his monastery garden from about 1857 to 1868. Sweet peas normally fertilize themselves, but Mendel artificially cross-fertilized a tall and a short variety, obtaining seeds which gave only tall plants. These when self-fertilized gave tall and short plants in the ratio of three to one. The short plants so produced always bred true, but only one out of three of the tall plants did the same, the other two producing tall and short plants again in the ratio of three to one. Mendel interpreted his results to mean that each plant possessed two character-determining factors for height, one derived from each parent. The factor for tallness dominated over the recessive shortness factor, so that all of the plants in the first generation after crossing were tall. When this generation was self-fertilized, the factors could be arranged to give in the offspring two factors for tallness, or two for shortness, or a tall and a short, or again a short and a tall. The first two combinations would breed true, giving only tall or short plants respectively, while the second two combinations would give tall and short plants in the ratio of three to one. Mendel's results supported a particle theory of inheritance, and he sent them to Nageli who had put forward such a theory. But Nageli paid little attention to Mendel's findings because he thought them 'empirical rather than rational', and perhaps also because he was more interested in the evolutionary factors which changed the species than the heredity factors governing their stability. Mendel in fact denied that there was any variability in the plant characteristics which he had studied, and to that degree the evolutionary mood of contemporary biological thought was alien to him. By the end of the century the ideas of progress and evolution had lost something of their earlier force, and it was then that Mendel's results were appreciated, finding their place in the current stress upon the stability of the species, 'the continuity of the germplasm', and upon discontinuous mutations where change occurred.

From 1901-4 various biologists, notably Sutton and Montgomery in America and de Vries and Boveri on the continent, pointed out that the behaviour of Mendel's factors corresponded to that of the chromosomes during the production and union of the egg and sperm. It appeared therefore that the chromosomes carried the Mendelian unit factors which agreed with Weismann's theory that the chromosomes carried the germplasm units. If each chromosome of a pair contained a Mendelian factor, the associated character of an organism would be governed by two factors, as Mendel's findings required. The splitting of the chromosome pair during the formation of the sexual cells would leave only one factor determining that character in the egg or in the sperm, but the union of the sexual cells would give two factors again in the fertilized egg, one from each parent. If the parents differed, the dominant factor of the two would determine the associated character in the offspring, while the recessive factor lay dormant, but unchanged, appearing in subsequent generations when united by chance with one of its own kind. Since the number of heredity factors was large compared with the number of chromosome pairs, it was predicted by Sutton that several factors should be linked together in one chromosome. In 1906 Bateson and Punnett discovered this phenomenon of factor linkage in the sweet pea, certain colour and pollen characteristics usually being inherited together, indicating an association between them.

The American school of geneticists which arose during the early years of the present century, consisting notably of Morgan, Muller, Bridges, and Sturtevant, found a particularly useful animal for the study of inheritance in the fruit fly. This insect had only four chromosomes, and it passed from egg to adult in twelve days. Using the fruit fly, it was found that characteristics usually inherited together became separated on occasion during the process of reproduction. Morgan in 1911 interpreted this fact to mean that the chromosome containing the associated factors, or genes as they were now called, had broken at a point between the localized positions of the genes determining those characteristics, each piece then uniting with a similar fragment from another broken chromosome. If the chance of such a break was equal along the length of a chromosome, a measure of the distance between two genes could be obtained from the frequency with which their associated characteristics became separated during reproduction, and maps showing the positions of the genes along a chromosome could be drawn up. Plough in 1917 demonstrated such a rearrangement, or 'crossing over', of two chromosomes and on this basis chromosome maps of the fruit fly and other organisms were constructed.

In 1933 it was found by Bridges and others that the larval salivary gland cells of the fruit fly and other insects possessed nuclei containing very large chromosomes. A series of bands were observed along the lengths of such chromosomes, which thus appeared to be made up of a number of disc-like units. At first these units were thought to be the actual genes, but

534 / A History of the Sciences

the effects produced by chromosome breakages suggested that the genes were much smaller than the chromosome bands. However, a gene could be associated with a particular band, as a new arrangement produced by a crossing over could be observed with the microscope in the giant chromosomes of the salivary gland cells, and the position of particular bands on a giant chromosome could be identified with the position of genes on the chromosome maps.

The crossing over and rearrangement of chromosomes helped to explain the mixing of genetic constitutions within a species. A set of characteristics associated with a single chromosome in a parent might be distributed over two in the immediate offspring, and still further separated and diffused in subsequent generations. Thus natural selection would have a great variety to work upon, selecting favourable combinations. No new characteristics could arise by this mechanism alone, though recessive genes would show themselves, and their associated characteristics would become general in the species if they were favourable. However, as earlier workers had found in other species, spontaneous changes were observed in the characteristics of the fruit fly, which were ascribed to changes or mutations in the structure of the associated genes. Most of the mutations, up to 90 per cent, were lethal and the remainder were recessive. In 1927 Muller found that X-rays increased the mutation rate in the fruit fly, and in the following year he discovered that elevated temperatures did also. These artificial mutations were of the same nature as those that appeared spontaneously, and it seemed possible that natural mutations were produced by radiation. However, Muller showed in 1930 that the amount of natural radiation on the surface of the earth was as much as five hundred times too small to account for the spontaneous mutation rate.

In the early days of modern genetics it was thought that the Mendelian factors, or genes, were quite individual and independent entities, their effects being unmodified by the immediate environment of the other genes on the same chromosome, or the larger environment of the nucleus, the cell, the organism as a whole, and the physical conditions in which it found itself, apart from the causal agents of mutation. However, Sturtevant found in 1925 that two genes producing the same characteristic had a much greater effect when they were adjacent on the same chromosome than when they were separated. Muller in 1935 pointed out that such a position effect would cause some undetectable chromosome rearrangements, such as the inversion of a small section, to produce the same phenomena as a gene mutation. In consequence, Goldschmidt suggested in 1938 that all supposed gene mutations were in fact chromosome rearrangement, and he proposed that the idea of gene determinants should be replaced by the conception that the arrangement of the chromosome as a whole determined a particular character. His suggestion was not widely adopted.

Mendel had thought of his factors as the determinants of the adult characteristics of organisms, such as the final height attained by his pea plants. Later it was appreciated that the characteristics of organisms depended not only on their genetic constitution but also on their environment. Bridges showed in 1928 that the eggs of the giant fruit fly developed into a normal sized insect if the larvae did not have an adequate food supply, demonstrating that the genes governed developmental processes rather than final characteristics. Such processes were studied by Goldschmidt, who measured the rate at which pigment was deposited in the skin of certain caterpillars under the control of mutants of the same gene, and by Beadle, who from about 1931 studied the genetically controlled biochemical processes of growth and metabolism in mutants of the fungus, *Neurospora*. In contrast to the normal fungus, the *Neurospora* mutants produced by X-rays could not build up their body materials from simple chemical substances, as they had lost one or more of the synthetic stages in the metabolic processes of the normal *Neurospora*. Thus the mutants had to be provided with the usual products of that stage or stages, and the chemical nature of the substances provided to a series of mutants which had lost a succession of synthetic stages then afforded an indication of the biochemical syntheses normally carried out by the fungus.

The conception that the genes and chromosomes governed the developmental processes of organisms, rather than their adult characteristics, offered the prospect that genetics would find a fruitful union with embryology. Hitherto, however, the two sciences have stood apart, for the factors governing the development of the individual organism which were discovered by the embryologists were found to reside in the cell material outside of the nucleus of the fertilized egg, the cytoplasm, not in the chromosomes of the nucleus as the geneticists suggested. Some embryologists, notably Boveri, Loeb, and Jenkinson, indeed suggested from about 1917 that the main characters of an organism, determining the phylum, the class, the order, the genus, and perhaps the species to which it belonged, were governed by factors in the cytoplasm of the fertilized egg, while only variety characters, such as the shortness or tallness of Mendel's peas, and perhaps species characters were determined by factors in the nucleus.

Such a view was based on the findings of the 'developmental mechanics' school of embryologists which arose towards the end of the nineteenth century. One of Haeckel's pupils, Roux, 1850-1924, pointed out in the 1880's that the embryologists up to and including his teacher had only described the process of embryological development, they had not elucidated its causes. To discover what he termed the mechanics of embryological development, Roux suggested that embryologists should adopt a new approach, turning away from description to experiment. Adopting such a principle, Roux himself discovered that the median plane of an animal with bilateral symmetry was governed by the first cleavage plane of the fertilized egg from which it came. The cleavage plane in turn was determined by the line of entry of the sperm into the egg, and the line of entry was already laid down before fertilization by the massing of the cytoplasm of the egg at the point where the sperm was to enter.

By orientating eggs in a variety of directions and by turning them over during the course of development, Roux showed further that the force of gravity had no influence upon the growth of the embryo, so that the development of an egg was determined by mechanical forces entirely within itself. Following Weismann, Roux supposed that such forces derived from the germplasm of the egg, and in 1888 he made an experiment which he thought to support this view. With a hot needle, Roux killed one of the two cells produced by the first division of a fertilized frog's egg, and he found that the remaining live cell developed into half an embryo. Roux supposed that the factors governing the developmental forces became divided with the fission of the egg, so that only a half embryo could develop from one of the daughter cells. However, Hertwig showed that if the dead cell were carefully removed, the remaining live cell developed into a normal embryo. Moreover, Driesch, 1867-1941, discovered in 1891 that all of the cells up to the fifth division of a fertilized sea-urchin's egg would generate complete embryos, though of small size, if separated one from the other. In consequence, Driesch reacted strongly against Roux's mechanical point of view, and held that there was an inner vital force or purpose within each egg which preserved the integrity of the embryo, and which could regenerate lost parts.

Further work indicated that both sets of discoveries held, fertilized eggs falling into two main divisions, the mosaic and the regulative, with many shades of gradation between them. The various parts of a mosaic egg were differentiated from the start, the separation of the two cells formed by the first division of the egg giving half embryos. The regulative eggs were not so differentiated in the beginning, a whole embryo being formed from a single cell of the first, second, and even the fifth division. In the latter case, whole embryos were also formed by separated cell groups from even later stages of development when the egg had divided into a sphere of segmented cells. Complete embryos were only formed if the sphere of cells was cut lengthways from pole to pole, abnormal embryos resulting from other sections. A group of cells containing one pole reformed into a sphere of segmented cells, but it failed to develop to the next stage, the cup-like gastrula. On the other hand, a group of cells from the other pole produced an abnormally large gastrula, while a combination of cells from both poles formed a normal gastrula.

Such discoveries suggested that in the case of regulative eggs the poles of the first-formed sphere of cells produced chemical substances which organized the later development of the embryo. These chemical organizers were studied notably by Spemann, 1869-1941, in Germany. During the 1920's he showed that in amphibian eggs there was a centre outside of the cell nucleus, and apparently unconnected with it, which produced the organizing substances. Working with the substance producing nerve tissue, termed the neural evocator, Spemann found that it was not at all specific in its actions, the evocator from frogs inducing the formation of nerve tissues in newts. In this case it was nerve tissue of the newt, not that of the frog, which was formed, indicating that the evocator only promoted the development of nerve tissue in general, while other factors determined its specific character. The organizers were found to be nonliving substances, since cell-free extracts produced the same effects as the organizer cells. In 1936 Needham and others showed that the neural evocator was perhaps a sterol, a group of organic compounds including the sex hormones and vitamin D. In the mosaic eggs, as in the regulative ones, the centres organizing the development of the embryo have been found to lie in the cytoplasm of the egg, though again the specific character of that development appears to be governed by other factors.

To account for the discoveries of the embryologists, the American geneticists, Morgan, Bridges, and Sturtevant, have suggested that the cytoplasm of the egg is controlled by the genes of the chromosomes in the nucleus, and that the cytoplasm in itself is of little importance either for the heredity or the evolution of the species. Some cases, involving variety characteristics governed by cytoplasmic factors, have been explained in this way in terms of the theory of genetics. In the development of certain snails, the form which the shell assumes, a right or left-handed spiral coil has been shown to be determined by the cytoplasm of the female egg and not at all by the male sperm, which consists almost entirely of nuclear material. Sturtevant indicated in 1923 that the cytoplasm of the egg was dependent upon the genetic constitution of the snail from which the egg came, the gene governing the cytoplasm which gave rise to right-handed spiral shells being dominant. Where character differences of specific rank are involved, as in the crossing of different species, the cytoplasm of the female egg appears to have a considerable influence upon the characteristics of the offspring, Goldschmidt, with insects, and Michaelis, with plants, showing in 1937 that the influence of the cytoplasm became more important the greater the difference between the species crossed. Such cases again were explicable in terms of the nucleus exercising a genetic control over the cytoplasm. However, more recently cases of cytoplasmic inheritance apparently independent of nuclear control have been discovered, the name plasmagenes being given to the character determinants which are handed down from one generation to another through the cytoplasm.

While the science of Mendelian genetics has stood apart from embryology for some time, it early found a place in the Darwinian theory of organic evolution. Darwin had observed, but not analysed, the variation in the organic world upon which the forces of natural selection acted, preserving the variants with new adaptive characteristics and deleting those without. Genetics provided such an analysis, distinguishing several sources of variation in the heredity of organisms. Variants could arise through the appearance of a new gene by mutation, or a new combination of existing genes, and by internal chromosome changes, such as the deletion, duplication, translocation, and inversion of parts, or changes involving chromosome sets, like the doubling up in the number of chromosomes because of the failure of the chromosome pairs to separate in the primary cell division leading to the production of the sexual cells. The various types of genetic change could give rise to different kinds of evolutionary change. The doubling up in the number of chromosomes produced a new species in one generation, as the new organism had a chromosome number different from that of the species from which it came. On the other hand, the effect of a mutant gene was buffered by the effects of the complex of other genes, so that the evolution by gene mutation might be as slow and as gradual as Darwin had believed the change of the species to be. Goldschmidt suggested in 1944 that gene mutations produced only what he termed 'microevolution', a development within a species adapting it to a changing environment. 'Macroevolution', or the change of one species into another, was due to a complete rearrangement of the chromosome material, changing the chromosome patterns and thus the biochemical reaction systems which they governed.

While Darwinism has been the theory of organic evolution generally received in the scientific world, Lamarckism has found some supporters. From 1920 to 1937 McDougall trained rats over a number of generations to avoid a route leading to an electric shock in favour of a route leading to food, finding that the number of training runs decreased in successive generations. More striking evidence for the Lamarckian view has been obtained from studies of the growth of bacteria and protozoa. From 1930 Jennings has shown that some protozoa adaptively modify themselves to alien environmental conditions, such as high temperatures and poisonous chemicals, the acquired modifications being inherited, as the organisms adapt back to their old conditions as slowly as they changed over to the new. From 1938 Hinshelwood has treated the growth of the bacterial cell as a chemical reaction, studying the kinetics of the process, and has obtained similar results.

In recent years the appearance of drug-resistant strains of bacteria has become a problem of considerable importance, the development of such bacteria being generally ascribed to the chance production of a mutant organism, which is viable in the presence of the drug, and which becomes dominant through natural selection. However, Hinshelwood has advanced reasons for believing bacteria can adaptively modify themselves to drugs, notably because nearly all of the bacteria in a diluted inoculum survive and produce colonies in a sub-lethal drug medium, not just one or two mutants. Critics suggest that each bacterium survives and reproduces long enough to form a drug-resistant mutant, but Hinshelwood suggests that such a view does not account for the quantitative features of the phenomenon, the rate of division of the bacterial cell being greatly reduced in drug media, though the rate of production of organic matter is considerable.

Lamarckism always has been a minority opinion with a rather heterodox flavour, a quality which became enhanced in the 1940's with the development of the somewhat similar Michurinist theory in the Soviet Union.

Chapter 43

The Theory of Relativity

THE EARLY modern scientists formulated theories in which they attempted to do away with the conception that there were special vantage points in the universe from which certain privileged observers viewed and governed cosmic events, a conception which had been prominent in the ancient and the medieval views of the world. They endeavoured to show that the moon and the planets were similar to the earth, and not the more perfect and powerful entities that the ancients had presumed them to be. Together with the Calvinists, the early modern scientists removed from the heavens the hierarchies of angelic spirits which were thought to propel the heavenly bodies round their courses and to observe and guide terrestrial events. Finally the scientists of the seventeenth century conceived of the heavens and the earth as identical in kind and as governed by the same power, interpreting the motions of the heavenly bodies in terms of the laws of mechanics which were obeyed on the earth. But the scientists of the sixteenth and seventeenth centuries retained the view that there was one particular privileged observation and control point in the universe, just as the Calvinists retained the conception that there was a single and absolute Ruler of the universe after they had rejected the old system of cosmic government through the hierarchies of angelic beings. For Copernicus the sun was the physical absolute monarch of the solar system, governing all that he surveyed. Kepler had a similar view, bringing together his science and theology by suggesting that the sun was a suitable dwelling place for the Deity. In the Newtonian system also, God was the single privileged observer and governor of the cosmos. The Deity, wrote Newton,

'endures for ever, and is everywhere present, and by existing always and everywhere, He constitutes duration and space. . . . (He is) a Being incorporeal, loving, intelligent, omnipresent, who in infinite space, as it were in his sensory, sees the things themselves intimately, and thoroughly perceives them, and comprehends them wholly by their immediate presence to Himself.'

Newton's view that there were such quantities as absolute space, time, and velocity, and that there was an etherial medium pervading the whole of absolute space, was therefore connected with the conception that there was a single privileged observer in the universe. Similarly, in all subsequent ether theories there was in principle one set of privileged systems and observers, namely, those that were at rest in the cosmic etherial medium, for in theory they could measure the absolute velocities of moving objects. Even the French philosophers of the eighteenth century, with their profound conviction of the mechanical uniformity of the world and its inhabitants, did not get away from the idea that there could be in principle a privileged observer of events. In a Newtonian vein of thought, Laplace conceived of a divine calculator, who knowing the positions and speeds of all the particles in the universe at any given moment, could know all that had happened in the past and predict what was to happen in the future. Thus the philosophical trend initiated by the early modern scientists was not carried to completion until Einstein made basic to physics the postulate that all positions in the universe were equivalent, all possible observers being able to obtain the same information about each other.

The decline of the ether theories dates from 1887 when Michelson, 1852-1931, and Morley, 1838-1923, in America measured the velocity of light along the line of the earth's motion through the ether and at right angles to it. The velocity of light in the two cases was found to be the same, a result which seemed to indicate, so Michelson thought, that the ether moved with the earth. However Lodge, 1851-1940, at London found in 1893 that light passed between two massive steel discs rotating at high speed was not altered in velocity, showing that the discs did not drag the ether with them. The aberration of light from the stars also suggested that the ether did not move with the earth, and thus Michelson and Morley's experiment led to the abandonment of the view that there was a material ether pervading space which carried the wave vibrations of light.

An explanation of the Michelson-Morley experiment was suggested by Fitzgerald in 1892. He pointed out that if matter were composed of electrically charged particles, as it seemed to be, the length of a rod at rest relative to the ether would be determined solely by the electrostatic equilibrium obtaining between the particles of the rod, whereas in motion relative to the ether the rod would contract, as the motion of its constituent charged particles would generate a magnetic field, altering the equilibrium spacings of those particles. Thus the apparatus used in the Michelson-Morley experiment would contract in length when orientated along the direction of the earth's motion by just the amount required to compensate for the reduced speed of light. Attempts were made to measure the Fitzgerald contraction, but they were not successful. A wire, for example, should contract in length and hence exhibit an increased conductivity, when directed along the earth's line of motion, but experiment showed that this was not the case. Such experiments showed that the Fitzgerald contraction in a moving system could not be measured by observers within that system, and thus they could not determine the absolute velocity of their system, the laws of optics and other electromagnetic phenomena being independent of absolute velocity.

It appeared therefore that the absolute velocity of a body was unknowable, and that the framework of absolute space and absolute time could be dispensed with, along with the ether which composed the material of that framework. Einstein, then at Berne, suggested in 1905 that the phenomena of physics could be more economically covered if it were assumed that the laws of nature were the same in all systems moving with uniform relative velocities, and that the velocity of light in empty space was always the same. Ritz in 1908 attempted to account for the Michelson-Morley experiment by supposing that the velocity of light depended upon the speed of its source, but de Sitter, 1872-1934 at Leiden pointed out in 1913 that if this were so then two stars rotating round one another would show an apparently anomalous movement, which was not observed. Einstein's assumption that the velocity of light was constant and independent of the velocities of the source and observer seemed to be justified therefore, and the Newtonian view that such velocities were additive had to be dropped now that there was no stationary ether to carry the wave vibrations of light.

According to Einstein any velocity compounded with that of light gave the same result, namely, the constant velocity of light in empty space. Thus the relative velocity of a body measured by an observer could never exceed the velocity of light, a body conceived of as moving faster than light being in principle undetectable by means of light signals. The Newtonian laws of mechanics suggested that the application of a force would cause an indefinite increase in the velocity and kinetic energy of a body, the mass component of the kinetic energy remaining constant whilst the velocity component increased. Einstein showed that the constant application of a force would increase the mass of a body when it approached the velocity of light, the force conferring energy in the form of mass upon the body, as it could no longer enhance the velocity component of the kinetic energy when the limit of the velocity of light was approached. It appeared therefore that energy was equivalent to mass, Einstein expressing the equivalency in the relationship that a quantity of energy is equal to the product of an amount of mass and the square of the velocity of light. This relationship was later verified by measuring the mass of electrons moving at high speed, when it was found that the expected increase in mass was equal to that observed.

Einstein's postulates showed furthermore that the Fitzgerald contraction was not a real physical change in bodies, but an appearance due to the relative motions of bodies. If the observers on a series of systems in motion relative to one another were equipped with identical measuring rods and clocks, then measurements using light signals made by one observer would appear to show that the rods of the other observers were shorter in the directions of relative motion than his own, and their clocks would appear to run more slowly than his own. These phenomena would be noted by all the observers, and the laws of nature would appear to be the same in all systems in relative motion. The measurements made by any given pair of observers would be completely symmetrical, in particular they would both ascribe the same relative velocity to the other. Thus there were no privileged observers, and no absolute space nor absolute time. The length of a rod would depend upon the relative velocity of the observer that measured it, so too would the time kept by a given clock. Furthermore no two observers in relative motion would observe two events as simultaneous unless the events occurred in the same place. Einstein's teacher, Minkowski, 1864-1909, showed in 1908 that whilst observers in uniform relative motion would not assign the same time and space separations to two observed events, they would assign the same value to the interval, a combination of space and time, between the two events. Minkowski found that if time were made into a kind of distance by multiplying it by the velocity of light, symbolized as C, then if the difference in time between two events is T for a particular observer, and the distance between them S, the value of $S^2 - C.T^2$ is the same for all observers. Minkowski termed the square root of $S^2 - C.T^2$ the space-time interval between two events, and since it is the same for all observers, it is an absolute quantity in four-dimensional space-time, whilst space and time separately are not.

In his *Special Theory of Relativity*, published in 1905, Einstein considered only systems in uniform relative motion, but in 1915 he went on to deal with accelerated motions in his *General Theory*. His starting point was the long known fact of Newtonian mechanics that the value assigned to the mass of a body for the purposes of studying its gravitational fall was exactly the same as the value used to investigate the acceleration of the body when acted upon by mechanical forces. It appeared therefore that gravitational fields and accelerations were equivalent, and Einstein sought to make them identical. He pointed out that an observer in a stationary lift might conclude from the fact that bodies fell to the floor with an accelerated velocity that the lift was in accelerated motion upwards, as though it were propelled by a mechanical force through empty space free from gravitational fields. Similarly if the lift were allowed to fall freely, the observer inside might think that he was in empty space, free from the action of any mechanical or gravitational forces, as bodies would remain suspended in mid-air. On the other hand, an observer outside would think that the lift was at rest in the first case, and falling with accelerated motion in the second. Thus, judgements of the accelerated movements of bodies are relative to the standpoint of the observer, and so too are those relating to forces; any gravitational field can be ascribed to a relative acceleration.

In order to connect up gravitational fields with relative accelerations, Einstein suggested that both might be explained geometrically if Minkowski's four-dimensional space-time were curved. For example, a billiard ball on a billiard table constructed with depressions round the holes would behave as though it were attracted towards the holes, moving with acceleration towards them. The motions of the billiard ball could be explained either in terms of the geometry of the billiard table, or in terms of the assumption that the holes were centres of an attractive force acting at a distance. Similarly gravitational motions would be explained either in terms of actions at a distance, or in terms of the curved geometry of space-time. In his *General Principle of Relativity*, published in 1915, Einstein made the postulates: that the laws of nature were the same for all observers moving in any manner relative to one another; that the geometry of space-time was non-Euclidian; that all gravitational motions take place along the shortest paths in space-time; and that the curvature of a given region of space-time was dependent upon the amount of matter in that region. Combining these postulates, Einstein selected as his model for space-time a restricted type of the non-Euclidian geometry invented by Bernard Riemann,1826-66. The properties of Einstein's model were such that all observers moving relatively within it were symmetrical and equivalent one to the other, and that it gave geodesics, that is, paths corresponding to straight lines in Euclidian geometry, which could be identified with motions in gravitational fields.

Developing his theory, Einstein showed that the shortest path in space-time round a weighty particle of matter would be an ellipse which rotated itself round the particle, whilst the Newtonian theory indicated that such a path should be a stationary ellipse. The French astronomer, Leverrier, 1811-77, had found that the elliptical orbit of the planet Mercury showed a rotation round the sun, and the amount of rotation he observed agreed closely with that calculated by Einstein. More important were Einstein's predictions regarding the influence of gravitational fields upon light. It had been known previously that accelerated motions affected light vibrations, but now that gravitational fields were connected up with relative accelerations it appeared that gravity should also influence light. Since light had energy, it should possess mass and so be deflected by gravitational fields, or rather, it should move through a curved path in regions of space-time bent by the presence of matter. Measurements of the bending of light from stars by the sun obtained during the solar eclipses of 1919 and 1922 were found to agree fairly closely with the theorctical values predicted by Einstein, through determinations made during the eclipses of 1929 and 1947 showed some divergencies. However, the results of the 1952 eclipse agreed well with the relativity theory. Einstein predicted further that the light emitted by atoms in a strong gravitational field should lose some of its energy whilst moving away from the field, the light becoming redder. If the gravitational field were regarded as a relative acceleration, the same red-shift would be produced by the accelerated recession of the light source. Such a shift towards the red was observed from 1923 to 1928 in the spectral lines given out by the iron, titanium, and cyanogen on the surface of the sun.

The general theory of relativity covers, so far, only the three cases described, the precession of the orbit of Mercury, the bending of light in gravitational fields, and the red-shift of light in gravitational fields, though these cases in themselves are of considerable importance, particularly the last two, which connect up for the first time gravitational and electromagnetic phenomena. It has been found difficult to apply relativity theory to rotatory motions, though Einstein in 1916 showed that rotations could be assimilated to some degree into the general theory. The traditional evidence for the rotation of the earth on its axis—the flattening of the earth at its poles, the tendency of a pendulum and a gyroscope to change their orientation relative to the earth, and so on—may be accounted for either on the assumption that the earth spins on its axis or on the assumption that the universe as a whole rotates round the earth. On the second assumption difficulties arise from the consideration that the tangential velocities of bodies would increase more and more the further they were placed from the earth, approaching to an ever increasing degree the velocity of light.

Chapter 44

The Quantum Theory and the Structure of the Atom

AT THE turn of the nineteenth century theories involving particles and discontinuous change appeared in a number of fields which had been dominated previously by the conception of the continuity of matter and of change. Such theories were developed notably, but by no means entirely, by the Germans, who had had a *penchant* for particle theories throughout the nineteenth century. The theory that the cell was the unit of all living creatures had been almost entirely a German development; so too was the theory that heredity operated by a particle mechanism in an autonomous germplasm, at least in its early speculative stages with Nageli and Weismann. The Germans, Fechner, Weber, Riemann, Kirchhoff and Clausius, thought of electricity in terms of charged particles, whilst the English, Faraday, Kelvin, Maxwell and Fitzgerald, considered electrical phenomena to be consequences of strain in a continuous ether: Maxwell, as we have seen, failed to develop the idea of particles of electricity which was implicit in his model of the ether. The German, Helmholtz, commenting upon Faraday's laws of electrolysis, said in 1881:

'Now the most startling result of Faraday's laws is perhaps this: if we accept the hypothesis that the elementary substances are composed of atoms, we cannot avoid concluding that electricity also, positive as well as negative, is divided into elementary portions which behave like atoms of electricity.'

The implication had not escaped Faraday, but he rejected it along with the atomic theory of matter, preferring the view that 'matter is everywhere present, and there is no intervening space unoccupied by it'.

Besides the study of electricity, another field in which particle theories were developed was the investigation of light, heat, and other electro-magnetic radiations, where the continuous wave theory had had a considerable success during the nineteenth century. The starting point was a problem encountered in the study of the light and heat emitted by black bodies, which give a continuous spectrum of radiation as opposed to the line spectra given by the chemical elements. Experiment showed that black bodies when heated to a given temperature emitted a maximum amount of radiation energy at a particular wavelength, the wavelength of the maximum decreasing as the temperature was increased. At laboratory temperatures the maximum was in the visible part of the spectrum, the body emitting first a red glow, then light that was orange, yellow, white, and finally blue, as the temperature was raised.

Such phenomena could not be accounted for by the wave theory of light. Lord Rayleigh showed in 1900 that if electromagnetic radiations were given out by natural vibrators, there would be no wavelength at which a maximum amount of energy would be emitted: the quantity of energy given out would increase indefinitely as the wavelength of the radiation decreased. A natural vibrator, such as a stretched string, had a number of modes of vibration. The first mode of vibration in the case of the string was that in which the length of the string equalled half a wavelength, and subsequent modes occurred when the length of the string equalled 2, 3, 4, 5 half wavelengths, and so on indefinitely, so that there were an infinite number of modes of vibration in the very short wavelengths. According to the principle of the equipartition of energy, each mode of vibration should have the same amount of energy, and thus most of the energy of black body radiation should be given out in the very short wavelengths, the ultra-violet and X-ray part of the spectrum instead of the visible. Other theoretical treatments of black body radiation met with no greater success, notably that published in 1896 by Wien, 1864-1928, who based his analysis on purely thermodynamic reasoning without the use of a model.

A solution to the problem was suggested in 1900 by Max Planck, 1858-1947, then at Berlin. He indicated that if black body radiation were given out discontinuously in quanta, in such a way that the energy of a quantum was proportional to the frequency of the radiation, the emission of the longer wavelengths towards the red end of the spectrum would be favoured at low temperatures as the energy of the quanta would be small, whilst at higher temperatures more energy would be available, favouring the emission of the larger quanta of the shorter wavelengths. In this way Planck accounted for the fact that a maximum amount of energy was emitted at a certain wavelength, and that this maximum shifted to the shorter wavelengths with an increase in temperature. Planck's quantum theory was used by Einstein in 1905 to explain the deviations from the Dulong and Petit rule of the constancy of atomic heats, and to account for the ejection of electrons from metals when exposed to light. He also suggested that light, and electromagnetic radiations in general, were propagated across space in the form of particles, or photons as they came to be termed, thus filling in to some degree the gap left by the rejection of the ether in his theory of relativity published in the same year.

The field in which the quantum theory found its most important application was the investigation of the structure of the atom. As we have seen, the laws of electrolysis observed when electricity was passed through salt solutions led Helmholtz to suggest in 1881 that electricity existed in the form of particles, Johnstone Stoney terming them electrons in 1891. The passage of electricity through gases at low pressures provided further evidence that electricity was particulate in character, and gave a method of studying the structure of the atom. When electricity was discharged between two plates in a tube containing a gas at a very low pressure, three kinds of rays were observed: cathode rays passing from the negative to the positive plate; positive rays travelling in the opposite direction; and X-rays, formed when cathode rays impinged upon matter, such as a metallic target.

The cathode rays were the first to be intensively studied. It was found that they travelled in straight lines, as objects placed in their path cast sharp shadows, and that they possessed considerable momentum, since they caused a light paddle-wheel to rotate. Cathode rays therefore seemed to consist of a stream of particles in rectilinear motion, and it was shown that such particles were charged as the rays were deflected by electric and magnetic fields. In 1897 J. J. Thomson, 1856-1940, at Cambridge measured the velocity of these particles by arranging an electric and a magnetic field so that the deflections they produced cancelled out. The force exerted by the magnetic field on the particles was dependent upon their velocity, whilst the force exerted by the electric field was not. In Thomson's experiment the ratio of the strengths of the electric and magnetic fields then gave the velocity of the particles, and once this was known the ratio of the charge to the mass of the particles could be determined from the magnetic or electric deflection alone.

Thomson showed that these particles were negatively charged, and that they were a common constituent of all matter, as any gas placed in his discharge tube gave cathode rays with the same ratio of charge to mass. From the electrolysis of acid solutions it was found that the ratio of the charge to the mass of hydrogen ions was about two thousand times smaller, from which Thomson concluded that the mass of a hydrogen ion was about two thousand times greater than that of a cathode ray particle, an electron, the charges on them being equal though opposite in sign. Thomson's deduction was confirmed when the charge on an electron was measured, notably by Millikan in America, who from 1913 to 1917 examined the motions of charged oil drops in opposing electrostatic and gravitational fields. He found that the lowest charge an oil drop picked up was the same as the lowest common multiple between higher charges which the drop could receive, and this he took to be the charge of a single electron. The amount of charge was the same as that on a hydrogen ion, though opposite in sign, indicating that hydrogen was 1836 times as heavy as an electron.

On this basis Thomson suggested in 1904 that the atoms of the chemical elements consisted of electrons held together by a sphere of positive charge. He thought that the electrons were responsible for the entire mass of the atom, so that there were, for example, 1836 electrons in an atom of hydrogen. However, the study of radioactivity soon led to the abandonment of Thomson's model of the atom. The radioactive elements were found to give out three kinds of rays similar to those produced by the gas discharge tube: firstly, positively charged alpha rays, consisting of doubly charged helium ions similar to positive rays; secondly, beta rays, consisting of electrons like the cathode rays, though moving somewhat faster; and thirdly, uncharged gamma rays, which were electromagnetic radiations, like X-rays, though of even shorter wavelength. Rutherford, 1871-1937, first at Manchester then at Cambridge, showed that alpha particles for the most part went straight through matter, but a few, about one in twenty thousand, experienced very large deflections, often doubling back on their course. Thus it appeared that the atom contained a powerful deflecting force, or nucleus, which was very small compared with the size of the atom as a whole. In 1911 Rutherford suggested that the atom consisted of a small positively charged nucleus containing most of the mass of the atom, surrounded by electrons moving in orbits round the nucleus, like the planets round the sun. The number of positive charges on the nucleus of a given elementary atom was equal to the ordinal number of that element in the periodic table, and so too was the number of orbital electrons to make the atom as a whole electrically neutral. Evidence for Rutherford's model of the atom was provided by the fact that the heavier elements scattered alpha particles to a greater extent than the lighter elements, presumably owing to their greater nuclear charge and mass. Furthermore, it was discovered that when a radioactive element gave out an alpha particle, which was doubly charged, the new element formed was two places back in the periodic table, indicating on Rutherford's theory that it had two less positive charges on its nucleus, whilst the emission of a beta particle, which had a single negative charge, gave a new element one place forward in the periodic table, that is, it had an extra positive charge on its nucleus. As the atomic weight of an element was roughly twice its atomic number, or nuclear charge, it was presumed that the nucleus of an atom was made up of hydrogen nuclei, or protons, equal in number to the atomic weight, and the number of electrons required to bring the overall charge on the nucleus back to the atomic number.

According to the classical view, the Rutherford model of the atom contained an inherent defect. The orbital electrons were charges moving in the electrostatic field of the nucleus, and so they should continuously emit radiation. The electrons should lose their own kinetic energy in the course of so doing, and thus they would gradually spiral into the nucleus. Niels Bohr, then working with Rutherford, pointed out in 1913 that the model could be saved by means of the new quantum theory. If radiation could not be emitted continuously, but only in definite quanta, it was legitimate to suppose that there were certain stable orbits in which electrons could move without loss of energy, radiation being emitted only when an electron jumped from one orbit to another. In this way it would be possible to account for the fact that the atomic spectra of the elements consist of sharp lines, and not continuous bands. The energy of the quantum, and therefore the frequency of the radiation emitted when an electron jumped from one orbit to another, would be determined by the difference between the kinetic energies of the electron in those two orbits. If the orbits were fixed, so too were the energy differences between them, and each line in the atomic spectra of the elements should correspond to a particular electronic transition between two such orbits. By this means Bohr was able to account for the atomic spectra of hydrogen quantitatively, typifying the main orbits between which an electron could jump by a quantum number, N, which could have the values, 1, 2, 3, 4, etc., a jump from the second to the first orbit giving one spectral line, from the third to the second another, and so on.

With instruments of higher resolving power the lines of the atomic spectra were observed to possess a fine structure. To account for this phenomenon, Sommerfeld at Munich in 1915 suggested that a main electron orbit of quantum number, N, was subdivided into subsidiary orbits, the number of which was governed by a subsidiary quantum number, K, which could have the values, 1, 2, 3, 4, . . . N. Sommerfeld pictured these subsidiary orbits as ellipses, varying in shape from a circle to an elongated ellipse. They would have slightly different energies, as an electron in an elongated elliptical orbit would increase in speed and mass when moving in the vicinity of the atomic nucleus, according to Einstein's theory of relativity, and would exhibit a precessional movement round the nucleus, like the orbit of mercury round the sun. Another subsidiary quantum number, M, which could have the values, $-(K-1) \ldots 0 \ldots +(K-1)$, had to be introduced to account for the splitting of the lines of the atomic spectra in a magnetic field, a phenomenon that had been discovered by Zeeman, 1865-1943, at Leiden in 1896. Finally a spin quantum number, S, which could have one of two opposite values, was suggested by Uhlenbeck and Goudsmit in 1925 to account for further multiplicities in the atomic spectra lines, on the assumption that an electron could spin on its axis in one of two opposite directions.

Pauli at Hamburg in 1925 formulated the principle that no two electrons in a given atom could have the same set of quantum numbers, a restriction which limited and defined the possible electronic structures of the elementary atoms. When the main quantum number, N, was equal to 1, the orbit so defined could only contain a single electron or two oppositely spinning electrons, as the subsidiary numbers, K and M, could only possess their minimum values. The first main orbit covered, therefore, the cases of hydrogen, with one orbital electron, and helium, with two, the first main orbit or shell then being filled. When the main quantum number, N, was equal to 2, the subsidiary number, K, could have the values of 1 or 2, and M the values 0 when K was equal to 1, and -1, 0, $+1$, when K was equal to 2. In the second main shell there were, therefore, four subsidiary orbits, each of which could accommodate two oppositely spinning electrons, thus covering the next eight elements, lithium, beryllium, boron, carbon, nitrogen, oxygen, fluorine and neon. In this way the quantum numbers derived from the study of the atomic spectra indicated the electronic structures that the elements possessed, providing an electronic classification of the elements which agreed closely with the earlier periodic classification.

The electronic classification of the elements drew attention to the fact that the inert gases, which possessed little or no tendency to form chemical compounds, occurred when an electronic shell or sub-shell was completed, as in the cases of helium and neon above. It appeared, therefore, that a completed shell or sub-shell was a particularly stable electronic structure, and precision was now given to the theory of chemical combination. From 1919 to 1921 Langmuir in America developed the theory that chemical elements combine in such a way that they achieve the stable electronic structure of the inert gases in their compounds. He suggested that the elementary atoms could do this in two ways; either by donating and receiving electrons, or by sharing electrons, the combinations being termed an electrovalency and a covalency respectively.

Sodium, for example, contains one more electron than the inert gas neon, whilst fluorine contains one less. When they combine to give the electrovalent sodium fluoride, the sodium atoms donate their extra electron to the fluorine atoms, so that both kinds achieve the stable electronic structure of neon. The nuclear charges remain constant, and thus the sodium atoms become positively charged and the fluorine atoms negatively charged, the electrostatic force between them holding them together. Two fluorine atoms, both of which are one electron short of the neon electronic structure, can form a different type of bond by sharing one of their electrons with the other atom, giving the covalent fluorine molecule. Here the two atoms are electrically neutral, and are rigidly held at a fixed bond distance by the two electrons exchanged between them, whereas the two electrically charged ions of electrovalent compounds can move independently of each other in certain circumstances, such as in aqueous solution. Later it was appreciated that pure electrovalencies and covalencies were rare, most bonds being combinations of both types, as many covalent compounds possessed a dipole moment, indicating that there was a separation of charge in their bonds, whilst the electronic shells of electrovalent ions were found to be polarized or distorted.

The Bohr atom provided the chemists with a useful model for the purposes of interpreting the structure of molecules and the course of chemical combination, but for the physicists it was less satisfactory, as it explained the atomic spectra of only the hydrogen type of atom, and even then did not account for the relative intensities of the spectral lines observed. One of Sommerfeld's pupils, Werner Heisenberg, suggested in 1925 that all mechanical models of the atom like that of Bohr should be abandoned, and an alternative approach adopted in which directly measurable quantities, such as the frequencies and the intensities of the spectral lines, were fitted immediately into mathematical equations. In this way Heisenberg was able to account for the Zeeman effect, which had raised difficulties in the earlier Bohr theory. Yet another approach stemmed from the prediction, made by Louis de Broglie in 1925, that matter as well as radiation should have both particle and wave properties. Later it was found that electrons, protons, and alpha particles had wave properties, giving diffraction patterns like light and X-rays, but more immediately Erwin Schrodinger in 1926 was able to cover the same facts as Heisenberg on the assumption that electrons had a wave form. From this viewpoint Bohr's postulate that stable electron orbits were defined by the requirement that electrons must have angular momenta equal to an integral number times $h/2\pi$, where h is Planck's constant, connecting the size of an energy quantum of radiation with its frequency, was equivalent to the requirement that there should be an integral number of electron wavelengths round the circumference of a stable orbit.

However, in Schrodinger's scheme electrons no longer moved in definite orbits. His equations showed that the density of electric charge in the atom alternated in a wavelike form outwards from the nucleus, the peaks of the waves corresponding to what had been the orbits of the earlier Bohr atom. Schrodinger himself regarded the wave forms as representing the actual distribution of charge in the atom, but Born suggested in 1926 that the height of the wave at any point should be taken as a measure of the probability of finding an electron in that position. The view of Born indicated that an electron could not be located with complete accuracy; only the probability of finding an electron at a particular point could be determined. Heisenberg in 1927 showed that the momentum and energy of an electron were similarly indeterminate, the product of the momentum and position of an electron being uncertain to a degree that could never be less than $h/2\pi$, where h was Planck's constant. This 'Principle of Uncertainty' followed from the wave-particle duality of matter and radiation, and from the fact that the characteristics of objects were usually unavoidably altered during the course of experimentation. If the position of an electron were to be accurately measured, radiations of very small wavelengths would have to be used for the determination. But such radiations would possess quanta of high energy, and would alter the momentum and energy of the electron by impact. Similarly to measure the momentum of an electron, quanta of low energy would have to be used: the wavelengths of such quanta being large, the position of the electron would be correspondingly indeterminate. The observed wave-particle duality of matter and radiation also posed the problem as to what correspondence there was between particle properties and wave properties. Planck showed that the energy of a quantum was equal to the frequency of its radiation times the Planck constant h. De Broglie found that the velocity of a particle was equal to the group velocity of its associated waves, the group velocity of a train of waves being smaller than that of the individual waves, as can be seen sometimes in the case of water waves where the crest of a large beat wave runs slowly along a group of smaller waves. Later, in 1927, Sir Charles Darwin suggested that the two possible spin modes of a particle correspond to the two transverse components of a wave vibration which are separated by polarization.

In developing his model of the atom, Schrodinger was able to deduce quantitatively three of the four quantum numbers defining the structure of the atom, by considering the distribution of electrons around the nucleus in the three dimensions of space. The quantum mechanics so derived have been most fruitful, but they are non-relativistic, as Schrodinger did not consider the fourth dimension of time. Dirac in 1928 derived a relativistic quantum mechanics and in so doing obtained terms corresponding to the fourth quantum number, covering the spin of the electron, which had not emerged from Schrodinger's theory. Dirac also predicted the existence of a particle with a mass equal to that of the electron, though opposite in charge, namely, the positron which was later discovered. However, the relativistic wave mechanics are complex so that Dirac's theory, as yet, has found fewer applications than Schrodinger's.

The wave mechanics of Heisenberg and Schrodinger were designed to account for the behaviour of the outer electrons of the atom. Meanwhile the nucleus of the atom had been intensively studied, and ultimately attempts were made to apply quantum mechanics here, where the problems encountered presented greater difficulties. The study of the nucleus, which contains most of the mass of the atom, began with the development of a new method of ascertaining the atomic weights of the elements, which was based on the examination of the positive rays given by the passage of electricity through gases at low pressures. J. J. Thomson found in 1912 that different gases gave positive rays which behaved differently when deflected by electric and magnetic fields, so that they were not a common constituent of all matter like electrons. He determined the ratio of the charge to the mass of the positive ray particles, and thence found that the mass of the particles was equal to their atomic weights, which indicated that they were positively charged atoms of the gas. However, he discovered that in the case of neon there were two different kinds of particles, one of mass 20, and the other of mass 22 relative to hydrogen. The two forms were present in the ratio of about $10/1$, so that their average mass was $20 \cdot 2$, a figure in good agreement with the known atomic weight of neon, namely $20 \cdot 18$. Thus it appeared that neon was made up of two kinds of atoms, termed isotopes, which had different masses, but the same nuclear charge, and the same number of orbital electrons.

It seemed also that Prout's hypothesis, postulating that the atoms of the elements were made up of a whole number of hydrogen atoms, had some truth in it, since the atomic weights of two neon isotopes were exact multiples of the atomic weight of hydrogen. Subsequent work by Aston, 1877-1945, at Cambridge showed that the isotopes of the elements in general were not quite exact multiples of the atomic weight of hydrogen, or rather of one-sixteenth of the atomic weight of the oxygen isotope with mass number 16, which was taken as the standard reference substance. In general, Aston found in 1927, the very light and the very heavy elements, such as hydrogen and uranium, had slightly more mass than expected from Prout's hypothesis, whilst the elements of moderate atomic weight in the region of iron were somewhat lighter than expected. Aston termed these deviations the packing fractions of the elements, and he showed that a smooth curve could be drawn by plotting graphically the packing fractions against the mass number of the element, that is, the number of particles in the nucleus expected from Prout's hypothesis. The packing fraction was taken to be a measure of the stability of a given element, as the very heavy elements with a high packing fraction were radioactive and spontaneously decayed into light atoms with a lower packing fraction, losing mass in the form of energy in the course of so doing. On this basis the elements of moderate atomic weight in the region of iron were very stable, and so too were certain light elements that did not fit into Aston's curve, notably the isotopes of helium, carbon, and oxygen which had mass numbers that were multiples of four, indicating that the alpha particle was a stable unit, and that it might exist as such in heavier atomic nuclei.

At first the work of Thomson and Aston was thought to show that the atomic nucleus was made up of protons and electrons. The two neon isotopes, for example, would contain 22 protons and 12 electrons, and 20 protons and 10 electrons respectively, so that they would both have a nuclear charge of 10, but atomic weights of 22 and 20. From his alpha ray experiments, Rutherford showed in 1920 that the atomic nucleus was too small to accommodate all these particles, and he suggested that if the nucleus contained neutrons, particles of the same weight as protons but without a charge, it would be possible to dispense with the idea that there were electrons in the atomic nucleus. His pupil, Chadwick, in 1932 discovered that particles with an exceptionally long range were emitted when beryllium was bombarded with alpha particles. He found that these particles were not deflected by electric or magnetic fields, and so were uncharged, whilst they caused the ejection of protons from other elements, and thus had a mass of the same order as that of the proton. In the same year Anderson in America and Blackett in Britain discovered that cosmic rays and gamma rays gave rise to positrons, the positively charged counterparts of electrons, the existence of which had been predicted by Dirac at Cambridge in 1928. Heisenberg then suggested that the neutrons and protons in the atomic nucleus might be held together by a process of positron exchange, a proton losing a positron becoming a neutron, and a neutron acquiring a positron becoming a proton. Such a process would be similar to the interchange of two orbital electrons between two atoms in a covalent chemical bond. Heisenberg, with Dirac and Pauli, also applied the idea to electric fields, deriving the Coulomb inverse square law of electrostatic force on the assumption that the force between two charged particles resulted from the interchange of photons between them, one particle emitting a photon and the other absorbing it. However, in this case difficulties appeared, their calculations indicating that there should be other forces, which increased to infinity as the radius of the particles went to zero.

Another attempt to account for the binding forces of the atomic nucleus on the basis of the exchange idea was made by Yukawa of Japan in 1935. He suggested that protons and neutrons in the nucleus were held together by exchanging a type of meson which was a neutral particle with a rest mass some two hundred times as heavy as an electron. Yukawa's meson was discovered by Powell at Bristol in 1947, though other mesons were discovered as secondary products of cosmic rays by Anderson at Pasadena in 1938. Pauli in 1931 postulated the existence of another particle, the neutrino, or neutral electron, to account for the fact that electrons emitted by radioactive elements have a variable amount of energy. He suggested that neutrinos were emitted simultaneously, carrying away a variable complement of energy, so that the total energy given out in radioactive decay was a fixed quantity. Hitherto these particles have not been discovered experimentally. It is to be expected, however, that the neutrino would have enormous penetrating powers, and so would be difficult to detect. Pauli's principle, postulating that no two particles in a system can have the same set of quantum numbers, has been applied to the nucleus of the atom as well as to the orbital electrons. It has been suggested that there are two sets of energy levels in the atomic nucleus, one filled by neutrons and the other by protons, both obeying the quantum rules defining the energy levels of the orbital electrons. In the first of the levels or shells in the atomic nucleus there is room for four particles, two neutrons with opposite spins and two protons with opposite spins. Thus the first nuclear shell when complete gives the alpha particle or helium nucleus, which is a particularly stable structure. Further shells may be built up containing 8, 20, 50, 82 or 126 protons or neutrons when complete, which has helped to explain the stabilities and abundances of various atoms. For example, lead, with a nucleus containing 82 protons, is the stable end product of all of the three natural radioactive series, whilst the most abundant lead isotope is that with a nucleus containing 126 neutrons.

Perhaps the most spectacular field opened up by the discovery of the fundamental particles in the present century is the study of nuclear reactions. The radioactivity of the heavy elements was recognized to be a case of spontaneous atomic change by Rutherford and Soddy in 1902, and the first case of an artificial atomic transmutation was discovered by Rutherford in 1919, when he bombarded nitrogen with alpha particles and produced fast moving protons. From 1921 to 1924 Rutherford and Chadwick produced protons from all the elements up to calcium by alpha-ray bombardment, except in the cases of the stable elements, helium, carbon, and oxygen, which had atomic weights that were multiples of four. Blackett in 1925 showed that heavier elements were produced by this process, oxygen as well as hydrogen nuclei being formed when nitrogen was bombarded by alpha particles.

Besides alpha particles, the bombardment of atoms with other particles was investigated. In 1932 Cockcroft and Walton accelerated protons to high energies and bombarded lithium with them, producing two helium nuclei. In the following year Lawrence and Livingston developed the cyclotron for speeding up charged particles, and with it studied the effects produced by bombardments with heavy hydrogen nuclei, or deutrons, consisting of a proton and a neutron. After the discovery of the neutron in 1932, the effect of neutron bombardment was studied, notably by Fermi, then at Rome. He found that neutrons slowed down by transit through substances containing much hydrogen, such as paraffin-wax or water, were particularly effective in bringing about nuclear reactions. Upon bombarding uranium with neutrons, Fermi obtained in 1934 another radioactive element which he presumed to be heavier than uranium. This work was taken up by Hahn at Berlin who by 1937 had claimed the preparation of several trans-uranic elements with atomic numbers ranging from 93 to 96. However, the Joliot-Curies at Paris in 1938 pointed out that the radioactive characteristics of the substances produced by bombarding uranium with neutrons resembled those of much lighter radioactive elements, notably radio-lanthanum. Examining the matter afresh, Hahn and Meitner in 1939 found that lanthanum, and other elements of moderate atomic weight, were produced by the bombardment of uranium, indicating that the uranium nucleus had split into two. The high packing fraction of uranium suggested that much energy would be produced through the loss of mass when the uranium nucleus split into lighter fragments, whilst the high ratio of mass number to the nuclear charge of uranium pointed to the probable release of several neutrons during the process. Joliot-Curie in 1939 showed that neutrons were given out by disintegrating uranium, so that the fission of one uranium nucleus would bring about the break-up of others in the vicinity under suitable circumstances. Such circumstances could be controlled, as in the atomic pile, or uncontrolled, as in the atomic bomb. Fermi, now at Chicago, set up the first atomic pile in 1942, and others in America by 1945 had produced the bomb. The high packing fraction of hydrogen and other light elements indicated that energy might be obtained by converting them into heavier elements, and subsequently such a nuclear fusion was achieved in the hydrogen bomb, using the nuclear fission of uranium to generate the temperatures required.

Chapter 45

Astrophysics and Theories of World Structure

DURING THE nineteenth century the scope of observational astronomy was considerably enlarged by a number of technical advances, notably the building of ever larger reflecting telescopes, the examination of the spectra of the stars, and the introduction of photographic techniques. The application of the theory of black body radiation has allowed the surface temperatures of the stars to be estimated, while the theories of atomic physics have suggested nuclear reactions to account for those temperatures and the rates at which the stars emit radiations. In another direction the information provided by the new methods has given a quantitative precision to the view, put forward by William Herschel at the turn of the eighteenth century, that the Milky Way consists of a self-contained galaxy of stars, of which the sun is a member, and that the small white nebulae are similar star galaxies much further away. The sizes of the star galaxies, their relative velocities, and the distances which separate them, may now be calculated, providing values for the spatial dimensions of the known universe and a time scale for its history, though the time estimates in particular vary considerably according to the theoretical viewpoint of the calculator.

The first spatial estimates made beyond the confines of the solar system were the measurement of the distances of the nearest stars from their apparent movement during the course of the earth's annual orbit round the sun. Such a parallactic movement, separate from aberration effects, was first observed in 1832 by Henderson in South Africa, and it was used for the determination of stellar distances notably by Bessel, 1784-1864, at Konigsberg, and Struve, 1793-1864, at Pulkova. The stellar parallax method could only be used to ascertain the distances of the nearer stars, but in 1912 another method of far greater range was discovered. It has been found that the stars varying regularly in brightness were of two kinds; firstly, pairs of stars which periodically eclipsed one another, and secondly, pulsating stars, termed Cepheid variables, which were distinguished by the particular way in which their brightness varied with time. Miss Leavitt, at Harvard, found in 1912 that Cepheid variables with the same period had the same intrinsic brightness. Thus, once the distance of a close Cepheid variable had been measured by the stellar parallax method, the distances of all the others with the same period could be determined from their apparent brightnesses, by applying the law of the diminution of light intensity with the square of the distance. Another method, similar in principle, was developed shortly afterwards, based upon the assumption that stars with the same kind of spectrum had the same intrinsic brightnesses, so that their apparent brightnesses were a measure of their relative distances from the earth.

In 1918 the 100-inch reflecting telescope was set up at Mount Wilson, and with it Shapley used these methods of determining stellar distances to estimate the shape and size of our galaxy. He found that the galaxy was a flattened disc, the plane of which was that of the Milky Way, its diameter being about 300,000 times the distance which light travels in a year and its thickness about 10,000 such light years. In 1924 Hubble, also at Mount Wilson, examined the small white nebulae, many of which had been found to possess a spiral structure since 1845 when Lord Rosse in Ireland had observed the first case of this kind. Hubble succeeded in resolving some of the nearer spiral nebulae into individual stars, identifying Cepheid variables amongst them. He was thus able to show that the spiral nebulae were at distances of the order of a million light years away from the solar system, lying right outside the galactic system of the Milky Way. Hubble showed furthermore that the brightnesses of the nearer nebulae fell within a range from about half to twice the average, a grouping much more concentrated than that of the individual stars. Hence the distances of the fainter nebulae could be determined from their apparent brightnesses, Hubble finding in this way that the most distant of the visible nebulae were some 500 million light years away.

Another important line of investigation derived from the principle enunciated by the Austrian physicist, Doppler, 1803-

53, in 1845 that the movement of a source of wave-motions changed their frequencies. A velocity of recession relative to an observer decreased the frequency of the wave motions, in the case of light the radiation became redder, while a velocity of approach increased the frequency, light vibrations shifting towards the blue end of the spectrum. By means of Doppler's principle, Huggins, 1824-1910, at London, showed in 1868 that the star Sirius was receding from the solar system with a velocity of twenty-nine miles per second. In 1912 Slipher, at the Lowell observatory in America, found by the same means that the Andromeda nebula was approaching the solar system at a speed of about 125 miles per second, but by 1917 he had discovered that this case was exceptional, most of the nearer nebulae apparently moving away with a velocity of the order of 400 miles a second. The spectra of the more distant nebulae were examined in 1929 by Humason, at Mount Wilson, and he found that the light which they emitted became progressively redder the further away they were from the solar system. In 1930 Hubble propounded the law that the red-shift in the light given out by a given nebula was directly proportional to its distance from the solar system. On the basis of Doppler's principle, Hubble's law indicated that each nebula was receding with a velocity proportional to its distance from our galaxy, the most distant nebulae moving away with enormous velocities of the order of one-seventh of the speed of light.

If it were assumed that the velocities of the nebulae had been constant throughout past time, it appeared that the nebulae had been clustered together about 1,800 million years ago. Then the nebulae had begun to recede from one another, an event marking the beginnings of measurable cosmological time, and setting a limit to the age of the universe. However, Eddington suggested that the velocity of nebular recession had increased with time in such a manner that the upper limit to the past age of the universe lay between 10,000 and 90,000 million years. Others maintained that the red-shifts were not due to Doppler effects at all, the nebulae being stationary. Zwicky supposed that the red-shifts were due to the gravitational drag of inter-galactic matter upon the light passing through it, and MacMillan thought similarly that radiation gradually lost energy whilst travelling over the enormous inter-galactic spaces, effects which would be the more pronounced the more distant the source lay from the observer, as the Hubble law required. Milne indicated that the laws of the emission of radiation might change with time, so that light given out by the distant nebulae some 500 million years ago and observed on the earth today was necessarily redder than the corresponding radiations of the present time. From the fact that the stars appear to have equivalent amounts of energy, Jeans had calculated earlier that the upper limit to the past age of the universe was 10^{12} to 10^{13} years, on the assumption that there had been a random distribution of energy amongst the stars in the beginning.

Such estimates of the past duration of the universe varied considerably, as they hinged a great deal upon the particular cosmological theory adopted, upon whether it were assumed that the space of the universe was finite or infinite, Euclidian or non-Euclidian, expanding or static, and if expanding, upon the particular rate of expansion at a given epoch. Jean's estimate, for example, was made before the appearance of the general theory of relativity, which threw doubts upon the whole basis of his calculation. The cosmological theories of the twentieth century are as numerous and varied as the ether theories of the nineteenth century, of which they are in a sense the historical heirs. They both belong to a tradition which has sought to explain the phenomena of nature in terms of a cosmic continuum, a geometrical space-time in the first case and a mechanical all-pervading ether in the second. To this tradition belongs the work of Einstein who, putting an end to the invention of ether models with his theory of relativity, constructed the first of the world models of the present century.

In 1917 Einstein considered the problem as to whether the space and matter of the universe were finite or infinite. Adopting the view of Mach that the mass of a body was determined by the rest of the matter in the universe, Einstein pointed out that in an infinitude of matter each object would possess infinite mass and inertia. On the other hand, if the universe had a finite boundary in Euclidian space, the matter inside the universe would not be in equilibrium with the empty space outside, and such a world would not constitute a stable system. To overcome these difficulties, Einstein suggested that the universe might possess a finite volume but no finite boundaries, as would be the case if space were the three-dimensional analogue of the two-dimensional surface of a sphere, which has a definite area but no boundaries, that is, no edges to the area. All point areas on the surface of a sphere are symmetrical and equivalent to one another, and so too would be the three-dimensional analogues, the point volumes of spherical space. Thus the unit volumes of matter in spherical space would be all alike, and there would be no special boundary cases, like the particles at the edge of a finite universe in Euclidian space. Because of the absence of special boundary cases, all observers in Einstein's world would be equivalent, observing the same phenomena and obtaining the same information, as the special theory of relativity required.

According to the general theory of relativity, an aggregate of matter is associated with a local curvature of space, which shows itself as a gravitational field. The curvature of space therefore varies from point to point in the Einstein world, but it is possible to obtain the general average curvature of space by smoothing out the local curvatures, or what is equivalent, by averaging out the matter in the universe throughout its volume. The number of unit volumes of matter in the Einstein world thus determine the curvature radius of its spherical space, just as the number of unit areas making up the surface of a sphere determines its radius. Deriving an expression relating the average curvature of space to the amount of matter in the universe, Einstein then calculated the mass of the universe and the curvatures of its space on the assumption that the density of matter in our own local cluster of galaxies, the Milky Way, Andromeda, and others, was the same as the density of matter in the rest of the universe.

Einstein constructed his world model before the large redshifts of the distant nebulae had been observed. He assumed that the velocities of bodies in the universe were small compared with the speed of light, so that the spatial structure of his world did not vary with time. Indeed time was a separate dimension in the Einstein world, space-time being a four dimensional analogue of a cylinder, a combination of spherical space and linear time. Following Mach, Einstein had assumed that a particle in a universe empty of other matter would have no mass or inertia, but de Sitter showed in 1917 that this would not necessarily be the case if space and time were combined so that space-time was the four-dimensional analogue of a sphere. A test particle, such as a nebula, introduced into Einstein's world would remain at rest if it had no initial motion relative to an observer, but in de Sitter's world it would immediately recede with ever increasing velocity. De Sitter's world was full of motion but contained no matter, while Einstein's world was full of matter but contained no motion. Both were extreme models, and later they were considered to be possible representations of the initial beginning, or the final end, of cosmic evolution, but not models of the universe as it is.

The stability of the plenum in the Einstein world was dependent upon a balance between gravitational attraction and a cosmic repulsion, while the emptiness of the de Sitter world was due to a dominance of the cosmic repulsion. After the publication of Hubble's law, which was generally interpreted to mean that the universe was expanding, Eddington and his former pupil, Lemaitre, showed that the Einstein world was unstable and that it would either expand or contract if disturbed, depending upon whether the disturbance favoured the cosmic repulsion or gravity. If the Einstein world expanded, the nebulae would recede one from the other and ultimately the density of matter in the universe would become zero, producing the conditions of the de Sitter world. Einstein and de Sitter showed in 1932 that if the universe were expanding, as Hubble's law seemed to suggest, a finite amount of matter expanding in infinite Euclidian space was a possible model of the world. Hubble himself preferred a static Euclidian model of the universe, as calculations based upon the expanding models indicated that the amount of matter in the universe was about a thousand times greater than the quantity contained in the nebulae alone. Such an amount of matter, present in the form of inter-galactic dust, Hubble believed would absorb more light than was in fact lost in the inter-galactic spaces.

Several other world models were invented, their diversity following from the fact that three quantities were required to specify a world model while only two could be determined by observation. The three unknown quantities were the mass of the universe, the curvature of space, and a cosmological constant, which had been introduced by Einstein to cover Mach's requirement that the mass of a body was determined by the rest of the matter in the universe. It was this cosmological constant which gave rise in Einstein's theory to the cosmic repulsive force. The two quantities which could be determined by observation were the rate of expansion of the universe, measured by assuming that the red-shifts were Doppler effects, and the mean density of matter in the universe, ascertained by assuming that it was the same as the observed mean density of matter in our local group of galaxies. De Sitter in 1932 enumerated nine main types of possible world models, based upon the conditions that both the cosmological constant and the curvature of space could be either negative, zero, or positive.

In the absence of observational criteria to decide between the various possible world models, some cosmologists, notably Eddington, and to a lesser degree, Milne, endeavoured to derive theoretical systems concerning the nature of the universe from the basic procedures of the scientific method. Comparing the scientist to a fisherman who caught only those fishes which could not get through the meshes of his net, Eddington suggested that the characteristics of the method of science determined the content and the nature of scientific knowledge. The scientific method was not a mirror which reflected the nature of the world, but a kaleidoscope which determined by its structure the images perceived. Some critics suggested that the net of the scientific method could not determine the qualitative features of the natural phenomena 'caught' and studied, but Eddington countered this argument with the claim that science was concerned only with the quantitative and measurable aspects of phenomena, not the qualitative.

In the method of the physical sciences, Eddington regarded the measurement of lengths as fundamental. He pointed out that spatial estimations involved four determinations, the observation of the two ends of a standard measuring rod and the observation of the two ends of the length to be measured. Such a methodological procedure was basic to the physical sciences, and Eddington considered that it was connected with the fundamental notion that space-time has four dimensions. Developing his system, Eddington attempted to calculate the universal constants which are pure numbers, such as the ratio of the masses of the proton and the electron, without assuming any numbers determined experimentally. It had been found that the pure number universal constants of nature fell into three main groups, which had values of the order of 10^{80}, 10^{40}, or less than about 2.10^3. Eddington indicated that the first set was dependent upon the total number of particles in the universe, N, the second set upon the square root of N, whilst the third set was independent of N. Since N was finite, Eddington based his calculations on the theory of a finite world model, namely, an expanding version of Einstein's static model. He estimated N directly by dividing the mass of the Einstein world by the mass of the proton, and also more abstractly by assuming that the mass of an electron derived from its charge and the presence of all the other particles in the universe, obtaining a value of the order of 10^{79} in both cases.

Eddington took the measurement of length as basic in physics and interested himself in the pure numbers connected with the structural features of the universe, whilst Milne regarded the measurement of time as fundamental and concerned himself with the temporal movement of cosmic processes. Following Einstein, Milne adopted the view that all observers in the universe were equivalent and that light always had the same velocity for those observers. He then showed that the determination of lengths could be reduced to the measurement of the times taken for light to traverse those lengths, thus dispensing with the rigid measuring rods which Einstein and Eddington had assumed essential for the making of spatial estimations. The measurement of length therefore depended upon the measurement of time, and Milne indicated that, in general, theories concerning the spatial structure of the universe were dependent upon theories of the movement of time. He pointed out that in observing distant objects, such as the nebulae, we must allow for the fact that the further away they are placed the earlier in the history of the universe the light coming from them started on its journey. The light perceived today started out from the nearer nebulae about a million years ago and that from the distant nebulae some 500 million years ago, and thus we must correct for the time lag more and more the further out we go in order to obtain a contemporaneous picture of the universe. But the correction applied depends upon the time scale employed, so that it is possible to construct different world models using different time scales.

Milne indicated that two time scales were fundamental. There was one in which the vibrations of light provided the units of time, known as the T time scale, and another to which the scale of pendulum clocks approximated, termed the τ time scale. If the T scale were adopted, the red-shifts in the light coming from the nebulae implied that the nebulae were receding with velocities which were proportional to their distances from our galaxy, since the theory of relativity required that the time units marked by a moving time-keeper, in this case the source of light vibrations in a nebula, should lengthen in proportion to the velocity of the time-keeper relative to the observer. To account for the recession, Milne showed in 1932 that, if they were initially in a confined space, any collection of freely-moving objects with uniform velocities, as the nebulae might be supposed to be, would appear after a time to be receding from one another with velocities proportional to the distances between them, as those with the larger velocities would have moved further in the given time. Such an effect would be a manifestation of the second law of thermodynamics, the nebulae moving away from one another like a collection of gas molecules released in empty space. On the basis of the T scale, Hubble's law thus indicated that the nebulae had been confined together in a small region of space and had begun to recede from one another some 2,000 million years ago.

If the number of nebulae were finite, it appeared that observers on the fast-moving nebulae at the edge of the expanding universe would obtain a view of the world different from that of observers on the slower nebulae, which had not moved so far from the region of their origin. Thus, to preserve the principle that all observers in the universe are equivalent, there must be an infinite number of nebulae, so that observers on the nebulae at the limits of our vision can observe further nebulae which we cannot see. With an infinity of nebulae, all observers in the universe would perceive an apparently spherical world of radius CT, where C is the velocity of light and T the time that has elapsed since the origin of the nebular recession.

Eddington suggested that the constants of nature were fixed

for all time, their values being determined by the characteristics of Einstein's world model from which the universe had developed by expansion. Milne showed, however, that something changed with time no matter which of the two time scales were adopted. According to the theory of the T scale, the gravitational constant and Planck's constant increased with time, so that objects gradually became heavier and sub-atomic events more uncertain. On the τ scale the gravitational and Planck's constants remained fixed, but the number of light vibrations occurring in a time unit of this scale gradually increased, so that light from a given source became shorter in wavelength with time. From the point of view of the τ scale the duration of past time had been infinite and the nebulae were stationary, the red-shifts indicating that the light from the nebulae had originated at epochs when there were fewer light vibrations in a given unit of time.

Another theory postulating the infinite existence of the universe, that of 'continuous creation', was put forward in 1948 by the Cambridge mathematicians, Bondi, Gold, and, a little later, Hoyle, in order to overcome the difficulties imposed by the assumption that the universe had had a beginning, and a lifetime of only some 2,000 million years. Bondi, Gold, and Hoyle, extended into the dimension of time Einstein's principle of the equivalence of all observers in space, assuming that the universe presents to any observer the same aspect as it has had throughout all past time and will have in the future. This 'perfect cosmological principle', as it was termed, led them to suppose that matter was created continuously throughout the universe at the rate of 10^{-43} grammes per cubic centimetre per second, in order to make up for the thinning out of the average density of matter caused by the recession of the nebulae. Such a rate of creation was too small to be detected, though the total amount for the observable universe was very large, being equivalent to the birth of some 50,000 stars per second.

The various theories concerning the spatial and temporal structure of the universe proposed hitherto have embodied a proportion of theory to observation which is large, and it now seems that further progress in this field will be considerably aided by, if it is not attendant upon, the advance of empirical knowledge. The 200-inch reflecting telescope set up at Mount

Palomar in 1949, for example, has already given results indicating that the lifetime of the universe must be doubled to 4,000 million years, which has removed some of the difficulties that the 'continuous creation' theory set out to overcome. The new radio astronomy, which reaches even further into space than the Mount Palomar telescope, has established, notably at the hands of Ryle and his co-workers at Cambridge, that the number of radio sources in a given volume of space increases with the cube of the distance from our galaxy, a result contrary to the predictions of 'steady state' or 'continuous creation' types of theory, which indicate that the number of sources in a given volume should remain constant or even decrease as the distance is increased. Such developments of astronomical technique are likely to have their impact upon other branches of astrophysics, such as the study of the constitution of the stars, though here the laboratory investigation of terrestrial matter is also of service.

Modern theories of the constitution of the stars are based on the observation, made by Russel at Princeton in 1913, that the brightnesses of a large group of stars were proportional to their surface temperatures. These stars formed a single sequence, and Russel suggested that the sequence manifested the various stages of the general course of stellar evolution. Following the gravitational contraction theory proposed by Helmholtz, Russel held that a star initially consisted of a diffuse nebulous mass which contracted under the force of gravity, becoming hotter and denser, until a critical density was reached when contraction no longer sustained the amount of radiation emitted and the star gradually cooled down. According to this scheme, the stars should be red hot when large, becoming white and then blue as they contracted, and finally red again when they became small and cooled down.

However Adams, at Yerkes observatory, discovered in 1914 that the small star accompanying Sirius emitted the kind of light which was given out by white hot bodies, the star being a white dwarf, and not a red dwarf, as all small stars should be according to Russel's classification. A decade later Adams found that the light emitted by the small companion of Sirius exhibited a gravitational red-shift which was large, indicating that the star had an enormous mass even though it possessed a small volume. Adams' discovery supported the hypothesis, put forward by Eddington and Jeans a few years previously, that the outer electron shells of an atom were ionized off at the high temperatures and densities obtaining in the interior of a star, leaving the atom much reduced in size. The gravitational contraction of stars could proceed therefore much further than had been imagined previously, producing densities of matter vastly greater than those met with on the earth. The Indians, Kothari and Chandrasekhar, showed that bodies containing more matter than the planet Jupiter would generate gravitational pressures sufficing to crush the atoms in their interiors, so that the volume of such a body would actually decrease as its mass increased.

Eddington and others demonstrated that most stars do not possess great densities, as the contractional force of gravity is balanced by the outwards pressure of radiation which, varying with the fourth power of the temperature, attained high values inside the stars. Moreover, the electrons stripped from the atoms of the stellar interiors would behave like the molecules of a gas, generating a pressure at high temperatures which would oppose the gravitational pressure of the overlying matter. In the case of the sun with a surface temperature of 6,000 degrees, Eddington calculated that the internal temperatures must be of the order of 20 million degrees to produce the electronic and radiation pressures balancing the contractional forces of gravity. He showed furthermore that the total amount of radiation emitted by a star depended upon its mass, rather than upon its size, for the contraction of a star of given mass would bring about an increase in temperature which would compensate for the concomitant decrease in the surface area giving out radiation. On the basis of this theory, Eddington endeavoured to calculate the theoretical luminosity of the sun, on the assumption that the sun was composed largely of the elements with moderate atomic weights, such as iron and its neighbours in the periodic table. The theoretical figure was about a hundred times too large, from which Eddington concluded that there must be a great deal of hydrogen in the sun and other stars.

With the development of atomic physics, it became apparent that, at the high temperatures of stellar interiors, the thermal velocities of the atoms stripped of their electrons would be of the same order as the speed of the artificially accelerated particles used in the laboratory to study nuclear reactions. It seemed probable therefore that nuclear reactions occurred inside the stars, these reactions providing the energy emitted as radiation. In 1938 Bethe at Cornell and Weizsacker in Germany worked out a cycle of nuclear reactions, transforming hydrogen into helium, which could occur at the temperatures obtaining within the sun. They showed that the carbon isotope of mass 12 could react successively with three protons, forming nitrogen isotopes of increasing atomic weight, and that the union of a fourth proton with the nitrogen isotope of mass 15 would give an atom of helium and regenerate the original carbon isotope. An atom of helium was somewhat lighter than four protons, the mass lost being transformed into energy during the course of the reaction cycle. The rate at which hydrogen was converted into helium in a given star, and thus the rate at which radiation was emitted, depended upon the amount of carbon in that star. From the solar spectrum it was estimated that there was about 1 per cent carbon in the sun, and with this figure Bethe showed that the theoretical rate at which energy would be generated by the carbon-nitrogen cycle in the sun was equal to the observed rate at which the sun emitted radiation.

Eddington had estimated, from his relationship between the luminosity and the mass of a star, that the sun contained either 35 per cent or 90 per cent of hydrogen. Calculations based on Eddington's lower estimate gave the sun a future life span of 35,000 million years, assuming the present rate of emission of energy to be the average over that period. In 1947 Hoyle gave reasons for preferring Eddington's higher estimate, and he suggested that the sun and the stars generally could replenish their hydrogen stocks from the matter of interstellar space.

Hoyle has drawn attention to the importance of the highly dispersed matter existing in the interstellar and intergalactic spaces. Amongst other things, he suggests that the interstellar matter picked up by stars can lead to the formation of binary systems, the two stars forming such a system being forced together by the greater impact of the interstellar gas on their outermost sides than on their adjacent sides, so that the two stars converge and ultimately circle round one another. Lyttleton in 1936 advanced the hypothesis that the solar system might have been formed from such a binary system after a collision between an intruding star and the sun's companion, the sun's gravitational field capturing some of the remaining fragments, which formed the planets. Developing the star-pair theory of the origin of the solar system, Hoyle and Lyttleton have suggested more recently that the sun's companion might have been a small dense star that, spinning more rapidly the more it contracted, broke up ultimately into many fragments, some of which were captured by the sun's gravitational field. In support of this theory the average of the known distances between the stars of binary systems is of the same order as the distance between the sun and the bulk of the planetary material.

Chapter 46

Science and the National Movements in Italy and Germany

IN GERMANY and Italy the growth of the sciences, like other historical movements, have displayed a not dissimilar course of development during the modern period. The Germans and the Italians were the pioneers of modern science, reaching their first peak achievement with the work of Kepler and Galileo respectively in the early decades of the seventeenth century. But they did not sustain the effort, and almost two centuries were to elapse before they produced men of science who were at all comparable. During the late medieval period, the Germanic, and more particularly the Italian lands were the most cultivated regions of Europe, excelling especially in the practical techniques and the theoretical disciplines which were the twin roots of science. During the early modern period the men of these lands were favourably placed with regard to the development of science, and thus they were the pioneers, though they lost their leadership in the field of science when they lost it in others. The great geographical discoveries opened up opportunities which were the more effectively exploited by England, France, and Holland, and it was these lands that became the main centres of European endeavour, in science no less than other fields, England and France retaining their leadership in science right down to the mid or the late nineteenth century.

Meanwhile the Germans and the Italians, in most departments of life, adhered to traditions which had been laid down in the sixteenth century. Politically they remained divided up into a number of petty principalities, in contrast to the united nation states of Britain and France, whilst in science they retained an active interest but produced little that was novel. It is noteworthy that, of the ninety or so scientific journals founded before 1815, fifty-three were German, nine were Italian, fifteen were French, and eleven were English, whilst

America, Switzerland, Sweden, Holland, and Belgium had one each. For such a number of scientific journals to be founded, there must have been a considerable interest in science amongst the Italians, and especially the Germans, but it seems that this interest was not active enough to produce markedly novel advances. When the science of the Germans and the Italians showed signs of revival in the late eighteenth century, they took up, so to speak, the threads which had been dropped in the century before. Galvani and Volta continued the experimental traditions of the Florentine Academy of Experiments, while the nature-philosophers in Germany took up the speculative tradition of Copernicus and the early work of Kepler, exalting Kepler above Newton, both out of affinity of tradition and affinity of nationality.

In one important direction the Italians and the Germans did make a contribution during the bleak period of the eighteenth century; they developed a feeling and an understanding of historical development, which did not come easily to the English and the French with their static, mechanical view of the natural world and human society. Such an understanding appeared notably with the *New Science* of the Italian, Vico, 1668-1744, and it was developed especially by the school of thought to which the German nature-philosophers belonged. We have seen how the view that the natural world was a product of a general developmental process led the Germans to notable achievements in the sphere of embryology, the study of the development of the individual organism. However, it was a viewpoint which was not conducive to empirical scientific research at first, for the Germans were highly idealistic, creating a rich world of ideas to compensate for the poverty of their material existence. Hegel, the master of the German school, expressed the sentiment well when he wrote:

'The soul takes refuge in the realms of thought, and in opposition to the real world it creates a world of ideas. Philosophy begins with the decline of the real world: when she appears with her abstractions, painting grey on grey, then the freshness of youth and life is already gone, and her reconciliation is not one in reality but in an ideal world.'

In their world of ideas the German philosophers saw an historical process which had a quite definite destiny—the German way of life. In Hegel's system, after the Absolute Idea had engendered the dialectical logic within itself and moved throughout the natural world, it entered human history and epitomized itself in the German principle. 'Europe is the goal of all history,' he wrote. 'The Germanic principle is the reconciliation and solution of all contradictions.' Hegel was by no means alone in this sentiment: the nature-philosophers were great nationalists, and they looked forward to a time when the numerous German-speaking principalities would be united into a single, powerful Fatherland. The Napoleonic conquest of Germany was felt to be particularly humiliating, and when Napoleon was defeated at Leipzig in 1813, one of the more influential and practically-minded nature-philosophers, Lorenz Oken, whose scientific theories we have already met with, published a work entitled *New France, New Germany* in which he discussed how France might be rendered harmless and Germany reconstituted politically. To this end he suggested that the various German states be united. 'Any number of human beings that speak the same language form one people', he wrote, 'and must be held together by one and the same law.'

In 1817 Oken founded a scientific and literary journal called *Isis*, which became a vehicle for the dissemination of science and nationalist sentiment. Later in the same year he reported a student demonstration at Jena, the Wartburg Festival, held in commemoration of Luther's Reformation and the liberation of Germany at Leipzig. The demonstration was not only nationalistic but also liberal in sympathy, and for reporting it Oken was ejected from his chair at Jena. However, he continued to publish *Isis*, and in 1821 he conceived of a scheme for calling a national congress of German-speaking scientists and doctors. The purpose of such a congress was, Oken averred, 'the good of science and the well being and honour of the Fatherland'. Accordingly Oken sent out invitations to a large number of scientists and doctors, but he found it most difficult to organize his congress, as the German intellectuals, then perhaps the most national of the Germans in outlook, were still remarkably parochial in their views and habits. Commenting on a reply he had received from a certain Professor Goldfuss, Oken wrote:

'In this letter you see the true German. . . . Scruples with him as to purse, scruples as to the journey, scruples as to strange faces, scruples as to quarters, scruples as to the assembly rooms, scruples as to the governments.'

However, the first meeting of the association of German-speaking scientists and doctors, the Deutscher Naturforscher Versammlung as it was called, was held at Leipzig in 1822. Some twenty scientists who had published work attended, together with sixty guests. Subsequent meetings, which were held annually in one or other of the main German cities, gradually grew larger, some six hundred attending the 1828 meeting at Berlin, and a thousand the 1842 meeting at Mainz. Oken averred that these meetings were 'the spiritual symbol of the unity of the German people'. At the second congress he laid it down that members should adopt a 'lively and impromptu delivery in place of the painful reading aloud of written manuscripts', in order to make the congresses more popular. 'The German scholars were still the slave of their manuscripts', however, and Oken failed to secure his ruling. Indeed as late as 1832 a member addressed the congress in Latin.

At first the national science congresses were regarded with suspicion by the rulers of the German states, as they were both liberal and nationalist in tone. Members attending the first meeting at Leipzig refused to allow their names to be recorded, for fear their governments should find out. Metternich in Austria suggested, to Viennese scientists applying for passports, that it would be contrary to their own interests to go, with the result that the Austrian scientists were not represented until 1832, when the annual meeting was held in Vienna. However, quite early on the Prussian government saw that the national science congresses could become a controlled force for German unity, and they extended their patronage to the meetings from 1828 on. Thereafter the congresses came more and more under the control of the German governments, the state acting as host for a particular annual meeting appointing the president of the congress for that year and the secretary who organized and ran the meetings. At the Berlin meeting of 1828 Alexander von Humboldt, the geographer and explorer, was made president. In his presidential address he asked:

'How can our national unity be more forcibly expressed than by this association? . . . Beneath the protection of noble princes this assembly has yearly grown in interest and importance. Every element of disunion which difference of religion or form of government can occasion is here laid aside. Germany manifests herself, as it were, in her intellectual unity, and . . . this feeling of unity can never weaken in any of us the bonds that endear us to the religion, the constitution, and the laws of our country.'

Charles Babbage, professor of mathematics at Cambridge, attended the Berlin meeting of the German scientists in 1828, and upon his return he suggested the foundation of a similar institution in Britain. His suggestion led to the formation of the British Association for the Advancement of Science in 1831. Later came the All-Italy association of scientists in 1839, then the American Association for the Advancement of Science in 1848, and in 1872 the French Association. In Britain, France, and America the national question did not arise, the British and the American associations being formed to link together the scattered scientific activity in their countries, whilst the French association was formed to take science from Paris to the provinces. In Italy, however, the national question did arise, and the All-Italy science congresses had a character similar to those in Germany.

In Italy too, national sentiment had risen during the period of Napoleon's conquests, but, in contrast to Germany, the national movement in Italy was not directed against the French, for under Napoleon and his brother Italy had been united and reforms had been introduced. The Italian national movement was directed against Austria, who controlled or influenced the eight separate Italian states that were set up again in 1815, and the scientific expression of this movement crystallized round one of Napoleon's nephews, Carlo Bonaparte, 1803-57, who returned to Italy from America in 1837.

Like his uncle, Carlo Bonaparte was interested in science, publishing a work on American birds in 1832 and another on the fauna of Italy ten years later. National congresses of Italian scientists had been suggested by the Florentine liberal journal, *Antologia*, in 1831, but they became a reality only in 1839 when the first meeting was held at Pisa. This congress was called notably by Carlo Bonaparte, Paola Savi, professor of natural history at Pisa, Vincenzio Antinori, director of the Florence museum of physics, and his assistant, Giovanni Amici, who was one of the inventors of the achromatic microscope.

The first congress at Pisa was attended by some four hundred persons, and further congresses were held annually for the next nine years, the last at Venice in 1847 being attended by over one thousand seven hundred. As in Germany these science congresses were the first representation of national opinion. The Lombard liberal journal, *Rivista*, described the first congress as a 'cohort of illustrious men, of scholars come together from every part of the fair land where *si* is heard, to confer, to become reciprocally acquainted, to co-operate fraternally in the splendour and progress of science, and in the glory of the common country'. In keeping with the more empirical tradition of Italian science, more practical proposals for securing national unity were made at the Italian congresses than at the German ones. The building of railways was suggested to 'stitch up the boot of Italy'. Even forcible action was proposed. Carlo Bonaparte at the 1844 Milan congress spoke of 'freeing the Lombards from slavery', and at the final congress of 1847 the historian, Il Cantu, closed the section on geography and archaeology with the words that he would like 'to plunge a knife into the hearts of those who would not build Italian unity'.

In the following year came the abortive Italian risings, which the Austrians suppressed, and with them, the national scientific congresses. Italian scientists who had taken part in the risings fled abroad, one of the most notable being the chemist, Cannizzaro, 1826-1910, who went to Paris. Later, in 1860, he attended the Europe-wide conference of chemists at Karlsruhe, which had been called to decide upon some basis for the determination of atomic weights and valencies. Cannizzaro, at the conference, suggested that Italian science had long been neglected, his fellow countryman, Avogadro, having formulated a hypothesis some fifty years before which provided just the basis for which the chemists all over Europe were now seeking. As soon as the conference was over, Cannizzaro returned to Italy in order to join the rising of Garibaldi and his Thousand in Sicily, which, with a similar movement in northern Italy, led to the unification of the country in 1860. However, the annual scientific congresses did not start again immediately. One was held at Siena in 1862, another at Rome in 1873, and a third at Palermo in 1875, but it was 1907 before the annual congresses were resumed on a purely scientific footing as the Italian Association for the Advancement of Science.

Meanwhile in Germany the national scientific congresses were held uninterruptedly from 1822 on. At first the sentiment behind the congresses had been a harmonious combination of nationalism and liberalism, but the two became no longer compatible when the Prussian landowning aristocracy, the Junkers, began the task of uniting the German states in the 1860's. There was, therefore, a split into two camps, the nationalists endeavouring to build German national unity in conjunction with the Junkers, and the radical liberals, who saw their prime aim as the replacement of the Junker autocracy by a middle-class democracy. When this division was becoming acute in the 1860's, Darwinism came to Germany and was debated at the annual national science congresses. The radical liberal biologists, notably Haeckel, associated themselves, and were associated by their opponents, with the cause of defending Darwinism in Germany, and thus Darwin's theories raised a greater storm there than elsewhere.

In 1863 at the national congress of the German scientists held at Stettin, Haeckel's exposition of Darwinism was not unsympathetically received, even though he generalized the theory to cover himan society. Evolution, he said upon that occasion, is 'a natural law which no human power, neither the weapons of tyrants nor the curses of priests, can ever succeed in suppressing'. However, after the unification of Germany by Bismarck and the Franco-Prussian war, Haeckel was differently received. At the 1877 congress held at Munich he was taken to task by one of his former teachers, Virchow, for supporting what Virchow considered to be an unproved and mistaken theory. In his early days Virchow had been both a nationalist and a liberal, being dismissed from his post at the Charite Hospital, in Berlin, for supporting the abortive liberal revolution of 1848. In 1860 Virchow accepted Darwinism for a time, but later he changed his opinion. Virchow also abandoned something of this earlier liberalism, accepting, if not supporting, the Junker hegemony. In 1878 the teaching of Darwinism in the German secondary schools was forbidden by the Prussian minister of education, and Virchow supported the act from his seat in the Reichstag. At the Munich congress of the German scientists in 1877, Virchow declared that Darwinism was a socialistic doctrine, and indeed Darwin's theory did become associated with socialism in Germany, as radical liberalism declined and socialism took its place. The German socialists welcomed Haeckel's evolutionary monism, and he catered to this public by writing popular expositions of his views, such as *The Riddle of the Universe* (1899), which was widely read by the German working classes.

After Haeckel, the liberal-Darwinian tradition all but came to an end, and the main trend in German science became one of ever-increasing nationalism. In 1914 a manifesto was issued, signed by ninety-three eminent German scientists and scholars, including such men as Baeyer, Ehrlich, Haber, Ostwald, and Planck, which accused the contemporary British and French scientists of plagiarism, and claimed that the priority of German scientists to a number of discoveries had not been acknowledged outside of Germany. A number of eminent German-speaking scientists did not associate themselves with the manifesto, notably Einstein, but they were in a minority. After the First World War some German scientists, particularly Lenard and Stark, became extreme nationalists, advocating the characteristic doctrines of Hitler's national-socialist party, such as the inherent superiority of the German race and its attainments, and so on, even before that party was formed. Few German scientists went so far as Stark and Lenard, but when the national-socialist party came to power in 1933 the majority of them condoned its policies and actions, supporting the drive for a greater Germany, if not some of the methods adopted to attain that end. Some German scien-

ists, notably Max von Laue, remained untouched by nationalism throughout the period of the Third Reich, but again they
were in a minority.

From 1933 German nationalism began to weaken German
science, for the most part quite inadvertently. A basic conflict
existed between the national-socialist view that there was a
privileged race, and privileged persons within that race who
had a superior understanding given by intuition, and the ethos
of modern science which has always held that the majority
of men are equal and equivalent observers of nature, viewing
the same things and arriving at the same conclusions, given
the apparatus and adequate training. The postulate that all
observers in the universe were equivalent and symmetrical
was basic to Einstein's theory of relativity, which was especially condemned by Stark and others. The emphasis placed
upon privileged intuition, as opposed to communal observation and reason, led to a decline in the number of students
reading scientific subjects in the German universities. The decline was most marked in theoretical physics, the number of
German students reading mathematics with natural science
in the winter term of 1932-33 being 12,951, whilst there were
only 4,616 in the corresponding term of 1936-7. The decline
was less marked, though still large, in the more practical
subjects, such as engineering, where the numbers of German
students at the two periods were 14,477 and 7,649 respectively.
The quality of the education offered to those students also
declined, as the scientists who were considered to be politically
or racially unacceptable were ejected from their posts, and
they were replaced by men who were often selected primarily
for their national-socialist sympathies.

The smaller decline in the number of students reading practical scientific subjects reflected an empiricism which was encouraged in German science under the Third Reich. In 1937
Stark published an article in *Das Schwarze Korps,* and also
in the English journal, *Nature,* in which he contrasted the
pragmatic or practical science, which he favoured, against the
dogmatic or theoretical science, which he deplored, and
which he claimed was largely a Jewish product. Empiricism
was favoured in German science under the Third Reich because
it agreed with the glorification of the man of action, and be-

cause the theories of science tended to contradict the tenets of national-socialism. In physics there was the postulate of the equivalence of all observers, whilst the findings of the biologists and the anthropologists did not always agree with the German racial theory.

The spirit of national-socialism gradually infected the German scientists, and they began to see themselves as the great men and the leaders of science, instead of members of the community of science on an equal footing with others. There appeared a tendency towards self-glorification, both on the national level and the individual level. Lenard averred that it was he, and not Röntgen, who had discovered X-rays, whilst Gerlach stated, in connection with atomic research, as late as December 1944, 'I am convinced that we at present have still an important lead over America.' The same view was expressed in an official report on nuclear physics sent the year before to Goering, who was the titular head of the German research council. The leadership principle again did not encourage heads of research departments to adopt ideas put forward by their juniors. It appears that Heisenberg ignored a suggestion put forward by Houtermans in 1941, proposing the construction of a uranium pile to make elements heavier than uranium for use as atomic explosives.

The infusion of national-socialism into German science led to many things, such as the revival of astrological prognostication, and the pointless and cruel experiments performed on human beings by some German medical men, but the most important single historical consequence was the decline of fundamental research and its application in Germany, and with it the decline of her military power. The craftsmanship and the technical inventiveness of the German engineers remained high in the 1930's and '40's, as was manifest in their magnetic and acoustic mines, the V1 and V2, and their aircraft, but in other matters dependant upon fundamental science they had less success. Apart from Houtermans, who was ignored, no one in Germany appears to have thought of an atomic bomb composed of an element heavier than uranium, until very near the end of the Second World War. The possibility of developing a bomb composed of the light uranium isotope of mass 235 was suggested, but the idea was dropped, as the German scientists thought that the separation of the uranium isotopes was impossible. However, the separation was carried out in America, and so too was the preparation of elements heavier than uranium, by means of the uranium pile. The German scientists conceived only of a uranium pile for use as a source of energy, or as a bomb if very large. By mid-1945 they had not yet constructed such a pile, that is, they had not reached the stage attained in America by the end of 1942.

In the field of radiolocation the Germans did not get beyond the limit set by the use of conventional radio valves, namely, the generation of radio waves of the order of a metre or so in wavelength. In Britain the development of the magnetron gave waves of a few centimetres or so in length, affording much greater precision in the location of objects and less interference from extraneous sources. Again, the Germans did not appreciate the value of operational research, the use of scientific methods for ascertaining the most effective way of deploying limited military resources, a development which brought about a great saving of men and materials on the Allied side in the Second World War. The Germans underrated the value of science in general during the Second World War, and drafted many of their younger scientists into the armed forces. All in all it would transpire from the German experience that a society engendering and sustained by values which are antagonistic to those of science will become historically less and less effective in the modern world.

Chapter 47

Some Aspects of American and Soviet Science

THE TWO countries that have risen to prominence during the present century have scientific traditions which stood in marked contrast to one another during their formative periods. The primary contributions to science made by the Russians of the pre-Soviet times were theoretical in character —a non-Euclidian geometry, the definitive form of the periodic table, and the theory of conditioned reflexes; whilst from the Americans of the same period came mainly practical and applied forms of scientific activity—anaesthetics, the telephone, the aeroplane. Willard Gibbs, 1839-1903, at Yale, was here a notable exception, but his work was not appreciated in his own country, nor in Europe for some time. Today the science of both countries is more balanced, and little remains of the original contrast. American science has already lost much of its empirical flavour, and theoretical controversy of a philosophical or ideological character is a declining feature of Soviet science. National idiosyncrasies have been swamped in both cases by a tremendous growth of scientific activity, which has entailed the use of procedures and concepts which are universal.

In origin, the American scientific tradition derived from that of England, and some similarity remained between the two traditions for a considerable time. From the start, however, an emphasis was placed on the practical side of science in America, even more than in England. During the seventeenth century, we noted that there was something of an alliance between Puritanism and scientific activity in England, and we find that it was the Puritan states of America which contributed the greatest number of fellows to the Royal Society during the colonial period. Eleven of the American fellows came from New England, three from Pennsylvania, three from Virginia, and one from the Carolinas. Puritanism itself tended to give a practical orientation to science, useful applications of science being regarded as 'good works'. The first notable American scientific society illustrates in its name the utilitarian orientation of science in the New World: it was the *American Philosophical Society for Promoting Useful Knowledge*, which was founded by Benjamin Franklin, 1706-90, at Philadelphia in 1743.

With the movement for independence from Britain, which came to a head in the years 1775-82, there appeared in America a general intellectual awakening, and, in particular, a burst of scientific activity. The most important American scientist of the period was Benjamin Franklin, who found that lightning was electrical in character, and who based his invention of the lightning conductor upon the discovery. In England there was a controversy in 1780 as to whether a pointed rod or a rounded sphere made the better lightning conductor, a controversy which appears to have been not entirely unconnected with the fact that Franklin was a leading figure in the American independence movement. On the other side of the movement stood another notable American scientist, Count Rumford, though it seems that he did not become greatly interested in science until after he had emigrated to Europe. Rumford was the main founder of the Royal Institution in London, and, curiously enough, it was an Englishman, James Smithson, 1765-1828, who provided the means for the foundation of the corresponding Smithsonian Institution in Washington. Smithson was the natural son of the first Duke of Northumberland, who refused to acknowledge the relationship publicly, and in consequence Smithson developed an antipathy to the aristocracy in general, supporting the American movement for independence and the French Revolution. At his death in 1828 Smithson bequeathed his fortune to the American government, so that they might found an institution for 'the increase and diffusion of knowledge amongst men'. The American government obtained the money in 1838, and Joseph Henry, 1799-1878, was made the first director of the Smithsonian Institution in 1846. Henry noted that in America,

'though many excel in the application of science to the practical arts of life, few devote themselves to the continued labor and patient thought necessary to the discovery and development of new truths.'

Henry, therefore, endeavoured to make the Institution a centre of fundamental scientific research, though he was often called upon to undertake applied research, being appointed, for example, chief adviser to the Northern government on military inventions during the American Civil War.

A man who had considerable influence upon the formation of the American scientific tradition was Thomas Edison, 1847-1931. Henry Ford said of Edison that he

'definitely ended the distinction between the theoretical man of science and the practical man of science, so that today we think of scientific discoveries in connection with their possible present or future application to the needs of man. He took the old rule-of-thumb methods out of industry and substituted exact scientific knowledge, while, on the other hand, he directed scientific research into useful channels.'

Edison's 'invention factory', set up at Menlo Park in 1876, was the prototype of the large industrial research laboratories which accounted for the greater part of the resources devoted to scientific research in America before the Second World War. In these laboratories men more versed in fundamental science than Edison were employed, and they made advances of considerable importance, such as those, for example, of Irving Langmuir, who worked in the laboratories of the General Electrical Company from 1909 to 1950.

After the Second World War considerable concern was expressed in the United States concerning the one-sided character of American science. The atom bomb had been developed largely by technical skill and substantial resources, the fundamental scientific knowledge having been developed mainly in Europe, and the German experience seemed to indicate that a failure in the field of fundamental theory would lead to a political and military failure in the long run. At the request of President Roosevelt, Vannevar Bush, the war-time director of the Office of Scientific and Research Development, drew up an official report on the matter which was published in 1945.

'Our national pre-eminence in the fields of applied research and technology', wrote Bush, 'should not blind us to the truth that with respect to pure research,—the discovery of fundamental new knowledge and basic scientific principles,—America has occupied a second place. . . . In the next generation, technical advance and basic scientific discovery will be inseparable; a nation which borrows its basic knowledge will be hopelessly handicapped in the race for innovation.'

To remedy this state of affairs, Bush suggested the setting up of a National Research Foundation which was to be devoted to the pursuit of fundamental scientific research alone. Two years later, Steelman presented his report to the President of the United States on *Science and Public Policy*, which made the same points with more detailed documentation. The matter was debated in the American Congress, and in 1950 the National Science Foundation Act was passed, allocating funds for basic research. But even here the American practical tradition asserted itself, for a clause in the Act specified that the Foundation should initiate and support such applied researches of a military nature as might be requested by the Secretary of Defence.

The utilitarian character of the American scientific tradition is connected perhaps with the fact that, until recently, the United States was comparatively sparse in population and yet rich in natural resources. Such a situation placed a premium upon labour-saving devices, and upon methods of communicating and travelling over long distances, so that enquiry was orientated towards the attainment of these practical ends. Moreover, in pioneer communities generally, a high place has been accorded to the useful arts, and the man who was primarily a theoretician has been regarded as something of a failure. In the latter part of the nineteenth century, for example, there was a movement to replace Willard Gibbs at Yale, as his theoretical researches on thermodynamics and statistical mechanics seemed to be of little practical use. Nowadays, of course, a great deal of basic research is carried out in the United States, but the utilitarian tradition persists, colouring and giving shape to the most abstract and theoretical of disciplines—philosophy. A school of thought almost entirely

American in origin is that of philosophical pragmatism, which arose notably with Charles Peirce, 1839-1914, and William James, 1842-1910. In the field of the philosophy of science, the pragmatists suggest that the meaning of a scientific definition, conception, or theory, is nothing more than the set of operations which it entails. Thus 'length' would be defined by the series of practical operations involved in the making of spatial measurements. Such operations defining a conception are not necessarily physical: they can be pencil-and-paper operations, or purely mental, as in the case of some mathematical definitions, but they are conceived of essentially as *activities*, in contrast to the more static and contemplative European approach to the problem of what constitutes the meaning of a scientific idea or definition. The pragmatists in general tend to the view, as James put it, that 'An idea is "true" so long as to believe it is profitable to our lives.' 'The true is only the expedient in our way of thinking', wrote William James, 'in the long run, and on the whole of course.' Such a philosophy is essentially an explicit and generalized expression of the point of view of the pioneer and the practical man, whose primary concern with ideas is their utilitarian significance.

In contrast to the America of the nineteenth century, there was no shortage of labour in the old Russia, while the natural resources then known were closely controlled by government awarded monopolies, of a type similar to those of early Stuart England or eighteenth-century France. As in England and France at those earlier periods, such a state of affairs engendered a spirit of criticism and revolt, and the Russian intellectuals of the nineteenth century, like the French philosophers of the century before, used the theories of science to question the beliefs sustaining the society in which they lived. Amongst other things, they took up Darwinism, emphasizing a facet of Darwin's theory—the cooperative relationship between the organisms of a given species—which seemed to support their social theories. The zoologist, Kessler, of St. Petersburg published a work *On the Law of Mutual Aid* in 1880, and his followers, Prince Kropotkin and Novikov, collected materials illustrating intraspecific co-operation which were published two decades later. Perhaps it may be said then, that the Russian science of the nineteenth century had a theoretical and speculative character for much the same reasons as French science had during the eighteenth century: technical enterprise and the application of science were limited by the conditions of the period, whilst, at the same time, those conditions provoked theoretical criticisms, in which science became involved, and which gave shape to the characteristics of the science produced during that period. The factors giving rise to the theoretical-practical contrast between the characters of Russian and American science of the nineteenth century in fact resembled those leading to the similar contrast between the French and English science of the eighteenth century. Indeed, just as the American scientific tradition derived in the beginning from that of England, the Russian tradition came at first from that of France. The St. Petersburg Academy of Sciences, founded by Peter the Great in 1724, was modelled on the Paris Academy of Sciences, and various French savants, the most notable being Diderot, journeyed to St. Petersburg later in the eighteenth century. As we have seen, the St. Petersburg Academy was staffed for some time by foreign scientists, first the Swiss, Euler and the Bernoullis; then Germans, such as the embryologists Wolff and von Baer. The first notable Russian academician was Lomonosov, 1711-65, an anti-phlogiston theorist quoted by Lavoisier. Later, scientific societies were set up elsewhere, often in conjunction with the local university, as at Moscow, Kazan, Kharkov, Kiev, etc. From the far-distant Kazan came Lobachevsky, 1793-1856, who in 1830 developed the first comprehensive system of non-Euclidian geometry.

During the Russian Revolution, as during the French, the activities of the scientists were directed towards practical ends, whilst attempts were made to train more scientists, and so to increase the volume of scientific activity in the country. The disorders that followed the Russian Revolution were greater than those that followed the French, and such attempts did not reach fruition until the 1930's. The late president of the Soviet Academy of Sciences, Vavilov, stated in his short history of Russian science that the student body of the Soviet Union reached the pre-revolution figure of 112,-000, distributed over 91 universities and colleges, by the late 1920's, and the figure of 667,000, distributed over some 800 institutions, by 1941. An index to the amount of work turned out by the students so trained is provided by the number of Russian science periodicals publishing original research papers. According to Vavilov, there was one such periodical in the field of the physical sciences, with a circulation of about 200, before the revolution, whilst in 1948 there were five, each with a circulation of some 5,000. During the 1950's it became apparent that the Soviet scientific potential was considerable, rivalling if not exceeding that of America, and it became the practise to translate the main Soviet scientific periodicals into English in full. About the same time, or a little earlier, much of the scientific literature written in English was regularly translated into Russian in the Soviet Union.

The philosophy that was harnessed to the spirit of revolt in the old Russia, and which has influenced the science of the Soviet Union, namely dialectical materialism, originated with Karl Marx, 1818-83, and was first applied to scientific questions by his collaborator, Frederick Engels, 1820-95, in the *Dialectics of Nature*. The scientists of the Soviet Union took up this work of Engels in the 1920's, and for some time, notably 1945-55, controversies concerning the relationship between dialectical materialism and scientific theory were a conspicuous feature of Soviet science. The dialectical materialists hold that the theories of science should be based upon the following points of view. Firstly, the conception that there is a material universe, which existed before man appeared, and which would continue to exist without man's presence. Secondly, that all the objects and systems in the universe are interconnected by causal networks, which man can study and elucidate indefinitely by means of the scientific method, the knowledge so obtained gradually approximating more and more to the actual way that nature works without any limit. Thirdly, that the greatest insight into the systems of nature can be obtained by enquiring into their origins, the processes of their formation, and their historical evolution. Fourthly, that such processes should be found to consist of two primary elements, which oppose and interact with one another, providing the driving force of those processes. Fifthly, that each natural process is inherently self-limited and comes to an end when the inner driving conflict is annulled or resolved, but that this in itself prepares the way for the operation of another and quite different process. In other words, the quantitative movement of any process will ultimately bring about a qualitative change in the character of that process.

Not one of these tenets by itself is unique to dialectical materialism, though the particular combination of them is. The German nature-philosophers upheld the last three, but denied the validity of the first two, for they asserted that mind or spirit was the ultimate reality, nature being an externalised manifestation of the self-movement of spirit, in which alone there was a causal network of development. The mechanical philosophers of the nineteenth century, for the most part, upheld the first three tenets, but not the last two, maintaining in general that the driving forces of natural processes consisted of linear cause-and-effect chains, not complexes of interacting elements in which the distinction between cause and effect was obscured. However, nature-philosophy was abandoned during the nineteenth century, whilst doubts have been thrown upon the tenets of the mechanical philosophy in the present century. The school of thought stemming from Mach suggested that the postulate of an independent material world was a metaphysical superfluity, for only sensory perceptions were given directly to the scientist, and whilst these had a certain heuristic use when organised into laws and relationships, they gave no knowledge of the hypothetical material world in itself. Again, Heisenberg's Uncertainty Principle implied that there was a definite limit to the knowledge a scientist could obtain of the physical world: there would always be an irreducible uncertainty in the determination of the momentum and position of a subatomic particle. Finally, doubts have been cast upon the utility of the historical approach, notably by some western anthropologists and sociologists, who have suggested that the study of the static social structures of primitive or civilised communities is more informative than the investigation of their problematic historical evolution.

Western scientists of the classical tradition have been reluctant to accept these criticisms of the scientific viewpoint of the nineteenth century, and a number of them have adopted the same attitude as the Soviet scientists to questions which involve tenets common to both dialectical materialism and the mechanical philosophy. In the 1940's, Einstein in America and de Broglie in France, together with some physicists in the Soviet Union, expressed the view that physical events are fully determinate in principle, and that the Uncertainty Principle is merely indicative of the incompleteness of quantum mechanics in its present form. They argued that quantum mechanics as it stands today is essentially a method of dealing statistically with assemblies containing large numbers of physical systems, notably atoms or subatomic particles, so that it is illegitimate to make final statements concerning individual systems, or single particles, on the basis of that theory. The time taken for half of a given number of radioactive atoms to decompose, for example, can be determined with some precision, but the time at which a particular atom will decompose cannot be determined at all as yet, and it has been shown that this barrier to knowledge must always be insuperable, in principle, unless some of the essentials of the quantum mechanical theory are abandoned. Thus Einstein, de Broglie, and some of the Soviet physicists, notably Landau, held that quantum mechanics is a provisional theory, and that it will be found to be a limiting case of a wider generalisation in which the behaviour of a single atom will be more determinate and more explicable. To provide a basis for such a wider theory, pupils of these physicists have explored the scope of conceptual models in which particles appear as singularities in the electromagnetic field.

A similar though larger and more complex controversy arose in Soviet biology as to whether inherited changes in living organisms were determinate or indeterminate. From about 1930, the Soviet biologist, Lysenko, following the work of the plant breeder, Michurin, developed the theory that the heredity constitution was labile or 'shaken' at certain early stages of its growth, during which it was susceptible to changes induced by the environment, such changes adapting the plant, its seeds, and thus its offspring, to the new conditions. It has been generally known from the mid-nineteenth century that winter wheats, which normally do not come to fruition if sown in spring, could be made to mature if the seeds were wetted until they germinated, and then chilled before they were sown. Such 'vernalised' seeds could not be sown by ordinary methods, as they had germinated, but Lysenko showed in 1930 that if the wetting and chilling were reduced, the seeds were still vernalised, though they had not germinated, and so they could be sown by the normal agricultural methods. The winter wheats as ordinarily vernalised were in no wise changed. The seeds that they produced were winter wheats, requiring vernalisation if sown in the spring rather than in the winter. However, after much experimentation, Lysenko claimed that under certain critical vernalisation conditions, the plants were permanently changed into spring wheats, the seeds requiring no subsequent vernalisation.

Lysenko and his school, the Michurinists, then attacked the theories of the Mendelian geneticists, questioning their view that genetic changes in general were not adaptive to new environmental conditions. The theory of genetics and its application to agriculture were extensively discussed in the Soviet Union at a series of conferences held under the auspices of the Lenin Academy of Agricultural Sciences from 1934 on, the Michurinist school being dominant for a short period from 1948 when the Soviet school of Mendelian genetics was temporarily disbanded.

About the same time, theories which were generally accepted in other fields, in astronomy, chemistry, medicine, psychology, and anthropology, were criticised on the ground that they conflicted with the principles of dialectical materialism by Soviet scientists, giving rise to controversies both in the Soviet Union and in the west. However, the controversies soon lost much of their original impetus, for it came to be generally appreciated that the advance of economically fruitful applied science depended much more upon the development of new and more accurate fundamental theory in the laboratory and the field than upon the philosophical discussion of theories already well developed and no longer at the frontiers of knowledge.

Chapter 48

Science and History

IF WE wish to define what science has been and what it has accomplished historically, we find it difficult to formulate a definition which holds for all times and places. The sciences of the bronze-age civilizations differed markedly in character from those of the ancient Greeks, while Greek science possessed only some of the many-sided attributes displayed by science in the modern world. Behind the changing character of science throughout the ages, there has been an element of continuity, for the men of each period have developed and enlarged some aspects of the science bequeathed to them. Accordingly, we may perhaps say that science is a human activity developing an historically cumulative body of inter-related techniques, empirical knowledge, and theories, referring to the natural world. The American authority upon the history of science, Sarton, indeed considers that in this respect science is 'the only human activity which is truly cumulative and progressive'. But only part of science has been cumulative up to the present time, namely, its practical techniques and its empirical facts and laws. Judged by a long time scale, the theories of science have been ephemeral hitherto. The laws of levers and of the reflection of light, known to the Greeks, have become part of the permanent heritage of science, but the scientific theories of the Greeks are only of historical interest. Similarly, given a continuance of the present tempo of scientific activity, we can hardly suppose that any of the scientific theories of today will remain unmodified for long.

In the civilizations of the bronze age, mathematics and astronomy were largely utilitarian techniques used for keeping accounts, surveying, and calendar making. The sciences then did not differ greatly in character from the arts of the craftsmen, save that they were handed on through a written record, rather than by word of mouth. The ancient Greeks made an important advance in generalizing the discovery that empirically known facts belonging to a particular class could be theoretically demonstrated and shown to hold for all similar cases, as in the examples of the Pythagoras theorem or the laws of levers. The Greeks also used geometry to interpret theoretically their astronomical observations, so that empirical data now began to give a quantitative structure to cosmological theory. However, the dominant world systems of the Greeks were influenced by the conception that the heavenly bodies were superior to the earth, a notion which led them to prefer geocentric systems, notably the homocentric and the epicyclic, both of which, in their final forms, conflicted with facts known in antiquity. Moreover, the Greeks did not develop a consistent experimental method, though they made experiments on occasion; nor did they extend the application of science to new fields, except possibly military engineering and the making of general world maps.

In the subsequent civilizations of Rome, the Muslim world, and medieval Europe, science did not transcend the bounds set to it in Greek times, and its influence upon those civilizations was not large. During the modern period of history, however, science, and the forces promoting science, have developed an ever-increasing power of historical change. Experimental enquiry, together with the qualitative-inductive, and the quantitative-deductive methods discussed during the early decades of the seventeenth century, gradually found their appropriate place and application in all of the sciences. Applied first to mechanics and astronomy, they elucidated the workings of the solar system, and then to electricity, chemistry, biology, and other sciences, they rendered these subjects in turn more precise and fruitful. Such developments have helped to bring about a profound secularization of the human mind, science assuming, whether by sympathy or antipathy, a more and more important place in all general systems of thought, and, in the industrialized countries, colouring the views generally accepted concerning the nature of the universe and man's place in it. The applications of science too spread beyond the classical bounds of surveying and calendar making, first to navigation, and then to industry, agriculture, and medicine. The changes so wrought have done much to form the character of modern civilization, dissolving old traditions and old ways of life, so that when we speak of modern civilization spreading, say to the orient, we are thinking primarily of the spread of science and its applications.

The long-term consequences of the applications of science were not widely appreciated before the beginning of the present century. James Watt hardly could have foreseen the urban congestion that arose from the adoption of his steam engine in the factories, nor Faraday the relief of that congestion through the rise of the suburbs, which was aided by the application of his electrical researches to problems of public transportation and the transmission of industrial power. The inadvertent nature of the long-term changes brought about by science is perhaps most strikingly illustrated by the fact that they have begun to limit the realization of the values belonging to the period and the society which brought modern science into being. The individualism of men in modern times, and the value accorded to personal endeavour, have provided much of the driving force behind the development of modern science, both directly through the desire to make a personal exploration of nature and indirectly through the connection of science with the movements in which those values found an expression, such as the voyages of geographical discovery, the agrarian and industrial revolutions. But amongst other things, the development of the applications of science has more and more set a limit to the realization of those values. The steam engine, and the new textile machinery, ended the days of the individual hand-loom weaver, and subsequent developments made the technical units of industry even larger, drawing the individual into a composite organization, which circumscribed his activities. The electrical generating station at its inception served a large region, eliminating a number of individual steam engines, and later it was found to be more efficient when operated, as part of a national unit, through grid connections with other stations over a country as a whole. Finally, when atomic energy appeared, it was considered to be a development too precious and powerful for private use, and it has remained a state project in all countries, even in those where the value accorded to private endeavour is the most deeply rooted.

In the modern world science has led mainly to the secularization of thought and the development of utilitarian applications, but it has had some influence upon human values and standards of judgement. Attempts have been made by a number of scientists, particularly biologists, to derive an ethical code from the theory of evolution, but it is probable that the scientific method has had more influence upon human evaluations than any particular theory. The scientific method relies upon rational arguments rather than emotional appeals, and it suggests that empirical evidence should decide between rival viewpoints, practices which have become perhaps a little more general in personal relationships than they were a century or so ago. It is notable that the practice of settling differences of opinion by duelling began to decline in the early decades of the nineteenth century, 'the scientific century' as the men of the period termed it, and that the decline was most marked amongst the middle classes of Britain and France, the section of the world community which was then the most scientifically minded. The tendency of men imbued with the scientific attitude to adopt a rational and a humanistic point of view is illustrated by the opposition of the British scientists of the mid-nineteenth century to the policies of Eyre, the governor of Jamaica, who repressed a disturbance there in 1865 in a particularly arbitrary and barbaric fashion. Writing of the incident in his *Life of John Bright*, Trevelyan remarks:

'Except for Tyndal, the men of the finest scientific mind,—Darwin, Huxley, Mill, Leslie Stephen, Sir Charles Lyell,—ranged themselves on the side of law and humanity, whilst those whose cue it sometimes was to complain of the hardness of the scientific attitude to life,—Carlyle, Ruskin, Kingsley, Tennyson,—showed by their own conduct how prone sentimentalists are to inconsiderate worship of brute force and the "strong man".'

Another feature of the scientific method, which perhaps has had some influence upon human evaluations, is its dynamic and inventive quality. The scientific method is essentially a means of discovering new phenomena, and of formulating new theories, so that the sciences constitute ever-expanding systems of knowledge, old theories being overthrown constantly by new ones, so long as that method is practised. The American authority, Sarton, has written in this connection:

'Science always was revolutionary and heterodox; it is its very essence to be so; it ceases to be so only when it is asleep.'

Men with a feeling for this aspect of science tend to be moved by forward-looking values, and to be impatient of institutions which have much inertia. Joseph Priestley, we remember, 'saw reason to embrace what is generally called the heterodox side of almost every question'. Discoursing upon the relations between the Catholic Church and the sciences, Priestley remarked that, in the degree to which the Pope patronized science and polite literature,

'he was cherishing an enemy in disguise. And the English hierarchy (if there be any thing unsound in its constitution) has equal reason to tremble even at an air pump or an electrical machine.'

However, the influence of the scientific method upon the men who espouse it, for the most part, has been small. Scientists generally have adopted the values of the society to which they have belonged, even in the cases where those values have been detrimental to the advance of science, as in Germany under the Third Reich.

Similar considerations hold, to a greater or lesser degree, for the other changes brought about by science, and for the development of science itself. We cannot regard science as an entirely self-moving historical phenomenon, nor as a completely autonomous agent of historical change, even though it has a tradition and a momentum of its own. The development of science has only been one of a number of historical movements that have formed an interconnected complex, in which science until recently has been of minor force. The science of a given age has belonged, not only to its own tradition with its own methods, values, and accumulated knowledge, but also to its own historical period, in which other movements have made their own impact upon it. In comparatively static periods of history, such as the middle ages, science has not displayed a notable development, whilst in expansive periods science has often thrived. Moreover, within a given period, there have been fashions, hesitations, and abrupt changes in the development of science, which do not appear to have been due to internal causes. In the modern period of history we have that curious stagnation of science during the first half of the eighteenth century, which affected chemistry and optics in particular, and electricity to a smaller degree.

Such happenings indicate that scientific activity has been directed now into one channel, and then into another, and that upon occasion the forces promoting science were relaxed, or even reversed. In a general way, it may perhaps be said that the practical problems of a given historical period have had an influence upon the empirical enquiries undertaken by the scientists of the time, while the intellectual interests of the age have influenced the form in which scientific theories were expressed. Thus the geographical explorations of the sixteenth century stimulated the search for methods of determining the longitude, and promoted the study of astronomical and mechanical problems which such methods entailed. Similarly the theory of natural selection was influenced by the *laissez-faire* current of English thought during the nineteenth century, an influence which Darwin acknowledged indirectly by specifying his debt to the views of Malthus. The division, however, has not been rigid. Practical problems have stimulated the rise of new theories, as in the case of the theory of thermodynamics which arose in part from the study of steam engine problems, while intellectual currents have orientated empirical scientific enquiry into specific channels, as in the case of the romantic and historical German philosophy which promoted the study of embryology amongst the Germans of the late eighteenth and early nineteenth centuries.

In the past the forces promoting the growth of science were not consciously directed, and only the results which those forces produced were immediately apparent. Scientists at the turn of the seventeenth century appreciated 'the present languid state of natural philosophy', though the causes of that state remained obscure. In recent times, however, science has become more consciously and directly orientated into specific fields, the choice of which has passed more and more out of the hands of the scientists themselves. As scientific research grew more complex, the amateur tradition in science declined, and research became professionalized and externally directed, except in the academic sphere, through the setting up of research institutions governed by outside bodies, such as industrial firms and governmental ministries.

These bodies in the main have been concerned primarily with the applications of science, and for a number of decades now they have provided the bulk of the resources devoted to scientific research. At first they sought mainly the improvement of industry, agriculture, and medicine, as illustrated in Britain by the formation of the Department of Scientific and Industrial Research in 1917 and the Agricultural and Medical Research Councils a few years later. Subsequently, however, researches upon subjects of military interest have become more and more stressed. The Civil Estimates, and other publications, indicate that the monetary resources expended by the British government upon military research and development increased sixty-seven-fold, from the year 1936-7 to the year 1950-1, while over the same period the amounts devoted to research and development in the industrial field increased ten-fold, in medicine nine-fold, and agriculture eight-fold. By comparison, government expenditure on the universities, where most of the fundamental scientific research is still carried out, increased nearly six-fold over the same period. These figures are indicative of the general trends within the science of our time, and of the character which applied scientific activity is now assuming, countries which are industrially less advanced perhaps devoting a greater proportion of their resources to industrial research, and the more industrialized states allocating perhaps a greater proportion to researches of a military nature.

Such developments have had their effect upon fundamental science. They have created, for example, a greater demand for atomic scientists, and they have given an impetus to fundamental researches in the particular field of nuclear physics. They have also clothed those researches with a veil of secrecy, which hitherto has been alien to the scientific tradition. Again they have placed greater premiums upon the intellectual conformity of the scientist to the values and viewpoints of the dominant group within the particular society to which he belongs, a trend which has been accompanied by the association of some scientific theories with the one or the other of the two opposing ideologies of the mid-twentieth century.

Throughout history scientific theories have been favoured or opposed, apart from considerations based upon the criteria of the scientific method, according to the degree to which those theories have been congruent, or at variance, with the generally accepted beliefs of their time and place. Such judgements, and the actions based upon them, have been particularly conspicuous during those periods of history when two major movements of comparable strengths have stood in opposition to one another. During the period of the Protestant Reformation and the Catholic Counter-reformation, for example, the Copernican and Ptolemaic theories were often judged by criteria which were wholly external to the scientific method. A not dissimilar situation obtains in the mid-twentieth century, though the two movements be secular, and in the 1950's, for example, theories of genetics aroused passions akin to those inflamed by astronomical theories during the sixteenth and seventeenth centuries. It is a measure of the historical importance which science has assumed in modern times, however, that the scientific revolution made little or no contribution to the force of either the Reformation or the Counter-reformation, whilst it is generally recognized in our time that science has become one of the important determinants of the strength of any major historical movement during the twentieth century.

Bibliography

PRIMARY and secondary sources, together with some specialist articles, have been indicated in the subsidiary bibliographies covering individual chapters or groups of related chapters. Further articles may be consulted in the following journals: For the history of technology, *The Transactions of the Newcomen Society*, published since 1920. For the history of scientific and other theories, *The Journal of the History of Ideas*, published from 1940. For the history of the sciences in general there are the British journal, *The Annals of Science*, published from 1936, and the international journals, *Isis*, for shorter papers, published since 1913, and *Osiris*, for longer papers, published from 1936. The following books are some of the general works devoted to the entire history of a given science, or to the history of a number of sciences over a considerable period, and also some of the works dealing with the background to the development of science, or with the biographical details of individual scientists. Readers requiring further bibliographical information are referred to: G. Sarton, *A Guide to the History of Science*, Waltham, Massachusetts, 1952, and the critical bibliographies given periodically in *Isis*.

B. T. BELL, *Men of Mathematics*, London, 1937.

H. BUTTERFIELD, *The Origins of Modern Science*, London, 1949.

A. CASTIGLIONI, *History of Medicine*, New York, 1947.

H. CREW, *The Rise of Modern Physics*, Baltimore, 1928.

J. G. CROWTHER, *The Social Relations of Science*, London, 1941.

W. C. DAMPIER, (formerly Whetham), *A History of Science*, Cambridge, 1942. With M. Dampier Whetham (eds.), *Cambridge Readings in the Literature of Science*, Cambridge, 1924.

G. E. DANIEL, *A Hundred Years of Archaelogy*, London, 1950.

P. DOIG, *A Concise History of Astronomy*, London, 1950.

R. DUGAS, *Histoire de la Mechanique*, Neuchatel, 1950.

R. E. DICKINSON, and O. J. R. HOWARTH, *The Making of Geography*, Oxford, 1933.

J. C. FLUGEL, *A Hundred Years of Psychology*, London, 1933.

C. C. GILLISPIE, *The Edge of Objectivity*, Princeton, 1960.

J. C. GREGORY, *A Short History of Atomism*, London, 1931.

H. HOFFDING, *A History of Modern Philosophy*, trans. B. E. Meyer, London, 1900.

A. V. HOWARD, *Chamber's Dictionary of Scientists*, London, 1951.

S. D'IRSAY, *Histoire des Universités*, 2 vols. Paris, 1933-35.

F. A. LANGE, *The History of Materialism*, trans. E. C. Thomas, London, 1925.

J. R. PARTINGTON, *A Short History of Chemistry*, London, 1939.

T. K. PENNIMAN, *A Hundred Years of Anthropology*, London, 1951.

H. T. PLEDGE, *Science since 1500*, London, 1940.

J. H. RANDALL, Jr., *The Making of the Modern Mind*, Cambridge, Massachusetts, 1940.

A. REBIÈRE, *Pages Choisies des Savants Moderns*, Paris (n.d.).

E. ROLL, *A History of Economic Thought*, London, 1938.

P. ROUSSEAU, *Histoire de la Science*, Paris, 1945.

G. H. SABINE, *A History of Political Theory*, London, 1937.

C. J. SINGER, *The Evolution of Anatomy*, Oxford, 1925.
A Short History of Medicine, Oxford, 1928.
A Short History of Biology, Oxford, 1931.

D. E. SMITH, *History of Mathematics*, 2 vols. Boston, 1923.

D. J. STRUIK, *A Concise History of Mathematics*, New York, 1948.

A. D. WHITE, *A History of the Warfare of Science with Theology*, 2 vols. London, 1914.

(CHAPTER 1—6)

C. BAILEY (ed.), *The Legacy of Rome*, Oxford, 1935.

M. BERTHELOT, *Collection des Anciens Alchimists Grecs*, Paris, 1888.

V. GORDON CHILDE, *Man Makes Himself*, London, 1941.

M. R. COHEN and I. E. DRABKIN, *A Source Book in Greek Science*, New York, 1948.

B. FARRINGTON, *Greek Science*, 2 vols., Pelican, 1944 and 1949.

D. FLEMING, 'Galen on the Motions of the Blood', *Isis*, 1955, XLVI, 14.

H. and H. A. FRANKFORT, J. A. WILSON and T. JACOBSEN, *The Intellectual Adventure of Ancient Man*, Chicago, 1946.

K. FREEMAN, *Ancilla to the Pre-Socratic Philosophers*, Oxford, 1948.

S. GANDZ, 'Conflicting Interpretations of Babylonian Mathematics', *Isis*, 1940, XXXI, 405.

S. R. K. GLANVILLE, (ed.), *The Legacy of Egypt*, Oxford, 1942.

H. GOMPREZ, 'The Problems and Methods of early Greek Science', *Journal of the History of Ideas*, vol. IV, 1943.

T. HEATH, *Greek Astronomy*, London, 1932.

H. KELSEN, 'Causality and Retribution', *Philosophy of Science*, 1941, 533.

R. W. LIVINGSTONE (ed.), *The Legacy of Greece*, Oxford, 1922.

T. EAST LONES, 'Mechanics and Engineering from the time of Aristotle to that of Archimedes', *Trans. Newcomen Soc.*, vol. II, 1921-2.

A. LUCAS, *Ancient Egyptian Materials and Industries*, London, 1948.

R. MCKEON, 'Aristotle on the development and nature of the scientific method', *Journal of the History of Ideas*, vol. VIII, 1947.

A. NEUBURGER, *The Technical Arts and Sciences of the Ancients*, trans. H. L. Brose, London, 1930.

O. NEUGEBAUER, *The Exact Sciences in Antiquity*, Copenhagen, 1951.

J. R. PARTINGTON, *Origins and Development of Applied Chemistry*, London, 1935.

A. REYMOND, *History of the Sciences in Greco-Roman Antiquity*, trans. R. Gheury de Bray, London, 1927.

G. SARTON, *Introduction to the History of Science*, vol. I, Baltimore, 1927.

C. SINGER, *Greek Biology and Greek Medicine*, Oxford, 1922.

F. THUREAU-DANGIN, 'Sketch of a History of the Sexagesimal System', *Osiris*, 1939, VII, 95.

H. P. VOWLES, 'The Early Evolution of Power Engineering', *Isis*, 1932, XVII, 412.

(CHAPTERS 7-11)

T. W. ARNOLD and A. GUILLAUME (eds.), *The Legacy of Islam*, Oxford, 1931.

D. CAMPBELL, *Arabian Medicine and its Influence on the Middle Ages*, London, 1926.

T. F. CARTER, *The Invention of Printing in China and its Spread Westwards*, New York, 1925.

H. CHATLEY, 'Chinese Technology', *Trans. Newcomen Soc.*, 1941, XXII, 117.

M. CLAGETT, *The Science of Mechanics in the Middle Ages*, Madison, 1959.

S. R. DAS, 'Some Notes on Indian Astronomy', *Isis*, 1930, XIV, 388.

P. DUHEM, *Le Système du Monde*, 10 vols., Paris, 1913-57.

A. FORKE, *The World Conception of the Chinese*, London, 1924.

S. GANGULI, 'Notes on Indian Mathematics', *Isis*, 1929, XII, 132.

G. T. GARRATT (ed.), *The Legacy of India*, Oxford, 1937.

L. CARRINGTON GOODRICH and FENG CHIA-SHENG, 'The Early Development of Firearms in China', *Isis*, 1946, XXXVI, 114.

M. HASHIMOTO, *The Origin of the Compass*, Tokio, 1926 (*Isis*, 1930, XIV, 525).

C. HASKINS, 'Arabic Science in Western Europe', *Isis*, 1925, VII, 478.

C. H. HASKINS, *Studies in the History of Medieval Science*, Harvard, 1924.

A. J. HOPKINS, 'A Modern Theory of Alchemy', *Isis*, 1925, VII, 58.

LI CH'IAO-P'ING, *The Chemical Arts of Old China*, Easton, Pennsylvania, 1948.

M. MEYERHOF, 'Ibn Al-Nafis and his Theory of the Lesser Circulation', *Isis*, 1935, XXIII, 100.

A. MIELI, *La Science Arabe*, Leiden, 1938.

J. NEEDHAM, *Science and Civilisation in China*, vols. I, II, III, Cambridge, 1954, 1956, 1959.

CT. L. DES NOETTES, *L'Attelage: le Cheval de Selle a travers les Ages*, Paris, 1931. *De la Marine Antique à la Marine Moderne*, Paris, 1935.

D. L. O'LEARY, *How Greek Science passed to the Arabs*, London, 1948.

J. H. RANDALL, Jr., 'The Development of the Scientific Method in the School of Padua', *Journal of the History of Ideas*, vol. I, 1940.

P. C. RAY, 'The Chemical Knowledge of the Hindus of Old', *Isis*, 1919, II, 322.

G. SARTON, *Introduction to the History of Science*, vols. 1-5, Baltimore, 1927-48.

C. SINGER, *From Magic to Science*, London, 1928.

D. E. SMITH and L. C. KARPINSKI, *The Hindu-Arabic Numerals*, Boston and London, 1911.

F. S. TAYLOR, *The Alchemists*, New York, 1949.

L. THORNDIKE, *A History of Magic and Experimental Science*, 6 vols., New York, 1923-41.

VAN HEE, 'The Chou-Jen of Yuan Yuan', *Isis*, 1926, VIII, 103.

WANG LING, 'On the Invention and Use of Gunpowder and Firearms in China', *Isis*, 1947, XXXVII, 160.

L. WHITE, 'Technology and Invention in the Middle Ages', *Speculum*, 1940, XV, 141.

K. C. WONG and WU LIEN-TEH, *History of Chinese Medicine*, Shanghai, 1936.

(CHAPTERS 12-15)

A. ARMITAGE, *Copernicus the founder of Modern Astronomy*, London, 1938.

F. BACON, *Novum Organon*, trans. G. W. Kitchin, Oxford, 1855.

C. BAUMGARDT, *Johannes Kepler: Life and Letters*, London, 1952.

M. CASPAR, *Johannes Kepler*, Stuttgart, 1950.

Copernicus, *De Revolutionibus*, Book I, trans. J. F. Dobson and S. Brodetsky. *Occasional Notes of the Royal Astronomical Society*, vol. 2, No. 10, 1947.

Three Copernican Treatises, trans. E. Rosen, New York, 1939.

R. DESCARTES, *Discourse on Method, etc.*, Everyman, London, 1946.

J. L. E. DREYER, *History of the Planetary Systems from Thales to Kepler*, Cambridge, 1906.

B. FARRINGTON, *Francis Bacon, Philosopher of Industrial Science*, London, 1951.

J. A. GADE, *The Life and Times of Tycho Brahe*, Princeton, 1947.

GALILEO GALILEI, *Two New Sciences*, trans. H. Crew and A. de Salvio, New York, 1941.

A. GEWIRTZ, 'Experience and the non-mathematical in Descartes', *Journal of the History of Ideas*, vol. II, 1941.

William Gilbert of Colchester, *De Magnete*, trans. P. Fleury Mottelay, London, 1893.

F. R. JOHNSON, *Astronomical Thought in Renaissance England*, Baltimore, 1937.

A. KOYRÉ, *Etudes galiliennes*, Paris, 1939.

G. MCCOLLEY, 'The theory of the diurnal rotation of the earth', *Isis*, 1936, XXVI, 392.

G. SARTON, 'Simon Stevin of Bruges, 1548-1620', *Isis*, 1934, XXI, 241.

E. J. STRONG, *Procedures and Metaphysics*, Berkley, 1936.

R. SUTER, 'Biographical Sketch of William Gilbert and Colchester', *Osiris*, 1952, X, 368.

E. ZILSEL, 'The Origins of William Gilbert's Scientific Method', *Journal of the History of Ideas*, 1941, II, 1.

(CHAPTER 16)

RICHARD BAXTER, *The Certainty of a World of Spirits*, London, 1691.

THOMAS BROWNE, *Religio Medici,* Everyman, London, 1947.

JOHN CALVIN, *The Institutes of the Christian Religion,* trans. H. Beveridge, 2 vols., London, 1949. See particularly Book I and Book IV.

A. DE CANDOLLE, *Histoire des Sciences et des Savants depuis deux Siècles,* Genève, Basle, Lyon, 1873.

JOHN COTTON, *A Brief Exposition with Practical Observations upon the whole Book of Ecclesiastes,* London, 1654.

H. HAYDN, *The Counter-Renaissance,* New York, 1950.

H. W. JONES, 'Mid-Seventeenth Century Science: Some Polemics', *Osiris*, 1950, IX, 254.

R. F. JONES, *Ancients and Moderns*, St. Louis, 1936.

PERRY MILLER, 'The Marrow of Puritan Divinity', *Trans. Colonial Soc. Massachusetts,* 1935, XXXII, 247.

R. OVERTON, *Man's Mortallitie*, Amsterdam, 1643.

J. PELSENEER, 'L'Origine Protestante de la Science Moderne', *Lychnos*, 1946-7, 246.

T. SPRAT, *The History of the Royal Society of London,* London, 1667.

D. STIMSON, 'Puritanism and the New Philosophy in Seventeenth-Century England', *Bull. Inst. Hist. Med.,* 1935, III, 321.

E. M. W. TILLYARD, *The Elizabethan World Picture,* London, 1943.

J. WILKINS, *Discourse concerning a New World and Another Planet: in 2 Bookes,* 4th edn., London, 1683.

B. WILLEY, *The Seventeenth Century Background,* London, 1934.

T. WRIGHT, *An Original Theory, or New Hypothesis of the Universe,* London, 1750.

(CHAPTERS 17 and 18)

E. A. BURTT, *The Metaphysical Foundations of Modern Science,* London, 1924.

A. R. HALL and M. BOAS HALL, 'Newton's Theory of Matter', *Isis*, 1960, LI, 131.

L. T. MORE, *Isaac Newton, a Biography,* London, 1934.

I. NEWTON, *The Mathematical Principles of Natural Philosophy,* trans. A. Motte, 3 vols., London, 1803.

L. D. PATTERSON, 'Hooke's Gravitation Theory and its Influence on Newton', *Isis*, 1949, XL, 327, and 1950, XLI, 32.

J. PRIESTLEY, *History and present state of discoveries relating to vision, light, and colours,* London, 1772.

A. J. SNOW, *Matter and Gravity in Newton's Physical Philosophy,* London, 1926.

A. WOLFF, *A History of Science, Technology, and Philosophy during the Sixteenth and Seventeenth Centuries,* London, 1935.

(CHAPTERS 19 and 20)

R. H. BAINTON, 'Michael Servetus and the Pulmonary Transit of the Blood', *Bull. Hist. Med.,* 1951, XXV, 1.

H. P. BAYON, 'William Harvey: Physician and Biologist', *Annals of Science,* 1938, III, 59, 83, and 435, and 1939, IV, 65, and 329.

M. BOAS, 'The Establishment of the Mechanical Philosophy', *Osiris,* 1952, X, 412.

The Works of the Honourable Robert Boyle, ed. with a life of the author by Thomas Birch, 5 vols., London, 1744.

W. HARVEY, *Works,* trans. R. Willis, London, 1847.

J. B. VAN HELMONT, *Oriatrike,* trans. J. Chandler, London, 1662.

L. T. MORE, *The Life and Works of the Honourable Robert Boyle,* New York, 1944.

W. PAGEL, 'The Religious and Philosophical Aspects of van Helmont's Science and Medicine', *Supplement to the Bulletin of the History of Medicine,* No. 2, Baltimore, 1944. 'William Harvey and the Purpose of Circulation', *Isis,* 1951, XLII, 22.

The Hermetic and Alchemical Writings of Paracelsus, trans. A. E. Waite, 2 vols., London, 1894.

M. SERVETUS, *Christianismi Restitutio,* Nuremburg, 1790.

(CHAPTERS 21-42)

T. S. ASHTON, *Iron and Steel in the Industrial Revolution,* Manchester, 1924.

Centenary number of the *Memoirs of the Literary and Philosophical Society of Manchester,* third series, vol. IX, 1883.

H. C. BOLTON, 'The Lunar Society', *Trans. Birmingham and Midlands Institute,* Archeological Section, 1889, 79-94.

G. N. CLARKE, *Science and Social Welfare in the Age of Newton,* Oxford, 1937.

W. H. B. COURT, *The Rise of the Midland Industries, 1600-1838,* Oxford, 1938.

H. W. DICKINSON, *James Watt,* Cambridge, 1936.
Matthew Boulton, Cambridge, 1937.

E. H. HANKIN, 'The Intellectual Ability of the Quakers', *Science Progress,* 1921-2, XVI, 654.

B. HESSEN, 'The Social and Economic Roots of Newton's "Principia" ', in *Science at the Cross Roads,* London, 1931.

R. T. GOULD, *The Marine Chronometer, its History and Development*, London, 1923.

R. JENKINS, 'Savery, Newcomen, and the early history of the steam engine', *Trans. Newcomen Soc.*, Vols. II and IV, 1922-24. 'The heat engine idea in the Seventeenth Century', *ibid*, Vol. XVII, 1936-37.

F. R. JOHNSON, 'Gresham College', *Journal of the History of Ideas*, vol. I, 1940.

R. K. MERTON, 'Science and Technology in Seventeenth-Century England', *Osiris*, 1938, IV, 360.

J. U. NEF, *The Rise of the British Coal Industry*, 2 vols. London, 1932.

M. ORNSTEIN, *The Role of Scientific Societies in the Seventeenth Century*, Chicago, 1938.

A. RAISTRICK, *Quakers in Science and Industry*, London, 1950.

W. STEVENSON, *Historical sketch of the progress of discovery, navigation, and commerce from the earliest records to the beginning of the Nineteenth Century*, Edinburgh, 1824.

D. STIMSON, *Scientists and Amateurs: A History of the Royal Society*, London, 1949.

E. G. R. TAYLOR, *Tudor Geography, 1485-1583*, London, 1930.
Late Tudor and Early Stuart Geography, 1583-1650, London, 1934.

A. P. USHER, *A History of Mechanical Inventions*, London and New York, 1929.

J. WARD, *The Lives of the Gresham Professors*, London, 1740.

(CHAPTERS 25 and 26)

'Natural Philosophy through the Eighteenth Century, and Allied Topics', *Philosophical Magazine*, Commemoration number, July 1948.

P. BRUNET, *L'Introduction des Theories de Newton en France au 18me Siècle*, Paris, 1931.

P. J. HARTOG, 'Priestley and Lavoisier', *Annals of Science*, 1946-7, V, 1.

A. MELDRUM, *The Eighteenth-Century Revolution in Science*, Calcutta, 1933.

J. R. PARTINGTON and D. MCKIE, 'Historical Studies on the Phlogiston Theory', *Annals of Science*, 1937, II, 361, and 1938, III, 1, and 337.

I. TODHUNTER, *History of the Mathematical Theories of Attraction and the Figure of the Earth from the time of Newton to that of Laplace*, 2 vols., London, 1873.

J. H. WHITE, *The History of the Phlogiston Theory*, London, 1932.

A. WOLFF, *A History of Science, Technology, and Philosophy in the Eighteenth Century*, London, 1938.

(CHAPTERS 27 and 28)

C. L. BECKER, *The Heavenly City of the Eighteenth Century Philosophers*, New Haven, 1932.

J. B. BURY, *The Idea of Progress: an inquiry into its origin and growth*, London, 1920.

E. CASSIRER, *The Philosophy of the Enlightenment*, trans. F. C. A. Koelln and J. P. Pettegrove, Princeton, 1951.

H. DAUDIN, *De Linné à Lamarck: Methodes de classification et idée de serie en botanique et en zoologie, 1740-1790*, Paris, 1926.

P. HAZARD, *La Crise de la Conscience Européenne, 1680-1715*, Paris, 1935.
La Pensée Européenne au 18me siècle, Paris, 1946.

E. KRAUSE, *Erasmus Darwin*, trans. W. S. Dallas, with a preliminary notice by Charles Darwin, London, 1879.

J. B. LAMARCK, *Zoological Philosophy*, trans. H. Elliot, London, 1912.

A. LOVEJOY, *The Great Chain of Being*, Cambridge (Massachusetts), 1936.

D. MORNET, *Les Origines Intellectuelles de la Révolution Française*, Paris, 1947.

J. ROSENTHAL, 'The Attitude of Some Modern Rationalists to History', *Journal of the History of Ideas*, vol. IV, 1943.

B. WILLEY, *The Eighteenth-Century Background*, London, 1946.

E. ZILSEL, 'The Genesis of the Concept of Scientific Progress', *Journal of the History of Ideas*, vol. VI, 1945.

(CHAPTERS 29-32)

A. GODE-VON AESCH, *Natural Science in German Romanticism*, New York, 1941.

J. BOEHME, *The Signature of All Things*, Everyman, London, 1912.

G. CONGER, *Theories of Microcosm and Macrocosm in the History of Philosophy*, New York, 1922.

F. H. GARRISON, 'The Romantic Episode in German Medicine', *Bull. New York Academy of Medicine*, 1931, VII, 841.

J. NEEDHAM, *A History of Embryology*, Cambridge, 1934.

L. OKEN, *The Elements of Physio-Philosophy*, trans. A. Tulk, vol. 7 of the publications of the Ray Society, 1847.

W. PAGEL, 'The Speculative Basis of Modern Pathology', *Bull. Hist. Med.*, 1945, XVII, 1.

E. S. RUSSEL, *Form and Function: a Contribution to the History of Animal Morphology*, London, 1916.

O. TEMKIN, 'German Concepts of Ontogeny and History around 1800', *Bull. Hist. Med.*, 1950, XXIV, 227.

(CHAPTERS 33 and 34)

T. COWLES, 'Malthus, Darwin, and Bagehot: a Study in the Transference of a Concept', *Isis*, 1936, XXVI, 341.

C. DARWIN, *The Origin of Species*, 6th edn., Everyman, London, 1928.

F. DARWIN (ed.), *Charles Darwin: his Life told in an Autobiographical Chapter, and in a Selected Series of his Published Letters*, London, 1892.

P. GEDDES and J. A. THOMPSON, *Evolution*, London (n.d.). (The author is indebted to the Oxford University Press for permission to quote pp. 213-15 of this work.)

G. MCCONNAUGHTY, 'Darwin and Social Darwinism', *Osiris*, 1950, IX, 397.

E. RADL, *The History of Biological Theories*, trans. E. J. Hatfield, Oxford, 1930.

D. C. SOMERVELL, *English Thought in the Nineteenth Century*, London, 1936.

H. THOMAS, 'Geology and Contemporary Thought', *Annals of Science*, 1946-7, V, 325.

A. R. WALLACE, *My Life*, London, 1905.

K. A. VON ZITTEL, *History of Geology and Palaeontology*, trans. M. M. Ogilvie-Gordon, London, 1905.

(CHAPTER 35)

L. AUCOC, *L'Institut de France*, Paris, 1889.

The Edinburgh Journal of Science, vols. IV-VI, 1831-2.

O. J. R. HOWARTH, *The British Association for the Advancement of Science: a Retrospect, 1831-1931*, London, 1931.

B. JONES, *The Royal Institution*, London, 1871.

A. MAURY, *Les Académies d'autrefois*, Paris, 1864.

J. T. MERZ, *A History of European Thought in the Nineteenth Century*, vol. I, London, 1896.

(CHAPTERS 36-39)

E. CASSIRER, *The Problem of Knowledge: Philosophy, Science, and History since Hegel*, trans. W. H. Woglom and C. H. Hendel, New Haven, 1950.

A. FINDLAY, *A Hundred Years of Chemistry*, London, 1948.

E. MACH, *Die Principien der Warmlehre*, Leipzig, 1900.

J. T. MERZ, *A History of European Thought in the Nineteenth Century*, 4 vols., London, 1896-1914.

M. M. P. MUIR, *History of Chemical Theories and Laws*, New York, 1907.

E. T. WHITTAKER, *A History of Theories of the Ether and Electricity*, London, 1951.

W. WILSON, *A Hundred Years of Physics*, London, 1950.

(CHAPTERS 40 and 41)

C. A. BROWNE, *Source Book of Agricultural Chemistry*, Waltham, Massachusetts, 1944.

A. and N. CLOW, *The Chemical Revolution: a Contribution to Social Technology*, London, 1952.

J. G. CROWTHER, *British Scientists of the Nineteenth Century*, London, 1935.

P. DUNSHEATH (ed.), *A Century of Technology*, London, 1951.

B. LEPSIUS, *Deutschlands chemische Industrie, 1888-1913*, Berlin, 1914.

S. LILLEY, *Men, Machines, and History*, London, 1948.

E. J. RUSSELL, *British Agricultural Research: Rothamsted*, London, 1942.

(CHAPTERS 42-45)

A. EINSTEIN and L. INFELD, *The Evolution of Physics*, Cambridge, 1938.

P. G. FOTHERGILL, *Historical Aspects of Organic Evolution*, London, 1952.

F. A. LINDEMANN, *The Physical Significance of the Quantum Theory*, Oxford, 1932.

E. T. WHITTAKER, *From Euclid to Eddington*, Cambridge, 1949.

G. J. WHITROW, *The Structure of the Universe*, London, 1949.

(CHAPTER 46)

J. G. CROWTHER and R. WHIDDINGTON, *Science at War*, London, 1947.

S. A. GOUDSMIT, *ALSOS: the Failure in German Science*, London, 1947.

E. Y. HARTSHORNE, 'The German Universities', *Nature*, 1938, **142**, 175.

R. HINTON THOMAS, *Liberalism, Nationalism, and the German Intellectuals, 1822-1847*, Cambridge, 1951.

A. HORTIS, *Le Riunione degli Scienziati Italiani prima delle Guerre d'Indipendenza*, Citta de Castello, 1922.

K. SUDHOFF, *Hundert Jahre Deutscher Naturforscher-Versammlungen*, Leipzig, 1922.

(CHAPTER 47)

D. BOHM, *Causality and Chance in Modern Physics*, London, 1957.

M. BUNGE, *Causality. The Place of the Causal Principle in Modern Science*, Cambridge (Mass.), 1959.

J. G. CROWTHER, *Famous American Men of Science*, London, 1937.

A. EINSTEIN, *Out of my later years*, London, 1950.

P. H. OEHSER, *Sons of Science: the Story of the Smithsonian Institution and its Leaders*, New York, 1948.

The Situation in Biological Science, Moscow, 1949.

R. P. STEARNS, 'Colonial Fellows of the Royal Society, 1661-1788', *Osiris*, 1948, VIII, 73.

S. I. VAVILOV, *Soviet Science: Thirty Years*, Moscow, 1948.

(CHAPTER 48)

F. BARRY, *The Scientific Habit of Thought*, New York, 1927.

J. D. BERNAL, *The Social Function of Science*, London, 1946.

W. GELLHORN, *Security, Loyalty, and Science*, Ithaca, N.Y., 1950.

T. H. HUXLEY and J. HUXLEY, *Evolution and Ethics, 1893-1943*, London, 1947.

S. LILLEY, 'Social aspects of the history of science', *Archives Internationales d'Histoire des Sciences*, 1949, XXVIII, 378.

J. PRIESTLEY, *Experiments and Observations on Different Kinds of Air*, 3 vols., London, 1775-7, preface to vol. 1.

G. SARTON, *Introduction to the History of Science*, Vol. 1, Baltimore, 1927, introductory chapter.

C. H. WADDINGTON, *The Scientific Attitude*, Pelican, 1941, (ed.), *Science and Ethics*, London, 1942.

Index

Abbasids, 95, 102
Aberration of light, discovery of, 297
Abn Ahmad, 102
Absolute scale of temperature, 494
Absolute space, Einstein on, 543, 545
Absolute time, Einstein on, 543, 545
Abubacer, 100
Abulcasis, 100
Academia Secretorum Naturae, 262
Academies: 19th century, 439; *see also* under full name of various academies, as Paris Academy of Sciences
Academy, Plato's, 39
Accademia del Cimento, 262
Accademia dei Lincei, 262
Acceleration, in Einstein's theory, 545, 547
Adams, 574-75
Adaptation, Lamarck on, 385
Addison, Joseph, 327-28
Advancement of Learning (Bacon), 141
Aepinus, Franz, 474-75
Affinity, chemical, 457
Agassiz, Louis, 424
Agricultural chemistry, 516-17, 520-21
Agriculture: medieval Europe, 103-105; 19th-century progress, 513, 516-17, 520; stone age, 15
Air: Aristotle on, 42; Descartes on, 170; as element for Hindus, 93; 18th-century research, 310; Helmont on, 236; Mayow on, 240-41; phlogiston theory, 305; as primordial substance in Greece, 26, 27
Air pump, 275
Al-, in Arabic names, *see under* second element, as Battani, al-
Albert of Saxony, 119
Albiruni, 93, 99
Alchemy, 227-42; Alexandrian period, 64-67; Arab science, 97-98; Boehme and, 352-53; in China, 77, 84; Hindu, 92-93; medieval Europe, 115-16; Paracelsus' views on, 229-30, 233; primary matter and, 236
Alcmaeon of Croton, 31
Alcohol, in medieval Europe, 115-16, 117
Alexander the Great, 48
Alexandria, Museum at, 49, 67
Alexandrian Age, 47, 48-60; alchemy, 64

Algebra: in Babylonia, 18; Greek science, 49; Descartes on, 168; Hindu contributions to, 91, 93
Alizarin, 522
Alkali, 514
Alkaline earth metals, 457, 464
Alloys, 66
All-Italy Association, 582
Almagest, 112
Alpetrugius, 100
Alphabet, 24
Alpha particles, in atomic theory, 553, 556-57, 560, 562
Alpha rays, 552-53
Alphonsine Tables, 135
Alum, 513
Amber, 474
Ambrose of Milan, 67
American Association for the Advancement of Science, 582
Ampère, André Marie, 453, 477-78
Amphibia, 346
Anaesthesia, 581
Analytical Society, 443
Anarcharsis, 35
Anatomy: Alexandrian school, 56; Aristotle on, 44; cell theory, 387-92; in China, 78; in Egypt, 22; Galen on, 57; medieval Europe, 115; Pythagorean school, 31; in 17th century, 215
Anaxagoras, 35
Anaximander, 26-27
Anaximenes, 27
Andalusia, Arab science in, 100
Anderson, 560, 561
Anderson, John, 441
Andromeda nebula, 159, 566
Angelic beings: Dionysius on, 68-69; hierarchical concept, 179, 180, 181, 218; Sebonde on multiplicity of, 188-89
Aniline blues, 521
Animals: Plato on creation of, 39; *see also* Zoology
Annulates, 371
Anthrax, 518, 519
Anthropology, 447-48
Antiseptics, 518
Antitoxins, 520
Anyang Period, 37, 79
Apollonius of Citium, 57
Apollonius of Perga, 54
Apothecaries, 215, 229, 234, 513

phlogiston theory and, 304, 311; as products of sexual generation, 66; *see also* Calcination of metals

Meteorology, 64

Metonic cycle, 36, 80

Metre (measure), 436

Meyer, Lothar, 464

Michell, John, 298, 476

Michelson, Albert A., 542-43

Michurin school, 597, 598

Microbiology, applications in 19th century, 513-26

Microcosm, 351-52, 375

Micro-organisms, 520

Microscope, 364

Microscopy, 208

Middle Ages, technology and craft tradition in, 103-11

Middle East civilizations, 15-24

Miletian philosophers, 25-28

Military engineering, 50

Milky Way, 159, 299, 300, 564, 565

Millikan, Robert A., 552

Milne, 567, 570, 571-72, 573

Milne-Edwards, Henri, 424

Milton, John, 184

Mineral chemistry, 460-61

Minerals, 77, 334, 345

Mines, 271-74, 278

Minkowski, Hermann, 545

Mirror galvanometer, 510

Mitscherlich, Eilhard, 455

Mohammedans, *see* Arab science

Mohists, 74-75

Mohr, Friedrich, 490

Molecular structure, theory of, 463

Molluscs, 371, 381

Molyneux, William, 320

Momentum of electrons, 557

Monads, Leibniz on, 353-54

Mondino de Luzzi, 115

Monge, Gaspard, 436

Mongols, 101, 107

Montaigne, Michel E., 185

Montmor, Habert de, 264

Montmor Academy, 264

Montpellier society, 266

Moon: Alexandrian school, 52-55; Anaxagoras, 36; Aristotle, 42; Euler's theory, 293; Galileo, 160; Hindu astronomy, 91; Leplace, 294; tables, 270

Moray, Robert, 258

Moreland, Samuel, 273

Morgan, 533

Morley, 542-43

Moro, Anton, 396-98

Morphology, 361, 374-83, 426-27

Mortalists, 184-85

Morveau, 437

Mosaic egg, 537

Mosaic Flood, 396, 397

Mosander, Carl G., 464

Moseley, Henry, 466-67

Mo Ti, 74

Motion: Albert of Saxony, 119-20; Aristotle, 43; Carnot on perpetual, 489; concept in medieval Europe, 118-20; Descartes, 169, 170-71; Einstein's theory and, 544-45, 569; Galileo, 161, 163; Newton, 203; of projectiles, 150

Mount Palomar Observatory, 573-74

Mount Wilson Observatory, 565, 566

Müller, Johannes, 127, 245

Müller, 533, 535

Multiple proportions, law of, 452

Multiplication, 17

Murchison, Roderick, 407

Murdoch, William, 285

Muscovy Company, 249

Muslims, *see* Arab science

Musschenbroek, Pieter van, 475

Mutations, 429-30, 531, 534, 535, 539

Mycenaean civilization, 24

Mysticism, 231, 232, 351-53

Nadim, al-, 96

Nageli, Carl, 428-30, 530-31, 532

Napier, John, 252

Napoleon I, 438

Nasir ed-din, 102

Nationalism, science and, 280-81, 578-88, 585-88

National Research Foundation, 592

National Science Foundation Act, 592

National-socialism, science and, 585-88

Natural History (Buffon), 336

Natural History (Pliny), 62

Natural philosophy: abandonment of, in 19th century, 596; Bacon, 143, 316; cell theory, 389; Descartes, 165-66, 171-72, 173, 202, 316; 18th century, 279-81; French philosophers' view of nature, 349-50; Greek, 26, 35-47; Newton, 202-204; at Padua, 122; 17th century science and, 182; theory of light, 468; *see also* German natural philosophers; Iatrochemistry

Natural selection, 419, 422, 434, 534

Nature, laws of, 172, 173

Nautical almanacs, 245, 270

Navigation: exploration and, 243, 244, 246, 247-48; longitude problem, 269-71; magnetism experiments, 252; mathematics and, 250; medieval northern Europe, 106; Norman on, 138-39

Nebulae, 299, 300, 565, 566, 572

Nebula hypothesis, 295

Necessity, Plato's idea of, 38

Needham, John Turberville, 365

638 / Index

trans. H. L. Brose, London, 1930.
O. NEUGEBAUER, *The Exact Sciences in Antiquity*, Copenhagen, 1951.